# Selected Titles in This Series

# Essays in the History of Lie Groups and Algebraic Groups

History of Mathematics

Volume 21

# Essays in the History of Lie Groups and Algebraic Groups

**Armand Borel**

American Mathematical Society

London Mathematical Society

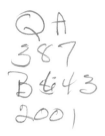
## Editorial Board

2000 *Mathematics Subject Classification.* Primary 01A55, 01A60, 17B45, 20–03, 20G15, 20G20, 22–03, 22E10, 22E46, 32M05, 32M15, 53C35, 57T15.

A list of photograph credits is included at the beginning of this volume.

---

**Library of Congress Cataloging-in-Publication Data**

Borel, Armand.
  Essays in the history of Lie groups and algebraic groups / Armand Borel.
    p. cm. — (History of mathematics, ISSN 0899-2428 ; v. 21)
  Includes bibliographical references and index.
  ISBN 0-8218-0288-7 (alk. paper)
  1. Lie groups—History.  2. Linear algebraic groups—History.  I. Title.  II. Series.

QA387.B643  2001
512′.55—dc21                                                                          2001018175

---

# Contents

# Introduction

This book consists of essays, some published previously, on topics belonging mainly to the first century of the history of Lie groups and algebraic groups. Partly written upon request, for various purposes, they do not aim at giving a comprehensive and exhaustive exposition. In order to put them in context, the first chapter attempts to sketch how they fit in the overall picture and complements them by some discussion of items not, or only briefly, touched upon elsewhere.

The "finite and continuous groups" of Sophus Lie were in fact analytic local groups of analytic transformations, as recalled in I, §1, but we shall deal almost exclusively with global aspects of Lie groups and algebraic groups. In retrospect, one can say that the passage from the local to the global was carried out in two ways, the transcendental or differential geometric one, highlighted by the contributions of A. Hurwitz, I. Schur, H. Weyl and É. Cartan, and the algebraico-geometric one, initiated in the 19th century mainly by L. Maurer, revived in the nineteen forties by C. Chevalley and E.R. Kolchin, and then developed by many others. Chapters II, III and IV pertain to the former, Chapters V, VI, VII and VIII to the latter.

Chapter II is the most elementary, devoted mostly to various proofs of the full reducibility of linear representations of $\mathbf{SL}_2(\mathbb{C})$.

Chapter III is more generally concerned with the work of Hermann Weyl on Lie groups, Lie algebras and invariant theory. It describes in particular his synthesis of Cartan's infinitesimal approach with the transcendental point of view initiated by A. Hurwitz (discussed in Chapter II), further developed by I. Schur, and its influence on É. Cartan, who was at the time noticing a remarkable connection between the theory of real simple Lie algebras and the study of a new class of Riemannian manifolds, later called symmetric spaces. This led Cartan to the work described in Chapter IV: first the building up of a theory of semisimple Lie groups and Riemannian symmetric spaces, in which both are remarkably intertwined, and then further developments which were all to have a far-reaching influence: generalization of the Peter-Weyl theorem to compact symmetric spaces, introduction of differential forms in algebraic topology, and bounded symmetric domains.

The remaining chapters are devoted to algebraic groups. Even though it was not a broadly recognized field in the 19th century, several, largely independent, contributions fit well under that heading, the most systematic being those of L. Maurer. They are surveyed in Chapter V.

The topic then fell into oblivion for almost half a century, and was taken up again in the 1940s, first by C. Chevalley and E.R. Kolchin. Their motivations were completely different: Chevalley wanted to develop and generalize the theory initiated by L. Maurer, while Kolchin's interest was mainly in the Picard-Vessiot Galois theory of homogeneous linear differential equations. Although he did not really need it, Kolchin started a theory over algebraically closed groundfields of arbitrary characteristic (while Chevalley was at first essentially bound to characteristic zero). After this pioneering work, the theory underwent spectacular developments in several directions. Chapter VI attempts to give an idea of the main steps. One can distinguish a first phase over algebraically closed groundfields, culminating with Chevalley's classification of algebraic simple groups (§§2, 3) and then the study of rationality properties over arbitrary fields (§4). Relationships with geometry, which in a way were there from the beginning, took a new prominence in the framework of Tits systems and Tits buildings (§5). From the thirties on, a rather persistent theme has been to what extent Lie group or, later, algebraic group properties can be read off the abstract group structure. Concretely, to what extent are abstract automorphisms described by Lie group or algebraic group automorphisms and field automorphisms? This is also a topic in which algebraic groups and geometry mix. It is discussed in §6. I have also included, as Chapters VII and VIII, two articles published earlier on C. Chevalley and E.R. Kolchin (slightly revised).

Chapter VI does not aim at completeness. The field of linear algebraic groups is still very active, and it was not my intention to cover the most recent developments. This, to me, is anyhow the purview of another type of publication. Without being strict about it, I have limited myself to work done (or at any rate well under way) during the first century of the theory (1873-1973). Following this rule, I have limited myself to some brief indications in VI, §7 on some very fruitful relations between the transcendental and algebraico-geometric points of view, woefully short in view of their growing importance.

When the editorial board for this series kindly suggested I contribute a volume to it, I felt that on one hand the papers underlying II, III, VII, VIII, with some minor modifications, belonged to it, and that, on the other hand, I could not contemplate rehashing them to fit them into a seamless narrative. I was even less tempted to do so in view of the forthcoming book by T. Hawkins: "The emergence of the theory of Lie groups", Springer 2000, which, among other things, includes a systematic, thorough exposition of much of the material in Chapters I, II, III. Thus from the start, as hinted by its title, this book was intended to be a rather heterogeneous collection of essays. It consists of four "old" (i.e. essentially published earlier) chapters and four "new" ones. This entails some overlap, especially between the old and the new, which I have not tried to suppress, preferring to let the old chapters keep the degree of autonomy initially intended. I still hope the book gives a good idea of the development of the topics it covers.

I am very grateful to T. Hawkins and T.A. Springer for corrections to, and remarks or questions on, earlier drafts, which led to a number of improvements and additions. As usual, deserved thanks are due to E. Gustafsson, who tirelessly typed into impeccable TeX rather unappealing typescripts, emanating from an old-fashioned typewriter, about to reach the status of an endangered species.

Finally, I am glad to thank the editorial staff of the AMS for suggestions and help in various aspects of the production of this book.

# Terminology for Classical Groups
# and Notation

I shall use mostly present-day notation, as indicated below, without further reference. Here, I also indicate the terminology used by Lie.

$$k \text{ is a commutative field}, \quad k^* = k - \{0\}$$

**1. $\mathbf{GL}_n(k)$** is the group of $n \times n$ invertible matrices with coefficients in $k$, and $\mathbf{SL}_n(k)$ its subgroup of matrices of determinant one.

Lie considered these groups mainly for $k = \mathbb{C}$, and I shall sometimes drop the ($\mathbb{C}$) in that case in Chapters I and V. These groups are now called respectively the general and special linear groups, but this is not Lie's terminology. For him, they are the general and special homogeneous linear groups. He reserves the term general linear group for our affine group $\mathrm{Aff}(k^n)$, i.e. the group of linear, not necessarily homogeneous, transformations of $n$-dimensional space. It maps canonically onto $\mathbf{GL}_n(k)$, and the inverse image of $\mathbf{SL}_n(k)$ is, for Lie, the special linear group (in $n$ variables).

If $V$ is a finite dimensional vector space over $k$ and no basis of $V$ is specified, we let $GL(V)$ be the group of (homogeneous) linear transformations of $V$.

**2.** The group $\mathbf{PGL}_n(k)$ is, by definition, the quotient $\mathbf{GL}_n(k)/k^*$ of $\mathbf{GL}_n(k)$ by its center, the group of dilations $x \mapsto k.x$ ($k \in k^*$). As usual, it may be identified with the group $\mathrm{Aut}(\mathbf{P}_{n-1}(k))$ of projective transformations of $(n-1)$-dimensional projective space $\mathbf{P}_{n-1}(k)$. For $k = \mathbb{C}$, it is called by Lie the general projective group.

**3.** For $k$ of characteristic not two, $\mathbf{O}_n(k)$ is the subgroup of $\mathbf{GL}_n(k)$ leaving the unit quadratic form $\sum_i x_i^2$ invariant, and $\mathbf{SO}_n(k)$ is its subgroup of elements of determinant one.

Let $k = \mathbb{C}$. Lie usually considers the image $\mathbf{PO}_n(k)$ of $\mathbf{O}_n(k)$ in $\mathbf{PGL}_n(k)$, viewed as the group of projective transformations leaving the standard non-degenerate hyperquadric invariant. For $n = 3$, it is the group of projective transformations leaving a non-degenerate conic invariant and was for some time called the "conic section group" (Kegelschnitt Gruppe).

If $F$ is a non-degenerate quadratic form on $k^n$, then $O(F)$ denotes the subgroup of $\mathbf{GL}_n(k)$ leaving $F$ invariant, and $SO(F)$ the subgroup of elements of determinant one in $O(F)$.

**4.** The symplectic group $\mathbf{Sp}_{2n}(k)$ is the subgroup of $\mathbf{GL}_{2n}(k)$ leaving invariant the standard non-degenerate antisymmetric bilinear form

(1) 
$$(x, y) = \sum_1^n \left( x_i . y_{n+i} - x_{n+i} \cdot y_i \right).$$

This terminology was introduced by H. Weyl. For $k = \mathbb{C}$, Lie calls the image $\mathbf{PSp}_{2n}(k)$ of $\mathbf{Sp}_{2n}(k)$ in $\mathrm{Aut}(\mathbf{P}_{2n-1}(k))$ the group of a non-degenerate linear complex.

Similarly, if $J$ is a non-degenerate antisymmetric bilinear form on $k^{2n}$, the subgroup of $\mathbf{GL}_{2n}(k)$ leaving $J$ invariant will be denoted $Sp(J)$. Since $J$ can always be put in the form (1) by a linear transformation, $Sp(J)$ is conjugate to $\mathbf{Sp}_{2n}(k)$ within $\mathbf{GL}_{2n}(k)$.

**5.** As usual,

$$\mathbf{U}_n = \{X \in \mathbf{GL}_n(\mathbb{C}), X.^t\bar{X} = 1\}, \qquad \mathbf{SU}_n = \{X \in \mathbf{SL}_n(\mathbb{C}), X.^t\bar{X} = 1\}$$

are the *unitary* and *special unitary* groups in $n$ variables.

**6.** Let me also recall some standard notation in group theory, also to be used often without further reference.

Let $G$ be a group, $A$ a subset. Then the normalizer $\mathcal{N}A$ or $\mathcal{N}_G A$ and centralizer $\mathcal{Z}A$ or $\mathcal{Z}_G A$ of $A$ in $G$ are defined by

$$\mathcal{N}_G A = \mathcal{N}A = \{g \in G | g.A.g^{-1} = A\},$$
$$\mathcal{Z}_G A = \mathcal{Z}A = \{g \in G | g.a = a.g\,(a \in A)\}.$$

The inner automorphism $x \mapsto g.x.g^{-1}\,(x \in G)$ is denoted by $i_g$, and its effect on $x$ is sometimes written $^g x$.

If $G$ is a Lie group with Lie algebra $\mathfrak{g}$, and $\mathfrak{a}$ is a subset of $\mathfrak{g}$, then similarly

$$\mathfrak{z}(\mathfrak{a}) = \mathfrak{z}_{\mathfrak{g}}(\mathfrak{a}) = \{X \in \mathfrak{g} | \,[X, \mathfrak{a}] = 0\}$$

and

$$\mathcal{N}(\mathfrak{a}) = \mathcal{N}_G(\mathfrak{a}) = \{g \in G | \mathrm{Ad}\, g(\mathfrak{a}) = \mathfrak{a}\},$$
$$\mathcal{Z}\mathfrak{a} = \mathcal{Z}_G(\mathfrak{a}) = \{g \in G | .\mathrm{Ad}\, g(X) = X\,(X \in \mathfrak{a})\},$$

where Ad refers to the adjoint representation, which associates to $g$ the differential at the identity of $i_g$.

# Photo Credits

The American Mathematical Society gratefully acknowledges the kindness of the people and institutions that granted the following photographic permissions:

Photographs of **Richard Brauer** (p. 86) and **Eduard Study** (p. 94) are the property of the American Mathematical Society.

Photographs of **Élie Cartan** (cover and p. 60) and **Claude Chevalley** (cover and p. 148) courtesy of Archives de l'Académie des Sciences, Paris.

Photograph of **Georges de Rham** (p. 85) courtesy of the Archives of Georges de Rham at the University of Lausanne.

Photograph of **Gino Fano** (p. 22) courtesy Dipartimento di Matematica, Università di Torino.

Photograph of **Hans Freudenthal** (p. 137) courtesy of Susan Freudenthal-Litteø and Mirjam Freudenthal.

Photographs of **Adolf Hurwitz** (p. 13), **Sophus Lie** (cover and p. 2), and **Emile Picard** (p. 100) courtesy of Staatliche Museen zu Berlin, Kunstbibliothek.

Photograph of **Ellis R. Kolchin** (cover and p. 166) courtesy of Mrs. Kate Kolchin.

Photograph of **Ludwig Maurer** (p. 102) courtesy of the Universitätsbibliothek Tübingen.

Photograph of **E. H. Moore** (p. 12) courtesy of the University of Chicago Archives.

Photograph of **J. F. Ritt** (p. 168) courtesy of the National Academy of Sciences of the U.S.A.

Photograph of **Issai Schur** (p. 32) courtesy of Mathematisches Forschungsinstitut Oberwolfach.

Photograph of **B. L. van der Waerden** (p. 135) courtesy of the University Museum of Groningen, The Netherlands.

Photograph of **André Weil** (p. 123) courtesy of Sylvie Weil.

Photograph of **Hermann Weyl** (cover and p. 30) courtesy of the Institute for Advanced Study.

# CHAPTER I

# Overview

## §1. Lie's theory

The theory of "finite and continuous groups" was developed by the Norwegian mathematician Sophus Lie from 1873 on and, after many publications, codified in the 3-volume treatise written in collaboration with F. Engel [LE]. A historical account of these beginnings, and of the genesis of the theory, may be found in [B] and, in much greater detail, in several publications of T. Hawkins, notably [H1] to [H4], so I shall be rather brief.

**1.1.** The notion of one-parameter group generated by "infinitely many repetitions" of an infinitesimal transformation was Lie's starting point. Consider first the linear case. Let $A \in \mathbf{M}_n(\mathbb{C})$. It generates the one-parameter group consisting of the transformations

$$(1) \qquad e^{tA} = \sum_{n \geq 0} \frac{t^n . A^n}{n!} \qquad (t \in \mathbb{C}).$$

The series is absolutely convergent for all $t$, and it satisfies the relation

$$(2) \qquad e^{sA} . e^{tA} = e^{(s+t)A} \qquad (s, t \in \mathbb{C}).$$

It is clear that a vector $Z \in \mathbb{C}^n$ is fixed under all transformations $e^{tA}$ if and only if it is annihilated by $A$. The search for fixed points for the one-parameter group is then reduced to a problem in linear algebra.

More generally, a vector field $X$ on $\mathbb{C}^n$, written locally as $X = \sum_i X_i(x) . \frac{\partial}{\partial x_i}$, yields differential equations

$$(3) \qquad \frac{dx_i}{dt} = X_i(x_1, \dots, x_n) \qquad (i = 1, \dots, m).$$

If $x(x_o, t) = \{x(x_o, t)_1, \dots, x(x_o, t)_n\}$ denotes the integral curve starting at $x_o$ for $t = 0$, then we have, in analogy with (1),

$$(4) \qquad x(x_0, t)_i = \sum_{j \geq 0} \frac{X^j x_i}{j!} := \left( e^{tX}(x_o) \right)_i$$

and, for suitably small values of $s, t$,

$$(5) \qquad e^{sX}(e^{tX}(x_o)) = e^{(s+t)X}(x_o).$$

Again the fixed points of these transformations are the zeroes of the vector field. In our first example, the components of the vector field defined by $A$ are $X_i = \sum_j a_{ij} x_j$, where the $a_{ij}$ are the coefficients of $A$.

Sophus Lie

Among the works done by Lie during the period 1869–1873, three seem particularly significant for the genesis of his theory. First was his collaboration with Felix Klein (1869-71) and their discussions pertaining to what became Klein's Erlanger Programm (1872). In two Comptes Rendus notes [KL1], Klein and Lie studied the orbits of some continuous commutative groups of projective transformations, one-parameter groups first, but then also 2- or 3-dimensional ones. In particular, they considered the orbits of the group of projective transformations of the 3-dimensional projective space $\mathbf{P}_3(\mathbb{C})$ which leave fixed the four vertices of a tetrahedron, i.e., up to conjugation, the projective transformations defined by the group of diagonal matrices in $\mathbf{GL}_4(\mathbb{C})$.

Around that time (1870-1) Lie also realized that some classical methods of integration of various ordinary differential equations could be unified by noting that they relied on the existence of a one-parameter group of transformations leaving the equation invariant. Some examples are the subject matter of §7 in [KL3], the contents of which are, according to Klein (p. 415 in vol. 1 of his Collected Papers), exclusively due to Lie.

However, all this is commutative. But, in Fall 1873, Lie outlined a general theory of not necessarily commutative groups. It seems that his work on Jacobi's integration theory and his interpretation of the Poisson bracket as the effect of an infinitesimal transformation on a function played a decisive role in this shift, by allowing him to gain experience in the handling of families of non-commuting infinitesimal transformations which occurred naturally.

Since I view this as an important step in the birth of the theory, I want to give a sketch of it. However, this will not be used nor referred to in the later chapters (and can therefore be skipped).

**1.2.** Let $(x_i, p_i)$ $(i = 1, \ldots, n)$ be coordinates in the cotangent space $T^*\mathbb{C}^n$ to $\mathbb{C}^n$.

The Poisson bracket $(u, v)$ of two (holomorphic) functions on $T^*\mathbb{C}^n$ is by definition

$$(5) \qquad (u, v) = \sum_i \frac{\partial u}{\partial p_i} \frac{\partial v}{\partial x_i} - \frac{\partial u}{\partial x_i} \frac{\partial v}{\partial p_i}.$$

It is antisymmetric and satisfies the Jacobi identity

$$(6) \qquad \big((u, v), w\big) + \big((w, u), v\big) + \big((v, w), u\big) = 0.$$

Lie associated to $u$ the infinitesimal transformation

$$(7) \qquad D_u = \sum_i \frac{\partial u}{\partial p_i} \frac{\partial}{\partial x_i} - \frac{\partial u}{\partial x_i} \frac{\partial}{\partial p_i};$$

hence

$$(8) \qquad (u, v) = D_u(v),$$

and the Jacobi identity is equivalent to

$$D_u \circ D_v - D_v \circ D_u := [D_u, D_v] = D_{(u,v)}.$$

Now let $F$ be a holomorphic function on $T^*\mathbb{C}^n$. Consider the partial differential equation in the unknown function $z$

$$(9) \qquad F\left(x_1, \ldots, x_n, \frac{\partial z}{\partial x_1}, \ldots, \frac{\partial z}{\partial x_n}\right) = 0.$$

Jacobi considered functions $u$ such that $(F, u) = 0$. The existence of such a function allows one to simplify the integration of (9). If $v$ is another such function, then so is $(u, v)$, by the Jacobi identity. Jacobi asserted that in the "generic case", given $u, v$ it is possible by iteration of Poisson brackets to find up to $n$ such functionally independent solutions; the more of them there are, the simpler the integration of (9) is. This then leads to the search for maximal sets of such functions, i.e. functions $u_1, \ldots, u_s$ such that

$$(10) \qquad (u_i, u_j) = \Omega_{ij}(u_1, \ldots, u_s),$$

where the $\Omega_{ij}$ are analytic functions in $s$ variables. In particular, if

$$(11) \qquad (u_i, u_j) = \sum_k c_{ij}{}^k(x_1, \ldots, x_n) u_k,$$

the $u_i$ span a Lie algebra (infinite dimensional if the $c_{ij}{}^k$ are not constant, and $s \geq 2$). Such sets of functions are called "groups" by Lie, and, later, "function groups".

Note that the equality $(F, u) = 0$ can also be written as $D_u(F) = 0$, so that, in agreement with Lie's idea about ordinary differential equations, the existence of infinitesimal transformations leaving (9) invariant helps to simplify its integration, a principle already mentioned above, which became a dominating motivation for

Lie. He later viewed his considerations on function groups as a precursor to his theory of finite and continuous groups.

**1.3.** It is a simple matter nowadays to define a real or complex Lie group as a real or complex analytic manifold $G$ endowed with a group structure such that the map $G \times G \to G$ given by $(x, y) \mapsto x.y^{-1}$ is analytic. But Lie could not say that, and this definition differs in two respects from his notion of "finite and continuous group". First, he considered only transformation groups. The notion of abstract group was not familiar at the time, and even later, when it had become more widespread, F. Klein had some misgivings about putting it in the foreground (Entwicklung der Mathematik I, 335-6), for fear of losing contact with applications. But Lie had two notions of isomorphism, one of transformation groups, which is recalled in IV, §5, and involves the space operated upon, and another one called "Gleichzusammensetzung", which focused attention on the law of composition; and he knew that each isomorphism class could be represented by the group acting on itself by left or right translations (the "parameter groups"), so the difference here is more of terminology than of substance. The second one is more important—his "groups" were local, so that, in present-day terminology, we would say that his basic notion was that of a germ of a transformation group.

Let $U$ be an open subset of $\mathbb{C}^n$ with coordinates $x_i$, and $V$ a neighborhood of the origin in $\mathbb{C}^p$ with coordinates $a_1, \ldots, a_p$. $G$ is a set of local transformations of $U$ parameterized by $V$. More precisely, we are given $n$ holomorphic functions $f_1, \ldots, f_n$ on $U \times V$. For fixed $a \in V$, the map

$$(12) \qquad\qquad g_a : x \mapsto f(x, a) \qquad (x \in U, a \in V),$$

where $f(x, a)$ has coordinates $f_i(x, a)$, is a local transformation of $U$, i.e. from some open subset of $U$ to another open subset depending on $a$. Moreover, there exist (holomorphic) functions $\varphi_i$ on $V \times V$ such that

$$(13) \qquad\qquad g_b\big(g_a(x)\big) = g_{\varphi(a,b)}(x),$$

whenever both sides are defined. Here

$$a = (a_1, \ldots, a_p), \quad b = (b_1, \ldots, b_p), \quad \varphi = (\varphi_1, \ldots, \varphi_p).$$

It is also assumed, although Lie did not do so initially, that $G$ contains the identity transformation, say for $a$ equal to the origin 0 in $\mathbb{C}^p$. Then, for $V$ small enough, $G$ also contains the inverse transformation to $a$. By taking first derivatives with respect to $a$ at the origin, for $b = 0$, Lie got infinitesimal transformations and showed that the one-parameter group generated by any of them belongs to $G$ (for sufficiently small values of the parameter). These infinitesimal transformations form a vector space that I shall denote by $\mathfrak{g}$. Then, by consideration of commutators and second derivatives, Lie showed that $\mathfrak{g}$ is stable under the bracket of infinitesimal transformations, i.e. $\mathfrak{g}$ is a Lie algebra in the usual terminology. This was not Lie's terminology; in fact, Lie and his contemporaries often did not make a clear terminological difference between groups (really local groups) and their Lie algebras.

The correspondence between Lie groups and Lie algebras was shown to go both ways. In particular, subalgebras (resp. ideals) of $\mathfrak{g}$ are the Lie algebras of Lie subgroups (resp. normal Lie subgroups). Also, any abstract Lie algebra is the Lie algebra of a group. This is expressed in three main theorems and their converses ([LE], III, Chapt. 25). Some remarks on the second one will be given in V, 4.6, 4.7.

## §2. Lie algebras

**2.1.** In principle, Lie's theory reduced problems on Lie groups, of an analytic nature, to algebraic problems on Lie algebras, and this led to a study of Lie algebras per se. However, this was not a direction which Lie was inclined to pursue, and it was left to other mathematicians. The first fundamental contributions are due to Wilhelm Killing, in four papers [K2], having as one of their main goals the classification of simple Lie algebras. When Killing started, he was aware of two infinite classes, the Lie algebras of the special linear group $\mathbf{SL}_n(\mathbb{C})$ and of the orthogonal groups $\mathbf{O}(n, \mathbb{C})$. At one point, it seemed probable to him that they were the only ones ([K1], letter 12, April 1886), but two years later he had obtained the classification, as we know it ([K1], letter 34, Febr. 1888; [K2 b]): four infinite classes (the classical Lie algebras) and only finitely many other Lie algebras, now called the exceptional Lie algebras and denoted $\mathbf{E}_6, \mathbf{E}_7, \mathbf{E}_8, \mathbf{F}_4, \mathbf{G}_2$, of respective dimensions 78, 133, 248, 52 and 14. Actually, he did not prove the existence, except for $\mathbf{G}_2$, but he described all possibilities for rank, dimension, root systems, and he found six such algebras, not noticing that two of them were isomorphic (of type $\mathbf{F}_4$). Even though his proofs were not complete, a fact that he deplored himself, this was an amazing achievement. See [H1] for a discussion of these contributions.

It has been remarked that, as far as terminology is concerned, posterity has not been kind to him: Cartan subalgebras, Weyl groups, fundamental reflections, roots, and the Coxeter transformation first appeared in his papers in some form. On the other hand, what now goes by his name, the "Killing form" seems to be a misnomer, and it may well be that I am the culprit. Cartan, Chevalley and Weyl never used this terminology. Once, J.J. Duistermaat and J.A.C. Kolk pointed out to me that, to their knowledge, its first occurrence is in a paper of mine (Sém. Bourbaki, Exp. 33, May 1951). I do not remember why I chose it, though I probably felt I was innovating, since it is between quotation marks. It is rather likely that discussions with some members of Bourbaki had influenced me, but I cannot blame it directly on Bourbaki, since "Killing form" appears in Bourbaki drafts only from 1952 on. It is true that Killing was the first to remark that the coefficients of the characteristic equation (of a regular semisimple element), i.e. the elementary symmetric functions of the roots, are invariant under the adjoint group, but he did not make much use of that remark and did not single out the sum of the squares of the roots, of which Élie Cartan made such fundamental use in his thesis (1894). It would be more correct to speak of the Cartan form.

Cartan's thesis [C] is the second fundamental contribution to Lie algebra theory. There Cartan notably gives a rigorous treatment of Killing's work and proves the existence of all the exceptional simple Lie algebras: rather he reduces the proof to the checking of many Jacobi identities, which are too numerous to be included and are left to the reader. A complete a priori proof was produced first, independently, by Harish-Chandra and Chevalley around 1948 (see VII for references). A basic result in Cartan's thesis is a criterion for a Lie algebra to be *semisimple* (i.e., a direct sum of (commuting) simple non-commutative subalgebras): namely, the Killing form is non-degenerate.

**2.2.** This period also saw the first studies of the universal enveloping algebra. Let $\mathfrak{g}$ be a (complex) Lie algebra. Recall that the universal enveloping algebra $U(\mathfrak{g})$

is, by definition, the quotient of the tensor algebra

$$T^*(\mathfrak{g}) = \bigoplus_{i \geq 0} T^i \mathfrak{g},$$

where $T^i \mathfrak{g}$ is the $i$-th tensor power of $\mathfrak{g}$, by the ideal spanned by the expressions $x \otimes y - y \otimes x - [x, y]$ $(x, y \in \mathfrak{g})$. It has a filtration $\{F_j U(\mathfrak{g})\}_{j \geq 0}$ by the images of the partial sums $\bigoplus_{i \leq j} T^i \mathfrak{g}$. The Poincaré-Birkhoff-Witt theorem asserts that the associated graded ring is the symmetric algebra $S(\mathfrak{g})$ of $\mathfrak{g}$.

All this was proved (in a different terminology) for the Lie algebra $\mathfrak{gl}_n$ of $\mathbf{GL}_n(\mathbb{C})$ in four papers in the 1880's, by Capelli, who moreover determined the center of $U(\mathfrak{g})$ and showed it is a polynomial algebra on $n$ explicitly given generators (see III, note 22). These papers are at the origin of the Capelli identity, which plays an important role in Weyl's book, "The classical groups", but it seems that Weyl read only the third one ([C] in III). At any rate, it is the only one he quotes, and all this was completely forgotten for almost a century. (As pointed out in Chapter III, I was made aware of it by C. Procesi.) In 1900 Poincaré studied $U(\mathfrak{g})$ in general, identified to the algebra of left invariant differential operators on a group $G$ with Lie algebra $\mathfrak{g}$, and more or less proved the Poincaré-Birkhoff-Witt theorem. In turn this was forgotten and rediscovered, with a rigorous proof in the case of Witt, much later. We refer to W. Schmid's paper [S] for a detailed discussion.

**2.3.** Let $\mathfrak{g}$, $G$ be as before and $X, Y \in \mathfrak{g}$. Consider a neighborhood $U$ of the origin in $\mathfrak{g}$ such that the exponential map is an isomorphism (of manifolds) onto a neighborhood $V$ of the identity in $G$. Given $X, Y \in U$ sufficiently small, there then exists a unique $Z \in U$ such that

$$e^X . e^Y = e^Z.$$

The Campbell-Hausdorff formula expresses $Z$ as a series of iterated brackets in $X, Y$, with universal coefficients. A first version appeared in 1906. Later on, after the convergence of the series was ascertained, this allowed one to define Lie groups over complete valued fields, in particular $p$-adic Lie groups, as well as Banach Lie groups.

**2.4.** Twenty years after his thesis, É. Cartan came back to the theory of semisimple Lie algebras in three important papers. In one he outlined a description of all irreducible representations of a complex simple Lie algebra, labeled by their highest weights. In the second one, he considered real simple Lie algebras, determined the real forms of a given complex Lie algebra $\mathfrak{g}$, i.e. the real Lie algebras $\mathfrak{g}_o$ such that $\mathfrak{g}_o \otimes_{\mathbb{R}} \mathbb{C}$ is isomorphic to $\mathfrak{g}$, and in the third one, he gave criteria under which the restriction to $\mathfrak{g}_o$ of an irreducible representation $\mathfrak{g}$ is orthogonal, symplectic, or neither. We refer to [H4].

The next fundamental advance in the theory of semisimple Lie algebras is due to Hermann Weyl, in the framework of a far-reaching combination of infinitesimal and global points of view, alluded to in 3.3 and discussed in Chapter III, and more extensively in [H4].

## §3. Globalizations

**3.1.** As recalled in §1, the original Lie theory was local. However, this did not prevent the early practitioners from considering global objects which they met naturally, in particular some classical groups. A first example of a global result is the discovery by Engel that the exponential mapping is surjective for

$$\mathbf{PSL}_2(\mathbb{C}) = \mathbf{SL}_2(\mathbb{C})/\{\pm 1\}$$

but not for $\mathbf{SL}_2(\mathbb{C})$. Thus two groups which are the same locally, i.e. from the point of view of Lie theory, behave globally differently. When a mathematician, Slocum, claimed to have found some difficulties or inconsistencies in Lie's theory, Engel pointed out in his review of his papers (Fortschritte der Mathematik, **31**, 1900, 148-150) that the author had overlooked the fact that the general theory proves only theorems valid in a neighborhood of a given point, and added, obviously with some irritation, that Lie cannot be required to repeat this each time. Of course, the facts that finite and continuous groups were usually local, sometimes global, and that group also occasionally meant Lie algebra, did not help to clarify matters.

**3.2.** Lie and his contemporaries were familiar with the classical groups, defined globally, of course, as well as with affine or projective varieties. Linear actions on affine varieties or actions by projective transformations of classical groups, pertaining notably to invariant theory, provided an abundant supply of global (algebraic) transformation groups, without even the necessity of a formal definition.

That such investigations had a global aspect not in the purview of the local Lie theory was clearly stated by E. Study in 1897, as final remarks (reproduced below in part and in free translation) in the paper where he proves the first and second main theorems of invariant theory for the orthogonal group [S2]. They were added to comply with a request of Lie to compare what Study had done with what Lie's methods allowed one to achieve:

> As Mr. Lie points out himself, his theory answers all questions on invariants, differential invariants, differential parameters, as well as on criteria about *analytic* invariance equivalence in a group.... But it does not follow that the Lie theory contains the *algebraic* theory of the invariants of the projective group or of its algebraic subgroups. It should be noted that these theories, still in the beginning stage, span the whole space, while the general theory of transformation groups with its "analytic" invariant concept of invariants and the corresponding equivalence criteria is limited to the consideration of a neighborhood of a point and, because of its generality, cannot do more....

Linear algebraic groups were studied from various points of view in the 19th century, over $\mathbb{C}$ (see Chapter V), and were then completely ignored for almost half a century. Their investigation was resumed in the 1940s and was developed much further, over arbitrary fields. This second phase is the subject matter of Chapter VI, completed by Chapters VII and VIII.

**3.3.** Another globalization for real or complex groups took place in the framework of manifolds, differential geometry and topology, initiated by A. Hurwitz, who introduced integration over a compact group and the "unitarian trick" to prove theorems on the complexification of these groups (see Chapter II). This was picked up by Schur (1922), then Weyl (1924) and Cartan, and led the latter to build up a theory of Riemannian symmetric spaces (1925-30). Some highlights of these developments are described in Chapters III and IV (see also [H4]).

# References for Chapter I

[B]     N. Bourbaki, *Note historique*, in Groupes et Algèbres de Lie, Chaps. 2, 3, Hermann, Paris, 1972, pp. 286–300.

[C]     É. Cartan, Sur la structure des groupes de transformations finis et continus, Thèse, Nony, Paris, 1894; Oeuvres Complètes I₁, 137–287.

[H1]    T. Hawkins, *Wilhelm Killing and the structure of Lie algebras*, Archive for History of Exact Sciences **26** (1982), 127–192.

[H2]    T. Hawkins, *Line geometry, differential equations and the birth of Lie's theory of groups*, in The History of Modern Mathematics (I. J. MacCleary and D. Rowe, eds.), Vol. 1, Academic Press, New York, 1988, 275–327.

[H3]    T. Hawkins, *Jacobi and the birth of Lie theory*, Archive for History of Exact Sciences **42** (1991), 187–278.

[H4]    T. Hawkins, The emergence of the theory of Lie groups: An essay on the history of mathematics, 1869–1926, Springer, 2000.

[K1]    W. Killing, Briefwechsel mit Friedrich Engel, Dokumente zur Geschichte der Mathematik **9**, Deutsche Math. Ver., Viehweg, 1997.

[K2]    W. Killing, *Die Zusammensetzung der stetigen endlichen Transformationsgruppen*, I, II, III, IV, Math. Annalen **31** (1886), 252–290; **33** (1888), 1–48; **34** (1889), 57–122; **36** (1890), 161–189.

[KL1]   F. Klein and S. Lie, *Sur une certaine famille de courbes et de surfaces*, C.R. Acad. Sci. Paris **70** (1870), 1222–6, 1275–9; Klein, Ges. Math. Abh. 1, 415–423; Lie, Ges. Abh. 1, 78–85.

[KL2]   F. Klein and S. Lie, *Ueber diejenigen ebenen Kurven, welche durch ein geschlossenes System von einfach unendlich vielen vertauschbaren linearen Transformationen in sich übergehen*, Math. Annalen **4**, 1871, 50–84; Klein, Ges. Math. Abh. 1, 424–459; Lie, Ges. Abh. 1, 229–285.

[LE]    S. Lie (with the collaboration of F. Engel), Theorie de Transformationsgruppen, 3 vols., Teubner, Leipzig, 1888, 1890, 1893.

[S]     W. Schmid, *Poincaré and Lie groups*, Bull. (N.S.) Amer. Math. Soc. **6**, 1982, 175–186.

[S2]    E. Study, *Ueber die Invarianten der projectiven Gruppe einer quadratischen Mannigfaltigkeit von nicht verschwindender Discriminante*, Ber. Verh. Kgl. Sächs. Akad. Wiss. Leipzig, Math.-Phys. Klasse **49**, 1897, 443–61.

CHAPTER II

# Full Reducibility and Invariants for $\mathbf{SL}_2(\mathbb{C})$

## §1. Full reducibility, 1890–96

**1.** Let $G$ be a group, $V$ a finite dimensional vector space over a commutative field $k$ (mostly $\mathbb{C}$ in this chapter), $n$ the dimension of $V$ and $\pi$ a representation of $G$ in $V$, i.e. a homomorphism $G \to GL(V)$ of $G$ into the group $GL(V)$ of invertible linear transformations of $V$. The choice of a basis of $V$ provides an isomorphism of $V$ with $k^n$, of $GL(V)$ with the group $\mathbf{GL}_n(k)$ of $n \times n$ invertible matrices with coefficients in $k$, and a realization of $\pi$ as a matrix representation:

$$g \mapsto \pi(g) = \big(\pi(g)_{i,j}\big)_{1 \leq i,\, j \leq n}.$$

Two main problems pertaining to this situation were considered already in the 19th century, in various special cases, for $k = \mathbb{C}$.

I) **Invariants.** Let $k[V]$ be the space of polynomials on $V$ with coefficients in $k$, and $k[V]_m$ ($m \in \mathbb{N}$) the space of homogeneous polynomials of degree $m$. The group $G$ acts via $\pi$ on $k[V]$ by the rule

$$g \circ P(v) = P\big(\pi(g)^{-1}.v\big) \qquad (v \in V,\, P \in k[V],\, g \in G)$$

leaving each $k[V]_m$ stable. (The argument on the right-hand side will usually be written $g^{-1}.v$ if there is no ambiguity about $\pi$.)

Let $k[V]^G$ be the space of polynomials which are invariant under $G$, i.e. which are constant on the orbits of $G$. It is an algebra over $k$, and the (first) problem of invariant theory is to know whether it is finitely generated, as a $k$-algebra.

In fact, this is only the first of a series of questions. If the answer is affirmative then, given a minimal generating set of $k[V]^G$, one may want to know whether the module of polynomial relations between its elements, the syzygies of the first kind, is itself finitely generated over the invariants and, if so, whether the relations between the relations, the syzygies of the second kind, also form a finitely generated module, and so on. Not going beyond the invariants themselves in this chapter, I shall not describe syzygies more precisely (see, e.g., [**P**] for more details).

II) **Full reducibility.** The representation $(\pi, V)$ is said to be reducible if there exists a $G$-invariant subspace $W \neq \{0\}, V$, irreducible otherwise, and fully or completely reducible if any $G$-invariant subspace has a $G$-invariant complement. If so, $V$ can be written as a direct sum of $G$-invariant irreducible subspaces. One is interested in groups having classes of fully reducible representations or in finding families of groups all of whose representations over a given $k$ are fully reducible.

Revised and augmented version of a paper with the same title published in l'Enseignement Math. (2) **44** (1998), 71-90.

In this chapter, I shall discuss the history of these two problems mainly for one group, the group $\mathbf{SL}_2(\mathbb{C})$ of $2 \times 2$ complex invertible matrices of determinant one, for $k = \mathbb{C}$ and holomorphic representations, i.e. representations in which the $\pi(g)_{i,j}$ in (1) are holomorphic functions in the entries of $g$. Occasionally, some remarks will be made on other groups, to put certain results in a more general context, or for historical reasons, but our main focus of attention will still be $\mathbf{SL}_2(\mathbb{C})$. Even so restricted, this history is surprisingly complicated, in part because the principal contributors were sometimes not aware of other work already done. In one case, it seems even that one of them had forgotten some of his own.

**2.** The irreducible representations of $\mathbf{SL}_2(\mathbb{C})$ were determined by Sophus Lie. As we know, there is for each $m \in \mathbb{N}$, up to equivalence, one representation of degree $m + 1$ in the space, to be denoted $V_m$, of homogeneous polynomials of degree $m$ on $\mathbb{C}^2$, acted upon via the identity representation of $\mathbf{SL}_2(\mathbb{C})$.

In fact, Lie formulated his result differently, more geometrically [**LE**]. For him, a representation is not linear, but projective, i.e. a homomorphism into the group of projective transformations of some complex projective space $\mathbf{P}_m(\mathbb{C})$. As usual, $\mathbf{P}_m(\mathbb{C})$ is viewed as the quotient of $\mathbb{C}^{m+1} - \{0\}$ by dilations. This identifies the group Aut $\mathbf{P}_m(\mathbb{C})$ of projective transformations of $\mathbf{P}_m(\mathbb{C})$ with the quotient $\mathbf{GL}_{m+1}(\mathbb{C})/\mathbb{C}^*$ of $\mathbf{GL}_{m+1}(\mathbb{C})$ by the non-zero multiples of the identity matrix, or also with $\mathbf{PSL}_{m+1}(\mathbb{C}) = \mathbf{SL}_{m+1}(\mathbb{C})/\text{center}$, i.e. modulo the group of multiples $c.\text{Id}$ of the identity matrix, where $c^{m+1} = 1$. Let $B$ be the group of upper triangular matrices in $\mathbf{SL}_2(\mathbb{C})$. The quotient $\mathbf{SL}_2(\mathbb{C})/B = C$ is a smooth complete rational curve, i.e. a copy of $\mathbf{P}_1(\mathbb{C})$. The group $B$ is solvable and connected; therefore, by Lie's theorem, it has a fixed point in any projective representation, and so, if this point is not fixed under $G$, its orbit is a copy of $C$. Lie looks for the cases where such a $C$ is "as curved as possible" ("möglichst gekrümmt"), meaning, not contained in a proper projective subspace. It is also a fact, implicitly assumed by Lie, that the action of $\mathbf{SL}_2(\mathbb{C})$ on such a curve is always induced by projective transformations of the ambient projective space. Therefore the search for smooth rational complete curves in projective spaces which are "as curved as possible", up to projective transformations, is tantamount to the classification of the irreducible holomorphic representations of $\mathbf{SL}_2(\mathbb{C})$ (linear or projective, there is no essential difference since $\mathbf{SL}_2(\mathbb{C})$ is simply connected), up to equivalence. Given $m \geq 1$, the smooth projective rational curves not contained in a proper projective subspace, of smallest degree (number of intersection points with a generic hyperplane), called rational normal curves, form in $\mathbf{P}_m(\mathbb{C})$ one family, operated upon transitively by Aut $\mathbf{P}_m(\mathbb{C})$, and the degree is $m$. The irreducible representations are those in which the $G$-orbit of a fixed point of $B$ has degree $m$. It is then the only closed orbit of $G$. In [**LE**], pp. 785-6, Lie reports that E. Study has proved the full reducibility of the representations of $\mathbf{SL}_2(\mathbb{C})$ (again, in an equivalent projective formulation which I do not recall here, but which will appear in §**13**), but he does not describe the proof because it is long, not quite correct (but this was fixed up by Engel), and simplifications are hoped for. He adds it is very likely to be true for representations of $\mathbf{SL}_n(\mathbb{C})$ for any $n \geq 2$. In fact, Study had made this conjecture in a letter to him, even more generally for semisimple groups, see V, §1.

**3.** In his thesis, Élie Cartan provides a proof of full reducibility [**Cr1**]. It is algebraic, dealing with Lie algebras, so it establishes in fact the full reducibility of the representations of the Lie algebra $\mathfrak{sl}_2(\mathbb{C})$ of $\mathbf{SL}_2(\mathbb{C})$, but this is equivalent. He does not state the theorem explicitly, however. The proof is embedded (pp. 100-2) in that of another theorem, due to F. Engel, to the effect that a non-solvable Lie algebra always contains a copy of $\mathfrak{sl}_2(\mathbb{C})$. But a statement is given at the beginning of Chapter VII (p. 116) with a reference to the passage just quoted for the proof.

In 1896, G. Fano, who knew about Study's theorem through [**LE**] and was surely not aware of Cartan's proof, maybe not even of Cartan's thesis, gave an entirely different one in the framework of algebraic geometry, using the properties of "rational normal scrolls" [**F**].

He first made two remarks of an algebraic nature which simplify the argument.

a) An induction on the length of a composition series shows that it suffices to carry out the proof when the space $E$ of the given representation contains one irreducible $G$-invariant subspace $F$ such that $E/F$ is also irreducible. In other words, since the $V_m$'s are the irreducible representations, up to equivalence, it suffices to consider the case of an exact sequence

$$(2) \qquad 0 \to V_m \to E \to V_n \to 0$$

(again, in projective language, see §**12**).

b) In (2), it may be assumed that $m \geq n$. If $m < n$, this is seen by going over to the contragredient representations.

$$(3) \qquad 0 \to V_n^* \to E^* \to V_m^* \to 0.$$

noting that $E$ is fully reducible if and only if $E^*$ is, and that for each $m$, the representation $V_m$ is self-contragredient. This also shows that it suffices to consider the case where $m \leq n$. In fact, this last reduction allows for a considerable simplification in Cartan's proof, whereas the reduction to $m \geq n$ is the one Fano uses (see §§**12** and **13** for more details).

## §2. Averaging. The invariant theorem

**4.** Another development came from a different source: the idea of averaging over a finite group. In 1896 it was shown that a finite group $G$ of linear transformations always leaves invariant a positive non-degenerate hermitian form. It was stated by A. Loewy without proof [**L**], and by E.H. Moore, who announced it at a meeting, communicated his proof to F. Klein, and published it later [**Mo**]. This argument is the now standard one (and Loewy stated later it was his, too): Starting from a positive non-degenerate hermitian form $H(\ ,\ )$ on $\mathbb{C}^n$, one considers the sum $H^o(\ ,\ )$ of its transforms under the elements of $G$:

$$(1) \qquad H^o(x,y) = \sum_{g \in G} H(g^{-1}.x,\ g^{-1}.y),$$

which Moore calls a *universal invariant* for $G$. It is obviously $G$-invariant and positive non-degenerate. This construction seems quite obvious, but Klein viewed it as interesting enough to make it the subject matter of a communication to the German Math. Soc. [**K**]. For Moore it was an application of a "well-known group theoretic process". In [**Lo**] and [**Mo**], this fact is used to show that a linear transformation of finite order is diagonalizable (which was known, but with more complicated proofs). A bit later, H. Maschke, a colleague of Moore at Chicago, made use of this universal invariant to establish the full reducibility of linear representations of a finite group [**Ma**]. The standard argument is of course to point out that if $V$ is a $G$-invariant

Elias H. Moore

subspace, then so is its orthogonal complement with respect to $H^o$. This is the gist of Maschke's proof, but it was presented in a rather complicated manner.

**5.** The idea of averaging was pushed much further by A. Hurwitz in a landmark paper [**Hu**]. He was interested in the invariant problem. He starts by saying it is well-known that one can construct invariants for a finite linear group by averaging, but he is concerned with certain infinite groups, specifically $\mathbf{SL}_n(\mathbb{C})$ and the special complex orthogonal group $\mathbf{SO}_n(\mathbb{C})$ $(n \geqq 2)$.

Hurwitz recalls first that if $G$ is a finite linear group acting on $\mathbb{C}^n$ and $P$ is a polynomial on $\mathbb{C}^n$, then the polynomial $P^\natural$ defined by

$$(2) \qquad P^\natural(x) = N^{-1} . \sum_{g \in G} P(g^{-1}.x) \qquad (x \in \mathbb{C}^n),$$

where $N$ is the order of $G$, is obviously invariant under $G$ (the factor $N^{-1}$ is inserted so that $P^\natural = P$ if $P$ is invariant). If now $G$ is infinite, the initial idea is to replace the summation in (2) by an integration, with respect to a measure invariant under translations. However, if the group is not compact (Hurwitz says bounded), this integral may well diverge. To surmount that difficulty, Hurwitz used a procedure which turned out later to be far-reaching, namely to integrate over a compact subgroup $G_u$, which insures convergence, but choosing it big enough so that invariance under $G_u$ implies invariance under the whole group, an argument

Adolf Hurwitz

later called the "unitarian trick" by H. Weyl [**W1**]. This is carried out for

$$\mathbf{SU}_n \subset \mathbf{SL}_n(\mathbb{C}) \quad \text{and} \quad \mathbf{SO}_n \subset \mathbf{SO}_n(\mathbb{C}).$$

I describe it for $G = \mathbf{SL}_2(\mathbb{C})$ and $G_u = \mathbf{SU}_2$. The latter is

$$(3) \qquad G_u = \mathbf{SU}_2 = \left\{ \begin{pmatrix} a & b \\ \bar{b} & \bar{a} \end{pmatrix}, a, b \in \mathbb{C}, |a|^2 + |b|^2 = 1 \right\}.$$

Write $a = x_1 + ix_2$, $b = x_3 + ix_4$, with the $x_i$ real. Then $\mathbf{SU}_2$ may be identified to the unit 3-sphere

$$(4) \qquad \mathbf{S}^3 = \{(x_1, \ldots, x_4) \in \mathbb{R}^4, x_1^2 + \cdots + x_4^2 = 1\}.$$

It can be parametrized by the Euler angles $\varphi, \psi, \theta$:

$$(5) \qquad \begin{aligned} x_1 &= \cos\psi . \cos\varphi . \cos\theta \\ x_2 &= \cos\psi . \cos\varphi . \sin\theta \\ x_3 &= \cos\psi . \sin\varphi \qquad (|\varphi|, |\psi| \le \pi/2, \theta \in [0, 2\pi]) \\ x_4 &= \sin\psi . \end{aligned}$$

The measure

$$(6) \qquad dv = \cos\psi . \cos\varphi . d\psi . d\varphi . d\theta$$

is then invariant under translations, and the volume of $\mathbb{S}^3$ with respect to $dv$ is $8\pi$. Let $\sigma : G \to \mathbf{GL}_N(\mathbb{C})$ be a holomorphic linear representation of $G$ and $P$ be a polynomial on $\mathbb{C}^N$. Integration on $G_u$ yields the polynomial $P^\natural$ given by

$$
(7) \qquad
\begin{aligned}
P^\natural(x) &= (8\pi)^{-1} \int_{G_u} P(g^{-1}.x) dv \\
&= (8\pi)^{-1} \int_{-\pi/2}^{\pi/2} \cos\psi . d\psi \int_{\pi/2}^{\pi/2} \sin\varphi . d\varphi \int_0^{2\pi} P(g^{-1}.x) d\theta
\end{aligned}
$$

$(x \in \mathbb{C}^N)$. It is invariant under the action of $G_u$, and the claim is that it is even invariant under $G$ itself. Given $x \in \mathbb{C}^N$, consider the function $\mu_x$ on $G$ given by

$$
(8) \qquad \mu_x(g) = P^\natural(g^{-1}.x) - P^\natural(x) \qquad (g \in G).
$$

It is holomorphic in the entries of $g$, and is identically zero for $g \in G_u$. This implies that it is identically zero on $G$. To establish this, it suffices to show that it is zero on a neighborhood $U$ of the identity. The tangent space to $G$ (resp. $G_u$) at the identity is the complex (resp. real) vector space $\mathfrak{g}$ (resp. $\mathfrak{g}_u$) of $2 \times 2$ complex (resp. skew-hermitian) matrices of trace zero. Take $U$ small enough so that it is the isomorphic image of a neighborhood $U_0$ of the origin in $\mathfrak{g}$ under the exponential mapping. Let $\tilde{\mu}_x$ be the pull-back of $\mu_g|U$ by the inverse mapping. Then $\tilde{\mu}_x$ is a holomorphic function on $U_o$ which is zero on $U_o \cap \mathfrak{g}_u$. But $\mathfrak{g}_u$ is a real form of $\mathfrak{g}$, i.e., as a real vector space, $\mathfrak{g}$ is the direct sum of $\mathfrak{g}_u$ and the space $i\mathfrak{g}_u$ of hermitian $2 \times 2$ matrices of trace zero. Hence $\tilde{\mu}_x$ is identically zero on $U_o$ and our assertion follows.

**6.** From this Hurwitz deduced that the algebra $\mathbb{C}[\mathbb{C}^n]^G$ (to be denoted $I_G$ to simplify notation) of invariant polynomials on $\mathbb{C}^N$ is finitely generated: The projector $P \mapsto P^\natural$ obviously satisfies the relation

$$
(1) \qquad (P.Q)^\natural = P.Q^\natural, \quad \text{if } P \text{ is } G\text{-invariant,}
$$

and it is linear and respects homogeneity. By Hilbert's basis theorem, the ideal $I$ generated by the elements of $I_G$ without constant term is finitely generated. Let $Q_i \in I_G$ $(1 \leqq i \leqq s)$ be a generating system of this ideal, which may be assumed to consist of homogeneous invariant elements of strictly positive degrees. Let $Q \in I_G$ be homogeneous. It certainly belongs to $I$. Therefore there exist homogeneous polynomials $A_1, \dots, A_s$ such that

$$
Q = \sum_{1 \leqq i \leqq s} Q_i . A_i.
$$

Then, we have, by (1),

$$
Q^\natural = Q = \sum_i Q_i . A_i^\natural.
$$

Since the $A_i^\natural$ have strictly lower degrees than $Q$, it follows by induction on the degree that $I_G$ is generated, as an algebra, by the $Q_i$.

**Remarks.** 1) In the 19th century, invariant theory was an area of great activity, even before the advent of Lie theory and representation theory. The above finiteness theorem was known earlier, though stated in a different way. It was proved for $\mathbf{SL}_2(C)$ first by P. Gordan ([G1], [G2]). In [Hi], D. Hilbert, after having established his basis theorem, proved the invariant theorem for $\mathbf{SL}_2(\mathbb{C})$ and then sketched its generalization to $\mathbf{SL}_n(\mathbb{C})$. For this he used a differential operator (the $\Omega$-process, also going back to earlier literature), which yields a linear map ♮: $\mathbb{C}[\mathbb{C}^n] \to I_G$ satisfying (1) above, so that, at this point, Hurwitz's argument is formally the same as Hilbert's. The chief novelty is that Hurwitz used integration rather than differentiation to define that operator and carry out the induction proof. To define the projector onto the invariants, it suffices that the representation of $G$ in $\mathbb{C}[\mathbb{C}^n]$ be fully reducible. This remark is contained in [W3], Chapter X, §7, pp. 300–303, based on an unpublished communication of M. Schiffer.

2) In analogy with Maschke's theorem, Hurwitz could have easily given a new proof of the full reducibility of the holomorphic representations of $\mathbf{SL}_2(\mathbb{C})$, and, more generally the first proof for $\mathbf{SL}_n(\mathbb{C})$ $(n \geqq 3)$ and $\mathbf{SO}_n(\mathbb{C})$ $(n \geqq 4)$. Indeed if, as in §4, $H(\ ,\ )$ is a positive non-degenerate hermitian form on $\mathbb{C}^N$, the form $H^o$ constructed as in (2), but using integration,

$$H^o(x,y) = \int_{G_u} H(g^{-1}.x, g^{-1}.y)dv \qquad (x,y \in \mathbb{C}^N),$$

is invariant under $G_u$ and still positive non-degenerate, whence follows the full reducibility of the (continuous) representations of $G_u$. It remains to show that every $G_u$-invariant subspace is $G$-invariant. Let $V$ be a $G_u$-invariant subspace, and $W$ its orthogonal complement with respect to $H^o$. Fix a basis $(f_1, \ldots, f_N)$ of $\mathbb{C}^N$ whose first $p = \dim V$ elements span $V$ and whose last $N - p$ span $W$. Then the matrix coefficients $\sigma(g)_{ij}$ $(i \leqq p, j > p)$ are holomorphic functions on $G$ which vanish on $G_u$ and hence, by the argument outlined previously, are identically zero on $G$. Therefore $V$ and $W$ are $G$-invariant, and full reducibility is proved.

**7.** I spoke of a "landmark paper". This is only by hindsight, because the paper was completely forgotten for about 25 years and, apparently, no specialist in Lie groups or Lie algebras was aware of it and had realized that a proof of Study's conjecture for $\mathbf{SL}_n(\mathbb{C})$ was at hand.

Meanwhile, progress was made on two fronts:

a) Character theory for complex representations of finite groups, orthogonality relations, etc. (Frobenius, Schur, Burnside, 1896-1906).

b) Construction of all irreducible representations of complex simple Lie algebras by Cartan ([**Cr2**], 1913).

In 1922, Schur discovered Hurwitz's paper and used it to extend the character theory a) to representations of $\mathbf{SU}_n$ or $\mathbf{SO}_n$ [**S2**]. Two years later, Weyl combined b) and the Hurwitz-Schur point of view to generalize it to all complex or compact semisimple groups [**W1**]. Until Weyl came on the scene, Schur was not aware of Cartan's work, nor Cartan of Schur's or Hurwitz's. Weyl also pointed out a gap in [**Cr2**]: the construction of irreducible representations makes implicit use of full reducibility, a problem Cartan had not considered at all there. At that point, as a proof, there was then only Weyl's generalization of Hurwitz and Schur, which was highly transcendental. Both Cartan and Weyl felt that an algebraic proof of such a purely algebraic statement was desirable, but viewed it as rather unlikely that one would be forthcoming. Cartan could have pointed out that in the case of $\mathbf{SL}_2(\mathbb{C})$ (or rather, its Lie algebra), one was contained in his thesis. The fact that he did not makes me think that he had forgotten about it (but not forever, though: it is again given in his book on spinors [**Cr3**]). See III for more details.

## §3. Algebraic peoofs of full reducibility

**8.** In comparing physicists and mathematicians it is often said that the physicists, unlike mathematicians, do not care that much for rigorous proofs. Here, we are dealing with the search for a second proof, in a different framework, of a theorem already established, a problem which would normally seem even less appealing to a physicist. In that case, however, it did attract one, H.B.G. Casimir, whose approach had its origin in the use of group representations in quantum mechanics. Since it involves $\mathbf{SO}_3$ or $\mathbf{SO}_3(\mathbb{C})$ rather than $\mathbf{SU}_2$ or $\mathbf{SL}_2(\mathbb{C})$, let me recall first that $\mathbf{SO}_3$ (resp. $\mathbf{SO}_3(\mathbb{C})$) is the quotient of $\mathbf{SU}_2$ (resp. $\mathbf{SL}_2(\mathbb{C})$) by its center, which consists of $\pm\mathrm{Id}$. In particular, $\mathfrak{sl}_2(\mathbb{C})$ may be viewed as the complexification of the Lie algebra $\mathfrak{so}_3$ of $\mathbf{SO}_3$, so that we can take as a basis of it the infinitesimal rotations $D_x, D_y, D_z$ around the three coordinate axes in $\mathbb{R}^3$, where $x, y, z$ are the coordinates:

$$(10) \quad D_x = \begin{pmatrix} 0 & 0 & 0 \\ 0 & 0 & 1 \\ 0 & -1 & 0 \end{pmatrix} \quad D_y = \begin{pmatrix} 0 & 0 & -1 \\ 0 & 0 & 0 \\ 1 & 0 & 0 \end{pmatrix} \quad D_z = \begin{pmatrix} 0 & -1 & 0 \\ +1 & 0 & 0 \\ 0 & 0 & 0 \end{pmatrix}.$$

Viewed as differential operators on functions, these transformations are

$$(11) \qquad D_x = y.\partial_z - z.\partial_y, \quad D_y = z.\partial_x - x.\partial_z, \quad D_z = y.\partial_y - x.\partial_x.$$

The application to quantum mechanics makes use of

$$L_x = i^{-1}.D_x, \quad L_y = i^{-1}.D_y, \quad L_z = i^{-1}.D_z$$

called the components of the moment of momentum, and of

$$(13) \qquad\qquad L^2 = L_x^2 + L_y^2 + L_z^2,$$

the square of the moment of momentum.

The decisive idea is to use $L^2$. It is a differential operator, also represented by minus the sum of the square of the matrices in (11). It belongs to the associative algebra of endomorphisms of $\mathbb{C}^2$ generated by $\mathfrak{sl}_2(\mathbb{C})$, a quotient of the so-called universal enveloping algebra of $\mathfrak{sl}_2(\mathbb{C})$, but not to the Lie algebra itself.

An elementary computation shows that $L^2$ commutes with the infinitesimal rotations, hence with $\mathfrak{sl}_2(\mathbb{C})$ itself. A linear representation $(\sigma, V)$ extends to one of the enveloping algebra, and in particular $\sigma(L^2)$ is defined. If $\sigma$ is irreducible, then $\sigma(L^2)$ is a scalar multiple of the identity (Schur's lemma). In the representation $V_n$ of degree $n+1$, this scalar is equal to $n(n+2)/4$. It characterizes the representation, up to equivalence.[1]

In order to prove full reducibility, Casimir notes that it suffices to consider the case of the exact sequence (2) in §4. Assume first that $m \neq n$, the main case. Then $\sigma(L^2)$ has two eigenvalues, $m(m+2)/4$ and $n(n+2)/4$. The eigenspace $W$ for the latter eigenvalue intersects $V_m$ only at the origin. Since $\sigma(L^2)$ commutes with $\sigma(\mathfrak{sl}_2(\mathbb{C}))$, the space $W$ is also invariant under $\mathbf{SL}_2(\mathbb{C})$. Its projection in $V_n$ is invariant and non-zero, hence equal to $V_n$, so $W$ is the sought-for complement

---

[1]In the physics literature and in [**W2**], the irreducible representations of $\mathbf{SL}_2(\mathbb{C})$ are parametrized by $(1/2)\mathbb{N}$. The representation $V_j$ there is our $V_{2j}$. It has degree $2j+1$, and the eigenvalue of $L^2$ is $j(j+1)$. It is a spin representation, i.e. non-trivial on the center, if and only if $j$ is a half-integer.

to $V_m$. If $m = n$, the existence of an invariant complement is proved by a rather elementary computation, sketched in 11.3.

**9.** An analog of $L^2$ had been introduced in 1931 by Casimir for any complex semisimple Lie algebra, later called the Casimir operator. Using it, B. van der Waerden generalized Casimir's argument to give the first algebraic general proof of the full reducibility of finite dimensional representations of complex semisimple Lie algebras [**CW**].

Later it was realized that the Casimir operator is an element in the center of the universal enveloping algebra (which generates it for $\mathfrak{sl}_2(\mathbb{C})$). The full center was investigated in the late forties by G. Racah, also a physicist, on the one hand, and by C. Chevalley and Harish-Chandra on the other, and became a powerful tool in studying the topology of compact Lie groups and infinite dimensional representations of semisimple Lie groups.

Racah's motivation was representation theory. From a physicist's point of view, the eigenvalue of $L^2$ gave a parametrization of an irreducible representation of $\mathbf{SL}_2(\mathbb{C})$ by a number with a physical meaning, whereas the highest weight had none. For higher dimensional groups, the eigenvalue of $L^2$ does not characterize the representation up to equivalence, which makes the general argument in [**CW**] quite complicated. Racah's idea was to search for more operators commuting with the Lie algebra ($r$ independent ones if $r$ is the rank of the Lie algebra), the eigenvalues of which would again characterize the irreducible representations. This would then allow one to treat the case of two inequivalent irreducible representations in a short exact sequence exactly as for $\mathbf{sl}_2(\mathbb{C})$, and considerably simplify the proof. At that time, however, mathematicians were not searching for a new algebraic proof, and this was not at all a motivation for Chevalley and Harish-Chandra.

**10.** The paper [**CW**] was followed shortly by two other algebraic proofs, one by R. Brauer [**Br**] and one based on a lemma of J.H.C. Whitehead, which is now best expressed in the framework of Lie algebra cohomology, and became the standard algebraic argument for a number of years.

In 1956, a new proof was published by P.K. Raševskiĭ [**R**]. Consider the group $\mathrm{Aff}(\mathbb{C}^N)$ of *affine* transformations of $\mathbb{C}^N$. It is the semidirect product of the group of translations by the group $\mathbf{GL}_N(\mathbb{C})$. Accordingly, its Lie algebra is the semidirect product $\mathfrak{s} \oplus \mathfrak{t}$ of the space of translations $\mathfrak{t}$ by the Lie algebra $\mathfrak{s}$ of $\mathbf{GL}_N(\mathbb{C})$. The new ingredient is the proof that any representation of a semisimple Lie algebra in the Lie algebra $\mathfrak{aff}(\mathbb{C}^N)$ of $\mathrm{Aff}(\mathbb{C}^N)$ leaves a point of $\mathbb{C}^N$ fixed, or, globally speaking, any complex semisimple group of affine transformations of $\mathbb{C}^N$ has a fixed point. Now let $\sigma$ be a representation of the complex semisimple Lie algebra $\mathfrak{g}$ in $\mathbb{C}^M$, and $V \subset \mathbb{C}^M$ an invariant subspace. Then the set of subspaces $W$ of $\mathbb{C}^M$ complementary to $V$ forms in a canonical way an affine space, with space of translations $\mathbb{C}^M/V$. It is operated upon naturally by $\sigma(\mathfrak{g})$. The existence of a fixed point implies the existence of a $\mathfrak{g}$-invariant complement to $V$, whence the full reducibility.

When N. Bourbaki was preparing Volume 1 of the book on Lie groups and Lie algebras, entirely devoted to Lie algebras, an algebraic proof was needed. The cohomological one was not really suitable, requiring as it did lots of preliminaries on cohomology of Lie algebras, which it did not seem appropriate to introduce at that early stage of the exposition. Then Bourbaki turned to Raševskiĭ's proof and made it somewhat more algebraic and self-contained. After the book was published in 1961, I stumbled across a copy of [**Br**], and realized that Bourbaki's argument was the same as that of [**Br**], another example of a paper overlooked for over 25 years, the knowledge of which would have saved Bourbaki some work.

**11.** This pretty much concludes my story, but as Poincaré once wrote, there are no problems which are completely solved, only problems which are more or less solved. Still considering $\mathbf{SL}_2$, one may ask about problems I and II for $\mathbf{SL}_2(k)$, where $k$ is an algebraically closed groundfield of positive characteristic $p$. It is well-known that full reducibility does not necessarily hold. Take for example $k$ of characteristic two and the representation $V_2$ of degree three on the homogeneous quadratic polynomials. It has $x^2, x.y$ and $y^2$ as a basis. In characteristic 2, we have the rule $(a + b)^2 = a^2 + b^2$, so the linear combinations of $x^2$ and $y^2$ are the squares of the linear forms, and form a two-dimensional invariant subspace $V$. A complementary subspace is of dimension one; therefore, if invariant, it would be acted on trivially by $\mathbf{SL}_2(k)$. However, $\mathbf{SL}_2(k)$ does not leave any non-zero element of $V_2$ stable, so $V_2$ is not fully reducible.

This did not rule out a positive answer to problem I, but to prove one, another approach would have to be devised. D. Mumford proposed a notion weaker than full reducibility, now called geometric reductivity: if $C$ is an invariant one-dimensional subspace, there exists a homogeneous $G$-invariant hypersurface not containing $C$ (in the case of full reducibility it could be a hyperplane). Then M. Nagata showed that this condition indeed implies the finite generation of the algebra of invariants. Later, geometric reductivity was proved by C.S. Seshadri for $\mathbf{SL}_2(k)$ and by W. Haboush in general.

Even over $\mathbb{C}$, the problems of full reducibility and of the determination of irreducible representations resurfaced not for $\mathbf{SL}_2(\mathbb{C})$, but for its generalization as a Kac-Moody Lie algebra, or for the deformation of its Lie algebra as a "quantum group". This has led to further problems and to more contacts with mathematical physics.

## §4. Appendix: More on some proofs of full reducibility

We give here more technical details on the proofs of full reducibility for $\mathfrak{sl}_2(\mathbb{C})$ or $\mathbf{SL}_2(\mathbb{C})$ due to Cartan, Fano and Casimir, assuming some familiarity with Lie algebras and algebraic geometry. We let $\mathfrak{g}$ stand for $\mathfrak{sl}_2(\mathbb{C})$.

**12. Lie algebra proof**.

**12.1.** Let

$$(1) \qquad h = \begin{bmatrix} 1 & 0 \\ 0 & -1 \end{bmatrix}, \quad e = \begin{bmatrix} 0 & 1 \\ 0 & 0 \end{bmatrix}, \quad f = \begin{bmatrix} 0 & 0 \\ -1 & 0 \end{bmatrix}$$

be the familiar basis of $\mathfrak{g}$. It satisfies the relations

$$(2) \qquad [h, e] = 2e, \qquad [h, f] = -2f, \qquad [e, f] = -h.$$

The elements $h, e, f$ define one-parameter subgroups

$$e^{th} = \begin{pmatrix} e^t & 0 \\ 0 & e^{-t} \end{pmatrix}, \quad e^{te} = \begin{pmatrix} 1 & t \\ 0 & 1 \end{pmatrix}, \quad e^{tf} = \begin{pmatrix} 1 & 0 \\ -t & 0 \end{pmatrix} \qquad (t \in \mathbb{R}).$$

By letting them act on functions of $x, y$ and taking the derivatives for $t = 0$, we get expressions of $h, e, f$ as differential operators, namely

$$(3) \qquad h = x.\partial_x - y.\partial_y, \quad e = x.\partial_y, \quad f = -y.\partial_x.$$

Let $E$ be a representation space for $\mathfrak{g}$ and $E_c$ $(c \in \mathbb{C})$ the eigenspace for $h$ with eigenvalue $c$. Then (2) implies

(4) $$e.E_c \subset E_{c+2}, \qquad f.E_c \subset E_{c-2}.$$

More generally, if $(h - c.I)^q.v = 0$ for some $q \geqq 1$, then

(5) $$(h - (c+2).I)^q.e.v = 0 = (h - (c-2).I)^q.f.v = 0.$$

**12.2.** We now consider $V_m$. It has a basis $x^{m-i}.y^i$ $(i = 0, \dots, m)$, and $x^{m-i}.y^i$ is an eigenvector for $h$, with eigenvalue $m - 2i$. Let

(1) $$v_{m-2i} = \binom{m}{i} x^{m-i}.y^i \qquad (i = 0, \dots, m).$$

The $v_{m-2i}$ form a basis of $V_m$, and we have

(2) $$h.v_{m-2i} = (m - 2i)v_{m-2i} \qquad (i = 0, \dots, m).$$

A simple computation, using 12.1(2), (3), yields

(3) $$f.v_{m-2i} = -(i+1)v_{m-2i-2},$$
(4) $$e.v_{m-2i} = (m - i + 1)v_{m-2i+2}$$

$(i = 0, \dots, m)$, with the understanding that

(5) $$v_{m+2} = v_{-m-2} = 0.$$

(3) and (4) imply

(6) $$f.e.v_{m-2i} = -i(m - i + 1)v_{m-2i},$$
(7) $$e.f.v_{m-2i} = (i + 1)(m - i)v_{m-2i}.$$

**Remarks.** (a). The eigenvalues of $h$ in $V_m$ are integers. By consideration of a Jordan-Hölder series, it follows that this is true for any finite dimensional representation.

(b) In $P(V_m)$ the rational normal curve occurring in Lie's description of the irreducible projective representations of $\mathbf{SL}_2(\mathbb{C})$ (see §2) is the orbit of the point representing the line spanned by $x^m$. This is also the unique fixed point in $P(V_m)$ of the group $U$ generated by $e$, i.e. the group of upper triangular unipotent (eigenvalues equal to one) matrices. It is therefore also the locus of the fixed points of the conjugates of $U$ in $G$, and each such conjugate has a unique fixed point in $P(V_m)$.

**12.3.** Note that 12.2(5) is a consequence of 12.2(4) and the commutation relations 12.1(1). A similar argument shows more generally that if $E$ is a representation of $\mathfrak{g}$ and $v \in E$ satisfies the conditions

(1) $$e.v = 0, \qquad h.v = c.v \quad (c \in \mathbb{C}),$$

then the elements $f^i.v$ $(i \geqq 0)$ span a finite dimensional $\mathfrak{g}$-submodule $F$. In particular, $\mathbb{C}.v$ is the eigenspace with eigenvalue $c$, and all other eigenvalues of $h$ in $F$ are of the form $c - q$ $(q \in \mathbb{N}, q \geq 1)$.

**12.4.** *First proof of full reducibility.* We use 12.3, which is contained in [**Cr1**], and the two remarks a) and b) of §**3**. This reduces the proof of full reducibility of a $\mathfrak{g}$-module $E$ to the case of a short exact sequence

(1) $$0 \to V_m \to E \xrightarrow{\pi} V_n \to 0 \qquad (m \leq n).$$

Let $m < n$. Then $h$ has an eigenvector $v \in E$ with eigenvalue $n$, which does not belong to $V_m$. It is annihilated by $e$, since there are no weights $> n$ in $V_m$ or $V_n$, and hence none in $E$. By 12.3, it generates a $\mathfrak{g}$-submodule distinct from $V_m$, which must therefore be a $\mathfrak{g}$-invariant complement to $V_m$.

Now let $m = n$. Let $\{v_{m-2i}\}$ $(i = 0, \ldots, m)$ be the basis of $V_m$, viewed as a subspace of $E$, constructed in 12.2. Let $v'_m$ be a vector which maps under $\pi$ onto the similar basis element of the quotient, and let

$$v'_{m-2i} = (-1)^i (i)^{-1} . f^i . v'_m$$

$v'_{m-2i}$ project onto the basis of $E/V_m$ defined in 12.2. There exists $a \in \mathbb{C}$ such that

(2) $$h.v'_m = m.v'_m + a.v_m.$$

We claim it suffices to show that $a = 0$. Indeed, in that case, 12.3 again implies that $v'_m$ generates a $\mathfrak{g}$-submodule distinct from $V_m$, hence a supplement to $V_m$.

It remains to prove that $a = 0$. We claim first that

(3) $$h.v'_{m-2i} = (m - 2i)v'_{m-2i} + a.v_{m-2i} \qquad (i = 0, \ldots, m).$$

For $i = 0$, this is (2). Assuming it is proved for $i$, we obtain (3) for $i+1$ by applying $f$ to both sides and using 12.1(2).

For $i \geq 1$, we have, by 12.1(2),

(4) $$i.e.v'_{m-2i} = -e.f.v'_{m-2i+2} = -f.e.v'_{m-2i+2} + a.v_{m-2i+2}.$$

By (3) and 12.2(6), this yields by induction on $i$ (noting that $e.v'_m = 0$ since $h$ does not have the eigenvalue $m + 2$)

(5) $$e.v'_{m-2i} = (m - i + 1).v'_{m-2i+2} + a.v_{m-2i+2}.$$

If we apply (5) for $i = m + 1$, we get $a(m + 1) = 0$, hence $a = 0$.

**Remark.** This last computation is contained in [**CW**] and also, unknown to the authors, in [**Cr1**]. As we saw, the proof for $m < n$ reduces immediately to 11.3, and by b) in §**3** it suffices to consider that case when $m \neq n$. A direct computation along the lines of the previous proof is longer if $m > n$ (see 12.2). Cartan did it even for a Jordan-Hölder series of any length, which leads to a rather complicated argument. By using his operator, Casimir did not have to make any distinction between the cases $m < n$ and $m > n$.

**12.5.** To give a better idea of Cartan's proof, we discuss the case $m > n$ directly, without reducing to $m < n$.

We let $v'_n$ and $v'_{n-2i}$ $(i \geq 0)$ be as before. Note first that if $n$ and $m$ have different parities, then $V_n$ and $V_m$ have no common eigenvalue for $h$. In particular, $h$ has no element of weight $n + 2$ in $E$, and the eigenspace for $n$ is one-dimensional, hence spanned by $v'_n$. Again, by 12.3, $v'_n$ generates a complementary $\mathfrak{g}$-module. So we assume that $m \equiv n \mod 2$. As before, the whole point is to find $v'_n$ satisfying the condition 12.3(1), for $c = n$.

As above, there is a constant $a$ such that

(1) $$h.v'_n = n.v'_n + a.v_n.$$

We want to prove $v'_n$ may be chosen so that $a = 0$. As in 12.4, we see that

(2) $$h.v'_{n-2i} = (n - 2i).v'_{n-2i} + a.v_{n-2i} \quad (i \geq 0).$$

The weights in $V_n$ run from $n$ to $-n$, so the projection of $fv'_{-n}$ in $V_n$ is zero and we have, for some constant $c$,

$$(3) \qquad\qquad f.v'_{-n} = c.v_{-n-2}.$$

Let $v''_n = v'_n - c.v_n$ and, following 12.2(4), define $v''_{n-2i}$ inductively by the relation

$$(4) \qquad\qquad v''_{n-2i} = -i.v''_{n-2i+2} \qquad (i = 1, \dots, n).$$

By induction on $i$, we see that

$$(5) \qquad\qquad h.v''_{n-2i} = (n - 2i)v''_{n-2i} + a.v_{n-2i} \qquad (i = 0, \dots, n)$$

and also, in view of (3), that

$$(6) \qquad\qquad f.v''_{-n} = f.v'_{-n} - c.f.v_{-n} = 0.$$

For $i = n$, the equality (5) gives

$$(7) \qquad\qquad h.v''_{-n} = -n.v''_{-n} + a.v_{-n}.$$

Now apply $f$ to both sides and recall that $f.h = h.f + 2f$. In view of (6) and 12.2(4) for $m = n$, we get

$$(8) \qquad\qquad a.(n + 1).v_{-n-2} = 0.$$

But $n < m$, so $v_{-n-2} \neq 0$, whence $a = 0$.

We may therefore assume that $v'_n$ is an eigenvector of $h$. There is no eigenspace for $h$ with eigenvalue $n + 2$ in $V'_n$; hence

$$(9) \qquad\qquad e.v'_n = b.v_{n+2} \qquad (b \in \mathbb{C}).$$

By 12.2(4), for $i = (m - n)/2$ (recall that $m \equiv n(2)$), we get

$$(10) \qquad\qquad e.v_n = \big((m + n + 2)/2\big).v_{n+2}.$$

Therefore

$$(11) \qquad\qquad w_n = v'_n - b\big((m + n + 2)/2\big)^{-1} v_n$$

satisfies the conditions

$$(12) \qquad\qquad h.w_n = n.w_n, \; e.w_n = 0,$$

so that, by 12.3, $\mathfrak{g}.w_n$ is a copy of $V_n$ complementary to $V_m$.

**13. Fano's proof**

Fano's proof deals with projective transformations and uses algebraic geometry. Given a finite dimensional vector space $F$ over $\mathbb{C}$, we let $P(F)$ denote the projective space of one-dimensional subspace of $F$. If $F$ is of dimension $n$, $P(F)$ is isomorphic to $\mathbf{P}_{n-1}(\mathbb{C})$.

**13.1.** The proof is contained in §§7, 8, 9 of [**F**]. §9 shows how to reduce it to the case considered in **3**, a), b), that is, to the case of a short exact sequence 12.4(1) with $m \geq n$, but expressed in projective language. Namely:

The space $P = P(E)$ contains a minimal irreducible invariant projective subspace $W = P(V_m)$ of dimension $m$, and the induced projective representation in the space $W'$ of projective $(m + 1)$-subspaces containing $F$ is irreducible.

The problem is then to find an invariant projective subspace $D$ not meeting $W$. Such a subspace necessarily has dimension $n$, and $P(E)$ is the join of $W$ and

Guido Fano

$D$. Moreover, by the remark b) in **3**, it may be assumed that $m \geq n$. Let us write $N$ for the dimension of $P$. Then $N = m + n + 1$ and $m \geq (N - 1)/2$.

As in 12.2(b), $U$ is the one-parameter subgroup of $G = \mathbf{SL}_2(\mathbb{C})$ generated by $e$. Its fixed point set is also the subspace $E^e$ of $E$ annihilated by $e$. Since $U$ is unipotent, any line invariant under $U$ is pointwise fixed, so that the projective subspace $P(E^U)$ associated to $E^U$ is also the fixed point set $P(E)^U$ of $U$ on $P(E)$. Similarly, it may be identified with the set $P(E)^e$ of zeros of the vector field on $P(E)$ defined by the action of $U$.

In §7 of [**F**], Fano proved that $P(E)^e$ is a projective line. I am not sure that I understand his argument, so I shall revert to the linear setup. As just pointed out, we have to show that $E^e$ is two-dimensional.

In $V_m$ and $V_n$ it is one-dimensional, so the exact sequence 12.4(1) shows that $\dim E^e \leqq 2$. As in §**3**, let $E^*$ be the contragredient representation to $E$. Then $E^{*^e}$ is the dual space to $E/eE$, so it is equivalent to prove that $\dim E^{*^e} = 2$. Therefore we may assume that $m \leq n$ (our assumption earlier, but not Fano's). Fix a vector $v' \in E$ projecting onto a highest weight vector in $V_n$. It is an eigenvector of $h$ if $m < n$, is annihilated by $(h - n.I)^2$ otherwise, and in both cases is annihilated by $e$ (see 12.1(4), (5)).

**13.2.** The next and main part of Fano's argument depends on some properties of the "rational normal scrolls", which we now recall (see [**GH**], pp. 522-527). Assume $N \geq 2$, and let $Z$ be a surface in $P$, not contained in any projective subspace. Then its degree is at least $N - 1$([**GH**], p. 173). Those of degree $N - 1$ have been classified, up to projective transformations ([**GH**], *loc.cit.*). Only one is not ruled, the Veronese embedding of $\mathbf{P}_2(\mathbb{C})$ in $\mathbf{P}_5(\mathbb{C})$.

The others are the *rational normal scrolls* $S_{a,b}$ $(a + b = N - 1)$, obtained in the following way: Fix two independent projective subspaces $A, B$ of dimensions $a, b$. Then $P = A * B$ is the join of $A$ and $B$. Let $C_A$ (resp. $C_B$) be a rational normal curve in $A$ (resp. $B$) and $\varphi : C_A \to C_B$ an isomorphism. Then $S_{a,b}$ is the space of the lines $D(x, \varphi(x))$ $(x \in C_A)$. If $a > 0$, but $b = 0$, then $C_B$ is a point, $\varphi$ maps $C_A$ onto a point and $S_{a,b}$ is the cone over $C_A$ with vertex $C_B$. It has a unique singular point, namely $C_B$, and this is the only case where $S_{a,b}$ is not smooth ([**GH**], p. 525).

A rational curve in $S_{a,b}$ which cuts every line $D(x, \varphi(x))$ in exactly one point is called a directrix. By construction $C_A$ (resp. $C_B$) is a directrix of degree $a$ (resp. $b$). The main result used by Fano is that if $a > b$, then $C_B$ is the unique directrix of degree $b$ ([**GH**], p. 525). Fano deduces this essentially from an earlier result of C. Segre [**Se**].

If $a = b$, then we may identify $A$ to $B$ by a map $\varphi$ which takes $C_A$ to $C_B$. It is clear that in that case $S_{a,b} = S_{a,a} = \mathbf{P}^1(\mathbb{C}) \times \mathbf{P}^1(\mathbb{C})$.

**13.3.** We now come back to the situation in 13.1. In $W$ there is exactly one rational normal curve $C$ stable under $G$. The zero set $P(E)^e$ of $e$ is a line (13.1) and $P(E)^e \cap C$ consists of one point, namely $W^e$. Let $Z$ be the set of transforms $g \cdot P(E)^e$ of $P(E)^e$ $(g \in G)$. Since $P(E)^e$ is stable under the upper triangular group $B$ and $G/B$ is complete (in fact a smooth rational curve), $Z$ is a projective subvariety, a $G$-stable ruled surface. We first dispose of a special case. The line $g.P(E)^e$ is the fixed point set of the subgroup ${}^gU = g.U.g^{-1}$, conjugate to $U$ by $g$. Assume that two distinct such lines have a common point. It would then be fixed by two distinct conjugates of $U$. But it is immediate that two such subgroups generate $G$, so that there would be a fixed point $D$ of $G$ in $P(E)$, necessarily outside $W$. Then $P(E)$ would be the join of $W$ and $D$, and we would be finished. From now on, we assume that the lines $g.P(E)^e$ either coincide or are disjoint. We want to prove that $Z$ has degree $N - 1$ in $P(E)$. First we claim that it is not contained in any hyperplane $Y$ of $P(E)$. Indeed, if it were, it would be contained in a $G$-stable proper subspace $U$, the intersection of the transforms of $Y$. The subspace $U$ would contain $W$ properly, which would contradict the irreducibility of the quotient representation in $P(V_n)$. The degree of $Z$ is therefore at least $N - 1$ ([GH], pp. 173-4). It remains to show that the degree of $Z$ is $\leqq (N - 1)$.

Let $C' \subset W'$ be the closed orbit of $G$ which plays the same role as $C$ in $W$. In particular, it has degree $n$. Let $Y$ be a generic hyperplane of $P(E)$ among those containing $W$. Viewed as a hyperplane in $W'$, it cuts $C'$ in $n$ distinct points $Q_i$ $(i = 1, \ldots, n)$. Let $U_i$ be the conjugate of $U$ which fixes $Q_i$ (see 12.2, (b)). The intersection $Z \cap Y$ is a (reducible) curve. We want to prove it has degree $N - 1$ in $Y$. We claim first that

(1) $$Y \cap Z = C \cup D_1 \cup \cdots \cup D_n \qquad (D_i = P(E)^{U_i}),$$

where the $D_i$ are disjoint projective lines, each intersecting $C$ at exactly one point.

First, by construction, $C \subset Z \cap Y$; in fact $C = W \cap Z \cap Y$. Let $x \in Z \cap Y$, $x \notin W$. It belongs to some line $D_g = g.P(E)^e$. The line $D_g$ also contains $g.W^e$, which belongs to $Z \cap Y$, too. Therefore $D_g \subset Y$, and of course $D_g \subset Z$; hence $Z \cap Y$ is the union of $C$ and some of the lines $D_g$. The line $D_g$ spans with $W$ a projective subspace of dimension equal to $\dim W + 1$, which represents a point of $W'$, fixed under $^gU$. It therefore belongs to $Y$ if and only $^gU$ is one of the $U_i$, i.e. if and only if $D_g$ is one of the $D_i$'s, and (1) follows.

Since $C$ has degree $m = \dim W$ in $W$, it follows that $Z \cap Y$ is a curve of degree $m + n$ in $Y$, and hence $Z$ is a surface of degree at most $m + n = N - 1$ in $P(E)$.

Thus $Z$ is a ruled surface, not contained in a hyperplane, of smallest possible degree. It is therefore a "rational normal scroll" (13.2). It is isomorphic to $S_{a,b}$, where $a = \dim W = m$ and $b = N - 1 - a = n$.

Recall that we have reduced ourselves to the case $a \geq b > 0$. Assume first that $a > b$. Then (see 13.2), $Z$ contains a *unique* directrix of degree $b$. It is a normal curve in a $b$-dimensional subspace, which must be invariant under $G$, since $Z$ is. This provides the complementary subspace $U$ to $W$.

Now let $m = n$. Then (13.2), $Z = C \times C'$ is a product of two copies of $\mathbf{P}_1(\mathbb{C})$, where $C$ is, as before, a $G$-stable rational normal curve in $W$ and $C' = P(E)^e$. The transforms $g.P(E)^e$ of $C'$ are the lines $\{c\} \times C'$ $(c \in C)$.

The lines $C_y = C \times \{y\}$ $(y \in C')$ are "directrices". We claim that they are all invariant under $G$. Clearly, the intersection number $C_y.C_z$ is zero if $y \neq z$ $(y, z \in C')$. Let $g \in G$. Since it is connected to the identity, we then have also $(g.C_y) \cdot C_z = 0$; therefore $g \cdot C_y \cap C_z = \emptyset$ unless $g.C_y = C_z$. Since $g.C_y$ must meet at least one $C_z$, we have $g.C_y = C_z$ for some $z$, and we see that $G$ permutes the set $F$ of curves $\{C_y\}$ $(y \in C')$. Each such curve contains a fixed point of $e$, hence of $U$. Therefore $C_y$ is stable under $U$. Now the subgroup $H$ of $G$ leaving each curve $C_y$ stable is a normal subgroup, which is $\neq \{1\}$ since it contains $U$. But $G$ is a simple Lie group, and therefore $H = G$, which proves our contention. Any curve $C_y$ is a rational normal curve in a subspace $W'_y$ which is necessarily $G$-stable. This provides infinitely many $G$-invariant subspaces and concludes the proof.

**Remark.** Let us compare the orders of the steps in the proofs of Cartan and Fano. In 12.4 and 12.5 the first item of business is to show that the action of $h$ on a certain $h$-stable two-dimensional subspace is diagonalisable. That space is $E^u$ in 12.4, and subsequently is shown to be $E^u$ in 12.5. Once a new eigenvector of $h$ annihilated by $e$ is found, 12.3 can be used. In Fano, the first step is to show that $E^u$ is two-dimensional, or rather, equivalently, that $P(E)^u$ is a projective line. There, the analogue of the first step of Cartan would be the existence of *two* fixed points on $P(E)^u$ of $h$, or of the group $H = \{e^{th}\}$ generated by $h$. One is $W^e$. In the generic case $m > n$, Fano's argument may also be viewed as a search for this second fixed point: it is the intersection of $P(E)^u$ with the (unique) directrix $C_B$. However, since the proof directly provides the $G$-orbit $C_B$ of that second fixed point, the argument is not phrased in that way.

**14.** To complete the list of full reducibility proofs for $\mathbf{SL}_2(\mathbb{C})$ at the turn of the century, I still have to mention another approach, valid for $\mathbf{GL}_n(\mathbb{C})$, but for a slightly less general class of representations, the polynomial ones. The definition is the obvious one: $f : \mathbf{GL}_n(\mathbb{C}) \to \mathbf{GL}_N(\mathbb{C})$ is polynomial if the entries of $f(g)$ are polynomials in the coefficients of $g$. In his thesis [S1] I. Schur determined the irreducible ones and their characters, and proved full reducibility. Unknown to him, he had been anticipated (except for the characters) by a Belgian mathematician, J. Deruyts, who expressed himself in a different language, in another framework, but his work was discovered, and explained in current terminology, by J.A. Green. See Chapter V in [Cu] for details and references. Note that, for a representation of

$\mathbf{GL}_n(\mathbb{C})$, to be polynomial is slightly more restrictive than to be holomorphic, but this is not so for $\mathbf{SL}_n(\mathbb{C})$. For $\mathbf{SL}_2(\mathbb{C})$, this already follows from Lie's discussion of its projective representations as recalled in V, §1, so that in fact there were still other proofs of full reducibility in that case available at the time. But Schur did not know about Lie-Study-Cartan and conversely. It is also true that any holomorphic representation of $\mathbf{SL}_n(\mathbb{C})$ is polynomial. (More generally, a connected complex semisimple Lie group $G$ has a unique structure of linear algebraic group such that all holomorphic finite dimensional linear representations are rational. To see this, one has first to know that $G$ has faithful linear representation, which follows from [W1] (see also IV.2.2). The entries of the finite dimensional holomorphic representations of $G$ then span a finitely generated algebra, which defines the algebraic structure.)

# References for Chapter II

[B]     D.J. Benson, Polynomial invariants of finite groups, London Math. Soc. Lecture Note Series **190**, Cambridge University Press, 1993.

[Bo]    N. Bourbaki, Groupes et Algèbres de Lie, Chaps. 7, 8, Hermann, Paris, 1975.

[Br]    R. Brauer, *Eine Bedingung für vollständige Reduzibilität von Darstellungen gewöhnlicher und infinitesimaler Gruppen*, Math. Zeitschr. **41** (1936), 330–339; Collected Papers, II, 462–471.

[Cr1]   É. Cartan, Sur la structure des groupes de transformation finis et continus, Thèse, Nony, Paris, 1894; Oeuvres Complètes, I₁, 137–287.

[Cr2]   ———, *Les groupes projectifs qui ne laissent invariante aucune multiplicité plane*, Bull. Soc. Math. France **41** (1913), 53–96; Oeuvres Completes, I₁, 355–398.

[Cr3]   ———, *Leçons sur la théorie des spineurs I, II*, Hermann, Paris, 1938.

[CW]   H.B.G. Casimir und B.L. van der Waerden, *Algebraischer Beweis der vollen Reduzibilität der Darstellungen halbeinfacher Liescher Gruppen*, Math. Annalen **111** (1935), 1–11.

[CE]    C. Chevalley and S. Eilenberg, *Cohomology of Lie groups and Lie algebras*, Trans. Amer. Math. Soc. **63** (1948), 85–124.

[Cu]    C.W. Curtis, Pioneers of representation theory: Frobenius, Burnside, Schur and Brauer, History of Mathematics **15**, Amer. Math. Soc., 1999.

[F]     G. Fano, *Sulle varietà algebriche con un gruppo continuo non integrabile di transformazioni proiettive in sè*, Mem. Reale Accad. Sci. Torino (2) **46** (1896), 187–218.

[G1]    P. Gordan, *Beweis, dass jede Covariante und Invariante einer binären Form eine ganze Function mit numerischen Coefficienten einer endlichen Anzahl solcher Formen ist*, J. Reine Angew. Math. **69** (1868), 323–354.

[G2]    ———, *Vorlesungen über Invariantentheorie*, written by G. Kerschensteiner, 2 vols., Teubner, Leipzig, 1885, 1887.

[GH]   P. Griffiths and J. Harris, Principles of algebraic geometry, Wiley, New York, 1978.

[Hi]    D. Hilbert, *Ueber die Theorie der algebraischen Formen*, Math. Annalen **36** (1890), 473–534; Ges. Werke II, 199–257.

[Hu]    A. Hurwitz, *Ueber die Erzeugung der Invarianten durch Integration*, Nachr. Kgl. Ges. Wiss. Göttingen, Math.-Phys, Klasse **1897**, 71–90; Math. Werke II, 546–564.

[K]     F. Klein, *Ueber einen Satz aus der Theorie der endlichen (diskontinuerlichen) Gruppen linearer Substitutionen beliebiger vieler Veränderlichen*, Jahresbericht Deutscher Math. Verein. **5** (1901), 57.

[LE]    S. Lie und F. Engel, Theorie der Transformationsgruppen III, Teubner, Leipzig, 1893.

[Lo]    A. Loewy, *Sur les formes quadratiques définies à indéterminées conjuguées de M. Hermite*, C.R. Acad. Sci, Paris **123** (1896), 168–171.

[Ma]    H. Maschke, *Beweis des Satzes, dass diejenigen endlichen linearen Substitutionensgruppen, in welchen einige durchgehende verschwindende Coefficienten auftreten, intransitiv sind*, Math. Annalen **52** (1899), 363–368.

[Mo]    E.H. Moore, *A universal invariant for finite groups of linear substitutions: with applications in the theory of the canonical form of a linear substitution of finite order*, Math. Annalen **50** (1898), 213–219.

[P]     V.L. Popov, Groups, generators, syzygies, and orbits in invariant theory, Transl. Math. Monographs **100**, Amer. Math. Soc., 1992.

[R]     P.K. Raševskiĭ, *On some fundamental theorems of the theory of Lie groups*, Uspehi Mat. Nauk. (N.S.) **8** (1953), no. 1, 3–20. (Russian)

[S1]    I. Schur, *Ueber eine Klasse von Matrizen, die sich einer gegebener Matrix zuordnen lassen*, Dissertation, Berlin, 1901; Ges. Abh., I, 1–73.

[S2]    ———, *Neue Anwendungen der Integralrechnung auf Probleme der Invariantentheorie 1. Mitteilung*, Sitzungsber. Preuss. Akad. Wiss. Berlin Phys.-Math. Kl. **1924**, 189–208; Ges. Abh., II, 440–459.

[Se]    G. Segre, *Sulle rigate razionali in uno spazio lineare qualunque*, Atti Reale Accad. Sci. Torino **19**, 1884, 355–373.

[W1]   H. Weyl, *Theorie der Darstellung kontinuierlicher halbeinfacher Gruppen durch lineare Transformationen, I, II, III und Nachtrag*, Math. Zeitschr. **23** (1925), 271–309; **24** (1926), 328–376, 377–395, 789–791; Ges. Abh., II, 543–647.

[W2]  _____ , Gruppentheorie und Quantenmechanik, S. Hirzel, Leipzig, 1928, 1931; English transl., Dutton, New York, 1932; reprinted, Dover, New York, 1949.

[W3]  _____ , The classical groups, their invariants and representations, Princeton University Press, 1939 and 1946.

# CHAPTER III

# Hermann Weyl and Lie Groups

## §1. First contacts with Lie groups

During the first thirteen years or so of his scientific career, H. Weyl was concerned mostly with analysis, function theory, differential geometry and relativity theory. His interest in representations of Lie groups or Lie algebras[1] and invariant theory grew out of problems raised by the mathematical underpinning of relativity theory:

> ... but for myself I can say that the wish to understand what really is the mathematical substance behind the formal apparatus of relativity theory led me to the study of representations and invariants of groups;...

[**W147**, p. 400]. This involvement, at first somewhat incidental, turned rapidly into a major one. Weyl soon mastered, furthered and combined existing techniques, and within two years announced in 1924 a number of basic contributions.

**1.** The first serious encounter with Lie group theory arose out of a problem considered in the 4th edition of "Raum, Zeit, Materie" [**WI**] on the nature of the metric in space-time, an analogue in this context of the Helmholtz-Lie space problem, which aims at characterizing the orthogonal group by means of some general mobility axioms. As you know, the mathematical framework of general relativity is a 4-manifold, say $M$, endowed with a Riemannian metric (of Lorentz type). The latter, in particular, assigns to $x \in M$ a quadratic form on the tangent space $T_x(M)$ at $M$. Riemann himself had already alluded to the possibility of considering metrics associated to biquadratic forms or more general functions. Weyl investigated, for manifolds of arbitrary dimension $n$, whether it was possible to start from a general notion of congruence, defined at each point by a closed subgroup $G$ of automorphisms of the tangent space at $x$, belonging to a given conjugacy class of closed subgroups of $\mathbf{GL}_n(\mathbb{R})$,[2] and deduce from some general axioms that it would be conjugate to the orthogonal group of a (possibly indefinite) metric. This would then prove that the notion of congruence was associated to a Riemannian metric of some index. The problem was reduced to showing that the complexification of the Lie algebra of $G$ was conjugate in $\mathfrak{gl}_n(\mathbb{C})$ to that of the complex orthogonal group. (Originally, it dealt with real Lie algebras. However, since only the quadratic character of the metric, and not the index, was at issue, it was sufficient to consider their complexifications.)[3] In [**WI**]. Weyl stated he could prove it for $n = 2, 3$ and,

---

Outgrowth of a Hermann Weyl Centenary Lecture, ETH, Zürich, November 7, 1985, published in *Hermann Weyl, 1885–1985*, Springer, 1986, pp. 53–82; reproduced here in a slightly modified form.

[1]The notes for this chapter are at the end of the chapter.

Hermann Weyl (right), talking with David Hilbert

shortly afterwards, published a general proof in [**W49**]. It was a rather delicate and long case-by-case argument, which Weyl himself likened to mathematical tightrope dancing ("mathematische Seiltänzerei").

É. Cartan read about the problem in the French translation of [**WI**], published in 1922, and lost no time in providing a general proof in the framework of his determination of the irreducible representations of simple Lie algebras [**C65**]. It was more general and more natural than Weyl's, even than Weyl's somewhat streamlined argument in [**WII**]. The comparison between them appears to have been a strong incentive for Weyl to delve into Cartan's work. Whether he had other reasons I do not know; at any rate he did so not long after, with considerable enthusiasm, as he was moved to write later on to Cartan (3.22.25):

> Seit der Bekanntschaft mit der allgemeinen Relativitätstheorie hat mich nichts so ergriffen und mit Begeisterung erfüllt wie das Studium Ihrer Arbeiten über die komtinuierlichen Gruppen.

[**WI**] makes ample use of tensor calculus. A student of Weyl considered the problem of showing that the usual tensors form a family characterised by some natural conditions. This eventually amounted to proving that all the Lie group homomorphisms of $\mathbf{GL}_n(\mathbb{R})^+$ into $\mathbf{GL}_n(\mathbb{R})$, where $\mathbf{GL}_n(\mathbb{R})^+$ is the group of elements in $\mathbf{GL}_n(\mathbb{R})$ with positive determinant, are compositions of inner automorphisms, passage to the contragredient, and sums of maps $M \mapsto |\det M|^\alpha$ $(\alpha \in \mathbb{R})$ [**Wn**].[4] About thirty years later, that former student told me that, at the time, Weyl was clearly broadening his interest in representations of semisimple Lie groups, and had suggested that he work further along those lines. He even felt he might have shared some of the excitement to come, had he done so, but he had preferred to go back to his main interest, analysis,[5] and became indeed a well-known analyst: I was talking about Alexander Weinstein.

Another push into Lie groups came, in a way, again from the tensor calculus in [**WI**], but for a completely different reason. In 1923 E. Study, a well-known expert in invariant theory for over thirty years, published a book on invariant theory [**St2**]. In a long foreword, he complained that invariant theory, in particular the so-called symbolic method to generate invariants, had been all but forgotten and that several mathematicians did less by other methods than would be possible using it. Among those was Weyl, identified by a quote, criticized for his treatment of tensors. Apparently Weyl was stung by this. This can be already be seen by the rather sharply worded footnote in his answer to Study, which makes up the first part of [**W60**], but it was also well remembered about 25 years later by one of his colleagues here at ETH at the time, M. Plancherel, who mentioned it to me then as an example of the extraordinary ability Weyl had, shared only by J. von Neumann among the mathematicians he had known, to get into a new subject and bring an important contribution to it within a few months. In fact, Weyl published two papers on invariant theory in 1924, [**W60**, I], [**W63**], which brings me to the achievements already alluded to.

For the sake of the discussion I shall divide them (and, more generally, Weyl's whole output in this area) into two parts: one concerned with linear representations of semisimple Lie groups, complex or compact, and semisimple Lie algebras, which operates with Lie algebra techniques and transcendental means, and is tied to the real or complex numbers; and one concerned with invariant theory and representations of classical groups, initiated by [**W60, W63**], in which Weyl wears an algebraist's hat. This will be convenient to me as an organisational principle, but is, of course, to some extent artificial and should not be construed as a sharp division. In later years, Weyl wrote a number of papers from both the transcendental-analytic and the algebraic points of view, but he remained actively interested for a longer time in the latter, and I shall discuss it later.

## §2. Representations of semisimple Lie groups and Lie algebras

**2.** At that time, *the* outstanding contribution to Lie algebras and their representations was the work of É. Cartan, which Weyl was discovering [**C5, C37**]. A second stimulus was provided by two papers of I. Schur [**Sc2**] about representations of the special orthogonal group $\mathbf{SO}_n$ (as well as of the full orthogonal group $\mathbf{O}_n$, but I shall limit myself to the former) and invariants for $\mathbf{SO}_n(\mathbb{C})$, in which he, in particular, extended the theory of characters and orthogonality relations known for

Issai Schur

finite groups by the work of Frobenius and Schur done at the turn of the century. For later reference also, let me recall briefly some features of the latter.

Let $G$ be a finite group, $\widehat{G}$ the set of equivalence classes of irreducible complex representations of $G$, and $F(G)$ the space of complex valued functions on $G$. On $F(G)$ there is a natural finite dimensional Hilbert space structure, with scalar product given by

$$(1) \qquad (f, g) = |G|^{-1} \sum_{g \in G} f(x)\bar{g}(x) \qquad (f, g \in F(G)),$$

where $|G|$ is the order of $G$. Given an irreducible representation $\pi \colon G \to \mathbf{GL}_n(\mathbb{C})$ by matrices $(a_j^i(g))$, let $V_\pi$ be the vector subspace of $F(G)$ generated by the coefficients $(a_j^i)$. It depends only on the equivalence class $[\pi] \in \widehat{G}$ of $\pi$ (and can be defined more intrinsically) and will also be denoted by $[\pi]$. Moreover, we have the orthogonal decomposition

$$(2) \qquad F(G) = \bigoplus_{\pi \in \widehat{G}} V_\pi,$$

and the $a_j^i$ form a basis of $V_\pi$ (orthogonal if $\pi$ is unitary). The space $F(G)$ is a $G \times G$ module via left and right translations, and the $V_\pi$ are the irreducible $G \times G$ submodules. If $E_\pi$ is a representation space for $\pi$, then $V_\pi \cong \operatorname{End} E_\pi$, and henceis isomorphic to $E_{\pi^*} \otimes E_\pi$ as a $G \times G$ module, where $\pi^*$ is the contragredient representation to $\pi$. As a $G$-module under right (or left) translations, $F(G)$ is the regular representation of $G$, and $V_\pi$ is isomorphic to the direct sum of $d_\pi$ copies of $\pi$ (or $\pi^*$), where $d_\pi$ is the degree of $\pi$.

Let $\chi_\pi$ be the character of $\pi$.[6] It belongs to the space

$$(3) \qquad C(G) = \{f \in F(g) | f(xyx^{-1}) = f(y) \ (x, y \in G)\}$$

of class functions on $G$; we have the orthogonality relations

$$(4) \qquad |G|^{-1}(\chi_\pi, \chi_{\pi'}) = \delta_{\pi, \pi'} \qquad (\pi, \pi; \in \widehat{G}),$$

and the $\chi_\pi$ $(\pi \in \widehat{G})$ form an orthogonal basis of $C(G)$.

A very simple consequence of the orthogonality relations is that the average over $G$ of the character of a given representation $\pi$ gives the dimension of the space of fixed vectors.

**3.** The key for the extension of these results to orthogonal groups was provided by a paper of Hurwitz [**Hu**]. The main concern there was the invariant problem for $\mathbf{SL}_n(\mathbb{C})$ and $\mathbf{SO}_n(\mathbb{C})$: given a finite dimensional holomorphic representation of one of these groups on a space $E$, show that the invariant polynomials on $E$ form a finitely generated algebra. The known averaging procedure for finite groups could not be directly applied to prove this, since $\mathbf{SO}_n(\mathbb{C})$ or $\mathbf{SL}_n(\mathbb{C})$ were not "bounded", but it could be applied to their "bounded" subgroups $\mathbf{SO}_n$ and $\mathbf{SU}_n$, and that was sufficient, since a holomorphic function on $\mathbf{SO}_n(\mathbb{C})$ (resp. $\mathbf{SL}_n(\mathbb{C})$) is completely determined by its restriction to $\mathbf{SO}_n$ (resp. $\mathbf{SU}_n$), in the same way as a holomorphic function on a connected open set of $\mathbb{C}$ is determined by its restriction to a line. This was the first instance of what Weyl first called the "unitary restriction" ("unitäre Beschränkung"), and later [**WVI**] the "unitarian trick" (see Chapter II).

I. Schur, whose initial motivation was also invariant theory, extended the orthogonality relations to $\mathbf{SO}_n$, drew the consequence about the average of a character to compute the dimension of the spaces of invariants in a representation of $\mathbf{SO}_n(\mathbb{C})$, using Hurwitz's integration device, and then went over the determination of all the continuous irreducible representations of $\mathbf{SO}_n$, their characters and dimensions. He also checked that these representations were in fact analytic and, even more, rational in the sense that their coefficients could be expressed as polynomials in the entries of the elements of $\mathbf{SO}_n$. Also full reducibility of finite dimensional representations could be established as in the finite group case, by the construction of an invariant positive non-degenerate hermitian form, where the averaging over the group was replaced by an integration.

**4.** Weyl was now ready to strike. He first extended Schur's method to $\mathbf{SL}_n(\mathbb{C})$ and the symplectic group $\mathbf{Sp}_{2n}(\mathbb{C})$ [**W62**] and then almost immediately afterwards combined the Hurwitz-Schur and the Cartan approaches in an extraordinary synthesis, announced first in the form of a letter to Schur [**W61**]. Until Weyl came on the scene, Cartan did not know about the work of Hurwitz and Schur, nor did Schur know about Cartan's, as can be seen from the introduction to [**Sc2**] in the latter case, and from a letter of Cartan to Weyl (3.1.25) in the former case. Schur expressed his admiration for Weyl's results, and shortly afterwards suggested that Weyl write them up and publish them in Math. Zeitschrift, which was soon done [**W68**].

In these papers Weyl first discussed separately the series of classical groups $\mathbf{SL}_n(\mathbb{C})$, $\mathbf{SO}_n(\mathbb{C})$, and $\mathbf{Sp}_{2n}(\mathbb{C})$, and then set up the general theory. A first main goal is to prove the full reducibility of the finite dimensional representations of a complex semisimple Lie algebra, a problem which Cartan had hardly alluded to in print before 1925. To be more precise, Cartan had in [**C37**] given in principle a construction of all irreducible representations of a given simple algebra[7] and had just not considered more general ones. However, Weyl pointed out that, as far as he could see, an argument of Cartan at one important point could be justified only if full reducibility were available.[8]

The key point was to show that the "unitary restriction" could be applied in the general situation. This was done in two steps: first Weyl showed that a given complex semisimple Lie algebra $\mathfrak{g}$ has a "compact real" form $\mathfrak{g}_u$, i.e., a real Lie subalgebra such that $\mathfrak{g} = \mathfrak{g}_u \otimes_{\mathbb{R}} \mathbb{C}$, on which the restriction of the Killing form is negative non-degenerate.[9] For instance, if $\mathfrak{g} = \mathfrak{sl}_n(\mathbb{C})$ one can take for $\mathfrak{g}_u$ the Lie algebra of $\mathbf{SU}_n$. To this effect, Weyl first had to outline the general theory of semisimple Lie algebras, for which the only sources until then were the papers of W. Killing and Cartan's thesis, all extremely hard to read, and then had to prove the existence of $\mathfrak{g}_u$ by a subtle argument using the constants of structure. Already this exposition, which among other things stressed the importance of a finite reflection group $(S)$, later called the Weyl group, was a landmark, and for many years it was the standard reference. But there it was really only preliminary material. Identify $\mathfrak{g}$ to a subalgebra of $\mathfrak{gl}(\mathfrak{g})$ by the adjoint representation, which is possible since $\mathfrak{g}$, being semisimple, is in particular centerless, and let $G^0$ be the complex subgroup of $GL(\mathfrak{g})$ with Lie algebra $\mathfrak{g}$. It leaves invariant the Killing form $K_{\mathfrak{g}}$ defined by $K_{\mathfrak{g}}(x, y) = \operatorname{tr}(\operatorname{ad} x \circ \operatorname{ad} y)$, which is non-degenerate by a result of Cartan. Then let $G^0_u$ be the real Lie subgroup of $G^0$ (viewed now as a real Lie group) with Lie algebra $\mathfrak{g}_u$. Since the restriction $K_u$ of $K_{\mathfrak{g}}$ to $\mathfrak{g}_u$ is negative non-degenerate, it can be viewed as a subgroup of the orthogonal group of $K_u$, hence is compact.[10].

This is a situation to which the Schur-Hurwitz device can be applied; therefore any finite dimensional representation $\pi$ of $\mathfrak{g}$ which integrates to a representation of $G^0$ is fully reducible. However, in general, a representation $\pi$ of $\mathfrak{g}$ will integrate to a representation not of $G^0$, but of some covering group $G_\pi$ of $G^0$. In the latter, there is a closed real analytic subgroup $G_{\pi,u}$ with Lie algebra $\mathfrak{g}_u$, and the Schur-Hurwitz method will be applicable only if $G_{\pi,u}$ is itself compact, i.e., is a finite covering of $G^0_u$, which is equivalent to $G_{\pi,u}$ having a finite center. This Weyl showed at one stroke for all possible $T_{\pi,u}$ by proving more generally that $G^0_u$ has a finite fundamental group, i.e. that its universal covering has finite center.[11] This first of all yields the full reducibility, but it also sets the stage to extend the character theory of Schur for $\mathbf{SO}_n$ to all compact semisimple groups. All this is by now so standard that a detailed summary is surely superfluous. I shall content myself with some remarks, mainly on some of the work which arose from it. Let $K$ be a compact connected semisimple Lie group and $T$ a maximal torus. It is unique up to conjugacy, and Weyl shows that it meets every conjugacy class. Its Lie algebra $\mathfrak{t}$ is a Cartan subalgebra of the Lie algebra $\mathfrak{k}$ of $K$. Its character group $X(T)$ is a free abelian group of rank equal to $l = \dim \mathfrak{t}$. A *root* is defined globally as a non-trivial character of $T$ in $\mathfrak{k}_{\mathbb{C}} = \mathfrak{k} \otimes_{\mathbb{R}} \mathbb{C}$ with respect to the adjoint representation of $K$. The Weyl group $W$ is the group of automorphisms of $X(T)$ induced by inner automorphisms of $K$. It is generated by the reflections $s_\alpha$, where $\alpha$ is a root and $s_\lambda$ the unique involution of $X(T)$ having a fixed point set of corank one, mapping $\alpha$ to $-\alpha$ and leaving stable the set of roots. Since $T$ meets every conjugacy class in $K$, it suffices to describe the restrictions to $T$ of the irreducible characters of $K$. At first these are finite sums of characters of $T$, i.e. trigonometric polynomials, invariant under $W$. To describe them, we consider the trigonometric sums

$$A(\lambda) = \sum_{w \in W} (\det w) w \cdot \lambda \qquad (\lambda \in X(T)).$$

The sum $A(\lambda)$ is skew invariant with respect to $W$, equal to zero if $\lambda$ is fixed under a reflection $s_\alpha$ in $W$, and depends only on the orbit $W \cdot \lambda$ of $\lambda$. Fix a closed convex cone $C$ in $X(T)_\mathbb{R} = X(T) \otimes_\mathbb{Z} \mathbb{R}$ which is a fundamental domain for $W$, call $\lambda$ dominant if it belongs to $C$, and say that a root $\alpha$ is positive if it lies in the half-space bounded by the fixed point set of $s_\alpha$ and containing $C$. Let $2\rho$ be the sum of the positive roots. We have Weyl's denominator formula

$$(5) \qquad A(\rho) = \prod_{\alpha > 0} (\alpha^{1/2} - \alpha^{1/2}).$$

Weyl showed that for every continuous class function $f$ on $K$ we have

$$(6) \qquad \int_K f(k)dk = \int_T f(t)\mu(t)dt,$$

where $dk$ (resp. $dt$) is the invariant measure on $K$ (resp. $T$) with mass 1 and

$$(7) \qquad \mu = |W|^{-1}|A(\rho)|^2 \qquad (|W| \text{ the order of } W),$$

a point which Schur had singled out in his praise of Weyl's results. From this and the orthogonality relations for characters of $K$ or of $T$, Weyl deduced that every irreducible character is of the form

$$(8) \qquad \chi_\pi = A(\lambda_\pi + \rho) \cdot A(\rho)^{-1},$$

where $\lambda_\pi$ is dominant.[12] He did not show, however, but derived from Cartan's work, that every dominant $\lambda$ occurs in this way.[13] I shall soon come back to this problem. He also deduced from (8) a formula for the degree $d^\circ \pi$ of $\pi$:

$$(9) \qquad d^\circ \pi = \prod_{\alpha > 0} \frac{\langle \rho + \lambda_{\pi,\alpha} \rangle}{\langle \rho, \alpha \rangle},$$

where $\alpha$ runs through the positive roots and $\langle \, , \, \rangle$ is a scalar product invariant under the Weyl group. Both (8) and (9) were not at all to be seen from Cartan's construction of irreducible representations.

## §3. Impact on É. Cartan

**5.** Among many things, these papers mark the birthdate of the systematic global theory of Lie groups. The original Lie theory, created in 1873, was in principle local, but during the first fifty years, global considerations were not ruled out, although the main theorems were local in character. However, a striking feature here was that algebraic statements were proved by global arguments, which moreover, seemed unavoidable at the time.[14] Weyl had not bothered to define the concepts of Lie group or of universal covering (the latter being already familiar to him in the context of Riemann surfaces).[15] He had just taken them for granted, but could of course lean on the examples of the classical groups, which had been known global objects even back in the early stages of Lie theory.

**6.** These papers had a profound impact on Cartan. He had first known Weyl's results through the announcements [**W61, W62**], in which the general case was only cursorily discussed. His first and very quick reaction [**C81**] was to show that, given the Hurwitz device, the use of "analysis situs" could be avoided by means of older results of his.[16] However, once the full papers were published, his outlook changed, and, from then on, the global point of view and analysis situs were foremost in his mind. He began to supplement his earlier work on Lie algebras with a systematic study of global properties of Lie groups. In [**C103, C113**] he developed the geometry of singular elements in a compact semisimple Lie group, and used the Weyl group systematically, to the extent even of deriving some basic properties of compact Lie groups or Lie algebras from results on Euclidean reflection groups. The scope of these investigations was further increased when he began to look from this point of view at a problem originating in differential geometry he was involved with at the time: the study of a class of Riemannian manifolds that he later (from [**C117**] onwards) called symmetric spaces. These spaces were originally, by definition, those in which the Riemannian curvature tensor is invariant under parallelism. They are locally Riemannian products of irreducible Riemannian spaces. Cartan had classified them [**C93, C94**] and had seen, with considerable astonishment, that this classification was essentially the same as that of the real forms of complex simple Lie algebras he had carried out ten years before [**C38**]. Up to that point, Cartan's investigations had been really local, although this was a tacit rather than an explicit assumption. In present day parlance we would say he had classified isomorphy classes of local irreducible symmetric spaces. However, he had soon recognized that these spaces were also characterized by the condition that the local symmetry at a point $x$, i.e. the local homeomorphism which flips the geodesics through $x$, is isometric [**C93**, Nr. 14], which of course led to his choice of terminology. There also Cartan systematically adopted a global point of view, put in the foreground a global version of this second condition (each point is an isolated fixed point of a global involutive isometry), and developed a theory of semisimple groups and symmetric spaces in which Lie group theory and differential geometry were beautifully combined (see, e.g., [**C107, C116**]). A particularly striking example is his proof of the existence and conjugacy of maximal compact subgroups by means of a fixed-point theorem in Riemannian geometry [**C116**]. Nowadays, all this has been streamlined, in the sense that group theoretical (resp. differential geometric) results have been given group theoretical (resp. differential geometric) proofs. Such an evolution is unavoidable and has many advantages, but sometimes loses some of the freshness and suggestive power of the original approach. It seems to me still fascinating to watch Cartan explore this new territory and display, as he once put it [**C105**]:

> toute la variété des problèmes que la Théorie des Groupes et la Géométrie, en s'appuyant mutuellement l'une sur l'autre, permettent d'aborder et de résoudre.

In [**C111**] he also points out that, since the set of singular elements in a compact semisimple group has codimension 3, Weyl's homotopy argument for the finiteness of the fundamental group (cf. [11]) can also be pushed to show that the second Betti number is always zero (in fact, what he sketches leads to a proof that the second homotopy group is zero). He then sets up a program to compute the Betti numbers of compact Lie groups or their homogeneous spaces by means of closed differential forms, conjecturing on that occasion the theorems which de Rham was soon going to prove. Using invariant differential forms, he reduced these computations to purely algebraic problems [**C118**]. This led later to the cohomology of Lie algebras.

See Chapter IV for more details on the contents of this section.

## §4. The Peter-Weyl theorem. Harmonic analysis

**7.** We now come back to the problem of showing that every dominant character $\lambda$ occurs in (8) as the highest weight of an irreducible representation. Already in a footnote to [**W61**], Weyl had stated that this "completeness' can be proved by transcendental means. In [**W68**, III, §4] he is more precise but also more circumspect: the problem is to decompose the regular representation of a compact group, but there are serious technical difficulties, and this method is maybe not really worth pursuing until simplifications are available since the result is known anyway from Cartan.[17] However, whatever difficulties there were, they were soon surmounted in the paper written jointly with his student F. Peter [**W73**], which not only proves the desired completeness, but is very broadly conceived and is to be viewed as the foundational paper for harmonic analysis on compact topological groups. In the simplest case, that of the circle group, this boiled down to the study of trigonometric series, but even there the group theoretical point of view was new.[18] We now refer for comparison to the earlier discussion of finite groups. Since $G$ is now a compact Lie group, the authors replace $F(G)$ by the space $L^2(G)$ of square integrable functions with respect to a fixed invariant measure. It defines, via right translations, a unitary representation of $G$. The completeness problem is to show that the algebraic direct sum $F$ of the spaces $V_\pi$, defined as above, with $\pi$ running through the equivalence classes of irreducible finite dimensional continuous representations, is dense in $L^2(G)$ (the Peter-Weyl theorem). Indeed, if that is the case, then the characters will form an orthonormal basis of the space of measurable class functions, and this will easily imply that any expression $A(\lambda + \rho)A(\rho)^{-1}$ ($\lambda \geq 0$) does occur as a character in the right-hand side of (8). To prove this density, the authors introduce the convolution algebra $C^*(G)$ of the continuous functions on $G$. They show that any finite dimensional representation $\pi$ also yields a representation of $C^*(G)$ by the formula

$$\pi(\alpha) = \int_G \alpha(x)\pi(x)dx$$

and view $\pi(\alpha)$ as a Fourier coefficient of $\alpha$. The convolution $\alpha*$ by $\alpha$ is an integral operator of Hilbert-Schmidt type, with kernel $k(x,y) = \alpha(x \cdot y^{-1})$, which is self-adjoint if $\alpha = \tilde{\alpha}$, where $\tilde{\alpha}(x) = \bar{\alpha}(x^{-1})$. This operator commutes with right translations; hence its eigenspaces are invariant under $G$ operating on the right. The authors consider in particular the operator associated to $\alpha * \tilde{\alpha}$. An extension of E. Schmidt's theory of eigenvalues and eigenspaces for such integral operators on an interval shows that $(\alpha * \tilde{\alpha})*$ has a non-zero eigenvalue if $\alpha \neq 0$ and that the corresponding eigenspace is necessarily finite dimensional. All such eigenspaces are then contained in $F$. Using the operators so associated to an "approximate identity" (a sequence of positive functions, whose supports tend to $\{1\}$ and whose integral is one), they show that every finite dimensional irreducible representation of $G$ occurs in this way (up to equivalence) and that every continuous function is a uniform limit of elements of $F$. This implies, in particular, that $F$ is dense. As a further consequence, the finite dimensional representations separate the elements of $G$, and the continuous class functions separate the conjugacy classes in $G$. Apart from some technical simplifications, such as the use of the spectral theorem for completely continuous operators (as done first in [**Wi1**, §21]), this is pretty much the way it is presented today, and it made a deep impression at the time it was published. Weyl himself viewed it as one of the most interesting and surprising applications of integral equations [**W80**, p. 196]. It was immediately extended to homogeneous spaces of compact Lie groups by Cartan, with emphasis on symmetric spaces [**C117**], thus supplying in particular a group theoretical framework to the theory of certain special functions, such as spherical harmonics, and then again by Weyl [**W98**].

**8.** Weyl had seen that the same approach would also yield the main approximation theorem in H. Bohr's theory of almost periodic functions on the line [**W71, W72**]. After Haar showed the existence of an invariant measure on any locally compact group, and noted that this allowed one to generalize [**W73**] to compact groups without further ado, J. von Neumann extended to groups S. Bochner's definition of Bohr's almost periodic functions (cf. [**Wi1**, §§33, 41]), and this led to what Weyl called the "culminating point of this trend of ideas" [**WVI**, p. 193], providing the natural domain of validity for the arguments of the Peter-Weyl theory, but he pointed out immediately its limitations by quoting a result of Freudenthal, to the effect that a group whose points are separated by almost periodic functions is the product of a compact group by the additive group of a vector space, i.e. one does not get more than the two cases initially considered. Therefore, an extension of this theory to other non-compact groups, in particular non-compact semisimple groups, would have to be based on quite different ideas, and Weyl never tried his hand at it. Still his work has exerted a significant influence on its development. First the obvious one: the character formula, the Peter-Weyl theorem, the use of a suitable convolution algebra of functions have been for all a pattern, a model. The results of Harish-Chandra on the discrete series, for instance, albeit much harder to prove, bear a considerable formal analogy with them. But, less obviously maybe, Weyl was also of help via his work on differential equations [**W8**], which gave Harish-Chandra a crucial hint in his quest for an explicit form of the Plancherel measure. In the simplest case, that of spherical functions for $\mathbf{SL}_2(\mathbb{R})$ (or real rank-one groups), the problem reduces essentially to the spectral theory of an ordinary differential equation on the line, with eigenfunctions depending on a real parameter $\lambda$. It was the reading of [**W8**] which suggested to Harish-Chandra that the measure should be the inverse of the square modulus of a function in $\lambda$ describing the asymptotic behaviour of the eigenfunctions [**HC**, II, p. 212], and I remember well from seminar lectures and conversations that he never lost sight of that principle, which is confirmed by his results in the general case as well.[19]

## §5. Group theory and quantum mechanics

**9.** Around 1927, Weyl got involved with the applications of group representations to quantum mechanics. His first paper [**W75**] contained notably some suggestions or heuristic arguments which also led to new developments in unitary representations of non-compact Lie groups, but of a quite different nature from those mentioned above. Weyl proposed that the spectral theorem should allow one to associate to an unbounded hermitian operator $A$ on a Hilbert space $H$ a one-parameter group $\{\exp itA\}$ $(t \in \mathbb{R})$ of unitary transformations of $H$ having $iA$ as an infinitesimal generator, as is well known to be the case in the finite dimensional case. Then, given operators satisfying the Heisenberg commutation relations

$$(10) \qquad [P_j, Q_k] = \delta_{jk}, \quad [P_j, P_k] = [Q_j, Q_k] = 0 \qquad (1 \leq j, k \leq n),$$

Weyl views these relations as defining a Lie algebra and considers the associated group $N$ of unitary transformations generated by the $\exp itU$, where $U$ runs through the real linear combinations of the $P_j$ and $Q_k$. To (9) there correspond commutator relations in $N$, which have since been known as the "Weyl form" or "integrated form" of (9) (and also prove that $N$ is a "Heisenberg group"). He then gives some heuristic arguments to prove the uniqueness (up to equivalence) of an irreducible unitary representation of $N$ with a given central character, out of which follows the uniqueness of the Schrödinger model for the canonical variables. Rigorous treatments of these suggestions led to the Stone theorem on one-parameter groups of unitary representations [**So**], which soon became a foundational result in unitary representations of non-compact Lie groups, and to the Stone-von Neumann uniqueness theorem [**Ne**], [**So**], itself a fundamental result and the source of many further developments (for all this see [**Ho4**] and [**M**]).

Weyl soon provided a systematic and impressive exposition in his book "Gruppentheorie und Quantenmechanik" [**WV**]. I shall not attempt to discuss its importance in physics and shall go on confining myself mostly to Lie groups. As far as Weyl was concerned, the main mathematical contribution stemming from it is the paper on spinors [**W105**], written jointly with R. Brauer. Infinitesimally, the spinor representations had already been described by Cartan in 1913 [**C37**], by their weights. But [**W105**] gave a global definition, based on the use of the Clifford algebra, itself suggested by Dirac's formulation of the equations for the electron. However, the most unexpected fall-out originated with a physicist, H. Casimir, and led to what was viewed at the time, erroneously, as the first algebraic proof of the complete reducibility theorem. In the representations of $\mathfrak{g} = \mathfrak{sl}_2(\mathbb{C})$, or equivalently $\mathfrak{so}_3(\mathbb{C})$, an important role in the quantum theoretic applications is played by a polynomial of second degree in the elements of $\mathfrak{g}$, which represents the "square of the magnitude of the moment of momentum" [**WV**], p. 156 (or p. 179 in the English version), the sum of the squares of the infinitesimal rotations around the coordinate axes. It commutes with all of $\mathfrak{g}$, and hence is given by a scalar in any irreducible representation: this yields an important quantum number $j(j+1)$, in the representation of degree $2j+1$ ($2j \in \mathbb{N}$). (See Chapter II, 8.) Casimir was struck by this commutation property and defined in 1931 an analogous operator for a general semisimple Lie algebra, called later on, and maybe in [**WIII**] for the first time, the Casimir operator, and indicated how it would allow one to derive the Peter-Weyl theorem from results about self-adjoint elliptic operators [**Cs**]. A year later, he noticed that in the case of $\mathfrak{sl}_2$, it could be used to give a purely algebraic proof of full reducibility, which was later extended to the general case by B. L. van der Waerden, using the general Casimir operator [**CW**]. As we shall soon see, Weyl was quite concerned at the time with finding algebraic proofs of results obtained first in a transcendental way but in a different context, and it seems that this problem was no longer of much interest to him,[20] although he had concluded his first announcement [**W62**] by saying that an algebraic proof would be desirable, and had suggested, at the time of the lectures [**WIII**], that it would be worthwhile to develop a purely algebraic theory of Lie algebras, valid at least over arbitrary fields of characteristic zero, a suggestion which was picked up by N. Jacobson (see [**J1**]) and had a considerable impact on his research interests.

## §6. Representations and invariants of classical groups

**10.** Although I have spoken at some length of the Math. Zeitschrift papers, I have not yet exhausted their content, and my survey has been incomplete on at least two counts. Making up for it will provide a bridge towards the more algebraic aspects of Weyl's work.

In Chapter I of [**W68**], devoted to the representations of $\mathbf{SL}_n(\mathbb{C})$ and $\mathbf{GL}_n(\mathbb{C})$, Weyl not only combined Hurwitz-Schur and Cartan, but also related the results to older ones going back to Schur's thesis [**Sc1**]: After having determined all the holomorphic irreducible representations of $\mathbf{SL}_n(\mathbb{C})$, he pointed out that the matrix coefficients are in fact polynomials in the matrix entries, and that these representations are the irreducible constituents of the representations of $\mathbf{SL}_n(\mathbb{C})$ in the tensor algebra over $\mathbb{C}^n$. They are therefore just the tensor spaces, described by means of symmetry conditions on the coefficients. In this he saw the "group theoretical foundation of tensor calculus," a point important enough for him to make it the title of that chapter and of the announcement [**W62**]. Moreover, Schur had given a direct algebraic construction of those, setting up the well-known correspondence with representations of the symmetric groups, stated in terms of Young diagrams, which also yielded an algebraic proof of full reducibility. This example of an algebraic treatment is one to which Weyl would come back repeatedly and which he would try to extend to other classical groups. It later became of even greater interest to him in view of its applications to quantum theory [**WV**, Ch. V].

The second point is "invariant theory". In broad terms its general problem is, given a group $G$ and a finite dimensional representation of $G$ in a vector space $V$, to study the polynomials on $V$ which are invariant under $G$ (or sometimes only semi-invariant, i.e. multiplied by a constant under the action of a group element). The questions which are usually asked are whether the ring of invariants is finitely generated (first main theorem), and, if so, whether the ideal of relations between elements in a generating set is finitely generated (second main theorem). In concrete situations, one will of course want an explicit presentation of the ring of invariants in terms of generators and relations. One may also look for the dimension of the space of homogeneous invariants of a given dimension (the "counting of the number of invariants"). Such a formulation, however, where $G$ and $V$ are general, emerged at a later stage of the theory, as an abstraction from the classical invariant theory, to which I shall come in a moment, which focuses on very specific instances of these questions.[21]

As remarked earlier, such problems were at the origin of the papers of Hurwitz [**Hu**] and Schur [**Sc2**], and it was to be expected they would also be very much in Weyl's mind, even independently of the Study incident. Indeed, he points out at the end of [**W68**] that the unitary restriction method now allows one to prove the first main theorem for all semisimple groups, which provides, for the first time, a natural group theoretical domain of validity for it. In addition, following Schur, he notes that the dimension of the space of fixed vectors in a given representation $\pi$ is given by integrating over $G$ the character $\chi_\pi$ of the representation. He pursues this in [**W69, W70**], where he also states that, more generally, the multiplicity of an irreducible representation $\sigma$ in $\pi$ is given by the integral $\int \chi_\pi(g)\overline{\chi_\sigma(g)}dg$, where $dg$ is the invariant measure with total mass 1, and applies this to a number of classical cases. Although he also stresses in [**W68**] the superiority of the integration method over the traditional procedures, based on differentiation operators of the kind of Cayley's $\Omega$-process, he soon became preoccupied with finding an algebraic framework for the main results of [**W68**] pertaining to classical group, which would encompass the classical invariant theory.

**11.** We have now come close to the two main themes of this second, more algebraic, part of Weyl's work, which culminates in his book on classical groups [**WVI**]:

> ...The task may be characterized precisely as follows: with respect to the assigned group of linear transformations in the underlying vector space, to decompose the space of tensors of given rank into its irreducible invariant subspaces... Such is the problem which forms one of the mainstays of this book, and in accordance with the algebraic approach its solution is sought for not only in the field of real numbers on which analysis and physics fight their battles, but in an arbitrary field of characteristic zero. However, I have made no attempt to include fields of prime characteristic.

After having briefly explained that the determination of representations logically precedes the search for algebraic invariants, he goes on to say:

> ...My second aim, then, is to give a modern introduction to the theory of invariants. It is high time for a rejuvenation of the classical invariant theory, which has fallen into an almost petrified state.

All this seems rather clear, but I dare say I am not the only one to have found the book rather difficult. According to the introduction, the program is first to decompose tensor representations and then to derive invariant theory. But Weyl apparently could not do this for all classical groups, algebraically and in that order, and so the itinerary between the two is more sinuous, starting in fact with invariant theory. It would seem also that, in spite of some occasional, rather pungent, comments on the symbolic method, Weyl had found invariant theory and some of its specific techniques of independent interest, since he gives them prominant billing in [**WVI**] and had also devoted to them a course here at ETH Zürich (of which a 13-page outline can be found in the Weyl Archives) and at the Institute for Advanced Study [**WIV**]. In addition, Weyl could not realize his program fully algebraically, and did not refrain from introducing and using the integration method, whether to realize his immediate goals or in its own right. This rather subtle interplay between various points of view does, of course, broaden the horizon of "the humble who want to learn" for whom the book is "primarily meant" [**WVI**, viii], but does not make it easier for them to get a clear picture of its organization. This somewhat tentative character may also have been felt by Weyl when he wrote [**W117**]:

> At present I have come to a certain end, or at least to a certain halting point, from which it seems profitable to look back upon the track so far pursued, and this is what I have tried to do in my recent book, *The Classical Groups, their Invariants and Representations*.

Reflecting upon the greater finality of the transcendental results as compared to the algebraic cones, with the added wisdom of fifty or so more years, one may observe that in the former case Weyl already had the natural framework and all the necessary tools at his disposal, but not so in the latter; in particular, the point of view of linear algebraic groups, or the use of the universal enveloping algebras of Lie algebras, which became essential, were not part of his vision of future developments, as described at the end of [**W117**] where he forecast

> ...a similar book dealing comprehensively with the representations and invariants of all semi-simple Lie algebras in an arbitrary characteristic.

I shall try to give some brief idea of the content of [**WVI**] and some of the developments to which it has led. Before that, however, I should backtrack and say something about the already often mentioned classical invariant theory and the earlier work of Weyl pertaining to it.

A typical example is the search for invariants of $p$ vectors for $G = \mathbf{SL}_n(\mathbb{C})$, $\mathbf{SO}_n(\mathbb{C})$, i.e. for polynomials in the coordinates $(x_{ij})$ $(i = 1, \ldots, n; j = 1, \ldots, p)$ of $p$ vectors $\mathbf{x}_j$ $(j = 1, \ldots, p)$ homogeneous in the coordinates of each $\mathbf{x}_j$, which are invariant under $G$. In other words, one is looking for the fixed vectors in the tensor product of $p$ copies of the polynomial algebra over $\mathbb{C}^n$. In the case of $\mathbf{SL}_n(\mathbb{C})$ or $\mathbf{GL}_n(\mathbb{C})$, one wants more generally the invariant of $p$ vectors and $q$ covectors $(\mathbf{y}_k)$ $(k = 1, \ldots, q)$. The first main theorem in this last case says that the ring of invariants is generated by the products $\langle \mathbf{x}_j, \mathbf{y}_k \rangle$, to which one should add the determinants in $n$ of the $p$ vectors or $q$ covectors in the case of $\mathbf{SL}_n$. For $\mathbf{O}_n$ there is no need to add covectors; the invariants of $p$ vectors are generated by the scalar products $(\mathbf{x}_j, \mathbf{x}_k)$. Also a generating set of relations between these elements was given. Let me now limit myself to the case of $p$ vectors. To prove such theorems one uses differentiation operators which commute with $G$, hence transform invariants to invariants, and which allow one to decrease the degree in certain variables and to carry out induction proofs. The first ones are the polarisation operators $D_{ik} = \sum_j x_{ji} \partial / \partial x_{jk}$. Another fundamental one is Cayley's operator

$$(11) \qquad \Omega = \sum (\mathrm{sng}\, \sigma) \frac{\partial^n}{\partial x_{\sigma(1),1} \cdots \partial x_{\sigma(n),n}},$$

where $\sigma$ runs through the symmetric group $\mathfrak{S}_n$ in $n$ letters. This operator commutes with the $D_{ij}$ for $i \neq j$, but not with the $D_{ii}$. Capelli [**Ca1**] showed that $H = \det(x_{ij}) \det(\partial / \partial x_{ij})$ (for $p = n$) does commute with all $D_{ij}$, and gave an expression for $H$ as a determinant in those, the so-called Capelli identity, which is the main formal tool in much of the theory.[22] There are of course many variants of this problem. One may, e.g., replace the identity representation of $G$ by a symmetric power, hence consider the invariants of $p$ homogeneous forms of a given degree. The "symbolic method" in principle reduces the form problem to the search for multilinear invariants, i.e. to the fixed points in tensor powers of the identity representation. In the nineteenth century, such procedures to generate invariants out of a given one were often used to try to check the validity of the first and second main theorems. In that task, they were discredited by Hilbert's work on invariant theory, which "almost kills the whole subject. But its life lingers on, however flickering, during the next decades" [**WVI**, pp. 27, 28].

In his answer to Study, in [**W60**], Weyl went back to Capelli, provided a new proof of Capelli's identity, and then established the first main theorem for $\mathbf{SL}_n$ and $\mathbf{SO}_n$, which were known, as well as for $\mathbf{Sp}_{2n}$, which had not been considered before. He also discussed the determination of invariants for some non-simple groups such as the group of euclidean motions or the subgroups of $\mathbf{SL}_n$, or $\mathbf{GL}_n$, leaving invariant a strictly increasing sequence of proper subspaces (now called parabolic subgroups, see [24]). [**W63**] centers on the use of the symbolic method.

**12.** After some contributions to the program outlined above, Weyl published his book [**WVI**]. It starts with an exposition of the first and second main theorems for invariants of vectors and covectors for the classical groups. A central point of the book, or at any rate of its algebraic part, is the double commutant theorem, proved first in [**W107**], which is the main principle on which Weyl organized the discussion of the decomposition of tensor representations. Let $A$ be a subalgebra of the algebra $\text{End}(E)$ of endomorphisms of a finite dimensional vector space $E$ over a field $K$ of characteristic zero, and let $A'$ be its commuting algebra in $\text{End}\,V$. Assume $V$ is a fully reducible $A$-module. Then $V$ is also fully reducible under $A'$, and $A$ is the commuting algebra of $A'$. In particular, $V$ decomposes into a direct sum of $A' \times A$ irreducible submodules. Any one of those is the tensor product $U_\sigma \otimes U'_\sigma$ of an irreducible $A$-module by an irreducible $A'$-module, whence we get a correspondence between some irreducible representations $\sigma$ of $A$ and $\sigma'$ of $A'$, which is particularly nice if neither $\sigma$ nor $\sigma'$ occurs twice in these pairs. The assumption of full reducibility under $A$ is in particular fulfilled if $A$ is the enveloping algebra of a finite group (even in positive characteristic prime to the order of that group).

The origin of this theorem, and its most perfect illustration, is the reciprocity between irreducible representations of the symmetric groups $\mathfrak{S}_p$ and the irreducible subspaces of the tensor representations of $\mathbf{GL}_n$, discussed first in Schur's thesis [**Sc1**] (for $K = \mathbb{C}$). Let $V$ be the tensor product of $p$ copies of $K^n$. It is operated upon by $\mathbf{GL}_n$, via the $p$-th tensor power of the identity representation, and by $\mathfrak{S}_p$, via the permutation of the factors; and these operations commute. Let $A'$ and $A$ be the corresponding enveloping algebras. Schur proved that $A'$ is the centralizer of $A$, and deduced full reducibility from this. He also showed that the correspondence $(\sigma, \sigma')$ is bijective and determines the characters of the $\sigma'$. After having discussed this case, Weyl went over to the orthogonal and symplectic groups. However, this has to be more roundabout since there is no finite group on the other side. Weyl had made the first steps in that direction in [**W96**]; here he availed himself also of some results of R. Brauer [**Br2**]. He considered simultaneously the enveloping algebra of the group under consideration in a tensor representation and its commuting algebra $A'$. None of them is a priori known to be fully reducible, and the interplay between information gained successively on each of them is rather subtle. An important fact is that the description of $A'$ is equivalent to the first main theorem. However, in the case of $\mathfrak{gl}_n$, this relationship can be used conversely to prove the first main theorem [**Br2**].

Weyl's next goal was the determination of the characters of the irreducible constituents of the tensor representations. Lacking an algebraic method valid for all classical groups, he turned to the transcendental one of [**W68**], based on integration. For $\mathbf{GL}_n$, however, algebraic treatments were available. I already mentioned one by Schur [**Sc1**]. Here Weyl followed a slightly earlier one due to Frobenius. He then went over to more general aspects of invariant theory, which he discussed from both the algebraic and the transcendental points of view.

A supplement to the main text in the second edition includes some complements published first in [**W122**]. In particular, it points out that a subalgebra of a matrix algebra over a subfield of $\mathbb{R}$ which is stable under transposition is fully reducible, which gives an algebraic proof of full reducibility in such cases.

**13.** [**WVI**] is obviously not an easy book.[23] Its results were not as spectacular and clear cut as those of the Math. Zeitschr. and Math. Annalen papers and, not surprisingly, it did not have such an immediate impact, but this was really only a question of time, and its influence has been felt more and more during these last fifteen years or so.

First of all, its treatment of classical invariant theory became the standard reference and made it available to potential users, whether specialists or not. This led to further applications but also to improvement of the theory, already over $\mathbb{C}$. As a first example, [**ABP**] provided a new proof of the determination of the tensor invariants for $\mathbf{O}_n$ directly from those for $\mathbf{GL}_n$, bypassing the use of the Capelli identity. The application the authors had in mind was to a new proof of the Atiyah-Singer theorem by the heat equation method. As a second example, M. Artin [**A**] was led to consider invariants of $r$ matrices and conjectured they would be generated by traces. This was soon deduced from the classical theory by Procesi [**P1**], who more generally described the invariants of $p$ vectors, $q$ covectors, and $r$ matrices, results which were then used to determine the rational Waldhausen $K$-theory of simply connected spaces [**DHS**].

As a second type of development, let me discuss some in which the restriction on the characteristic of the ground field was lifted, even beyond what Weyl could envisage. Progress was made almost simultaneously in two directions. On the one hand, the first main theorem was extended to all semisimple (even reductive) linear algebraic groups over an algebraically closed ground field $K$ of arbitrary characteristic.[24] In characteristic zero, this was essentially Weyl's theorem mentioned earlier (modulo a harmless reduction to $\mathbb{C}$). But this proof was based on the full reducibility, which is false in positive characteristic, so it could not be extended directly. Motivated by his geometric invariant theory, D. Mumford proposed a weaker notion, later called "geometric reductivity": Given a rational representation $G \to GL(V)$ and a pointwise fixed line $D \subset V$, there should exist a homogeneous hypersurface $W$ meeting $D$ only at the origin and invariant under $G$ (in the case of full reducibility, $W$ could be a hyperplane), and conjectured that every semisimple algebraic group would satisfy this condition. Soon after, M. Nagata showed that this would imply the first main theorem [**Na**]. The validity of the latter was then assured when W. Haboush proved geometric reducibility in general [**Ha**], a few years after Seshadri had established it for $\mathbf{GL}_2$.

On the other hand, by introducing new methods in combinatorics, [**DRS**] gave a characteristic-free proof of the first main theorem for $\mathbf{GL}_n$. This was quickly seized upon by de Concini and Procesi to extend the first and second main theorems to classical groups over a commutative ring $A$, subject only to the condition that a polynomial in $n$ indeterminates with coefficients in $A$ which is zero on $A^n$ is identically zero [**CP**]. In particular, this applies to any algebraically closed ground field, or to $\mathbb{Z}$.

## §7. Two later developments

**14.** At this point, although it is one more generation removed from Weyl, I feel it is natural to mention a further and very extensive characteristic-free theory, in which [**CP**] is an essential tool, the "standard monomial theory" of Seshadri et al. In its geometric form, it aims at giving canonical bases for spaces of sections of homogeneous line bundles on flag varieties.As far as I know, it does not yet work in full generality, but it does work in a vast class of cases. In characteristic zero, when it does, it allows one to give a canonical basis of an irreducible representation; more explicitly, it yields a combinatorial procedure to single out an irreducible representation with a given highest weight in the tensor product of fundamental representations in which Cartan looked for it (see [7]), assuming this is done for the fundamental representations which occur. This generalizes work by W. V. D. Hodge for $\mathbf{SL}_n$, who used it to give explicit equations for Schubert varieties and applied it to enumerative geometry, and it has similar applications and goals in the general case. See the survey [**Se**] by Seshadri.

As my final item, I now turn to a development of which [**WVI**], or more specifically the bicommutant theorem, has been a foster parent, namely R. Howe's theory of reductive pairs.

In the early seventies, some remarkable correspondences were set up between certain families of irreducible unitary representations of some pairs of real reductive groups. Knowing [**WVI**], Howe was led to think that the proper framework for such correspondence was a generalization of the situation of the bicommutant theorem: start from a real semisimple group $G$ and a so-called "reductive pair" $(H, H')$ in $G$, i.e. two reductive subgroups $H, H'$ each of which is the centralizer of the other. Then let $\pi$ be an irreducible representation of $G$, say first finite dimensional, to avoid analytical difficulties. Then it decomposes into irreducible $H \times H'$ submodules, each of which is a tensor product of irreducible representations $\sigma$ of $H$ and $\sigma'$ of $H'$, whence again we get some relationship between the irreducible representations of $H$ and $H'$. If, on the other hand, $\pi$ is unitary and infinite dimensional, then the decomposition into $H \times H'$ modules presents all kinds of difficulties. They become more manageable if one of the groups is compact. Also one may anyhow restrict one's attention to the closed irreducible subspaces. It turned out that the correspondences mentioned earlier could be given a natural explanation by viewing the two groups in question as a reductive pair in a symplectic group $\mathbf{Sp}_{2n}(\mathbb{R})$. or rather by going over a two-fold covering of $\mathbf{Sp}_{2n}(\mathbb{R})$ and taking for $\pi$ the so-called oscillator or metaplectic representation. The choice of that particular representation had been suggested by Weil's use of it, over local and global fields, in his group-theoretical treatment of $\theta$-functions [**Wi2**]. Accordingly, this principle has had many applications and variants for groups over local or global fields. In the latter case it leads to correspondences built from $\theta$-series between spaces of automorphic forms. This is one of the most fruitful principles in representation theory of, and automorphic forms for, classical groups, which has suggested many problems and has been confirmed by many special results. See [**Ho3**] for a survey of theorems and conjectures, and [**Ho5**] for further results. At this point, we seem again to be far removed from Weyl, and obviously such developments are not direct offsprings of his work. But the importance of [**WVI**] as one of the main influences on the genesis of this general principle has been stressed by Howe himself, who has also applied it to groups over finite fields [**Ho1**] and then to invariant theory itself [**Ho2**]. There Howe generalized the classical theory to the determination of "superinvariants," i.e. of invariants in the tensor products of symmetric and exterior powers, and proposed a general recasting of the whole theory from that point of view.[25]

With this I conclude my attempt to give an idea of Weyl's work on Lie groups and of its repercussions. As you can see, those were felt in a broad range of topics in analytical, differential geometric, topological or algebraic contexts and took many forms: general theorems or specific results on special cases, clear cut statements as well as less sharply delineated suggestions or guiding principles, mirroring the many-sidedness of Weyl's output and outlook.

Early in this chapter, I quoted from a letter to Cartan in which Weyl expressed his admiration for Cartan's work on continuous groups. He goes on to say, commenting on the results announced in [**W61**]:

> Meinen gegenwärtigen Anteil an dieser Theorie schätze ich gar nicht besonders moch; ich komme mir eigentlich nur vor wie der zufällige Treffpunkt von Ihnen und Schur.

This is of course putting it very mildly. Not only was much more than chance needed to produce such a synthesis, but Weyl had to be a meeting ground for, and to combine, nor only Schur and Cartan, but invariant theory, topology and functional analysis as well. At that time, no one else was conversant with all of these; in fact, except for Schur, hardly anybody was conversant with more than one. Although I limited myself to a rather sharply circumscribed and quantitatively minor part of Weyl's work, this already provides a demonstration of, a practical lesson in, the unity of mathematics, given to us by a man whose mind was indeed a meeting ground for most of mathematics and mathematical physics.

## Notes

[1] (29) As a rule, I shall use current terminology rather than that of the original papers, and content myself with some occasional historical remarks about the latter.

The term "Lie algebra" appeared first in [**WIII**], and was suggested by N. Jacobson, but the concept was present very early in the theory. Before [**WIII**], the usual terminology was "infinitesimal group," or sometimes just "group" (or "abstract group" in [**WII**]). Obviously, Weyl felt at the time more comfortable with the latter, since he reverts almost exclusively to it, after having introduced "Lie algebra" as an alternate to infinitesimal group in a formal definition, and later he uses only "subgroup" and "invariant subgroup" for the present-day "subalgebra" and "ideal."

"Lie group" was introduced by Cartan around 1930 (see, in particular, [**C128**]). For Lie (disregarding the distinction between local and global), and before [**C128**] they were the "finite and continuous groups." However, in [**C128**] Cartan used the latter expression for what we would call locally euclidean groups, and required twice differentiability for Lie groups.

[2] (29) We refer to pages xi–xii for some notation.

[3] (29) Cartan [**C65**] translated the problem into the formalism of affine connections he was developing at the time, which made Weyl's transition from half-philosophical considerations to the actual mathematical problem easier to grasp for some. In present day terminology, the problem can be stated geometrically as follows: Let $G$ be a closed subgroup of $\mathbf{SL}_n(\mathbb{R})$. Let $M$ be a smooth manifold of dimension $n$. Choose in a neighborhood $U$ of a point $x$ a trivialisation of the bundle $P$ of frames, i.e. $n$ smooth everywhere linearly independent vector fields $(e_n)$ (or, equivalent, as Weyl and Cartan express it, a set of $n$ everywhere linearly independent one-forms $\omega_i$). Let $Q$ be the fibre bundle over $U$ whose fiber at $y$ is the set of $G$-transforms of $\{e_i(y)\}$, where $\{e_i(y)\}$ is used to identify the tangent space $T_y(M)$ with $\mathbb{R}^n$. It is assumed that for every choice of the $e_i$'s, the principal $G$-bundle $Q$ has one and only one torsion-free affine connection (the existence is Axiom I and

the uniqueness is Axiom II). Then the complexification of the Lie algebra $\mathfrak{g}$ of $G$ is conjugate in $\mathfrak{gl}_n(\mathbb{C})$ to $\mathfrak{so}_n(\mathbb{C})$.

Algebraically, this amounts to the following problem. Let $G$ and $\mathfrak{g}$ be as before, and let

$$(C_{j,s}^i) \quad (1 \le i, j \le n' 1 \le s \le m = \dim G)$$

be a basis of $\mathfrak{g}$. Assume that for any set of $m \cdot n^2$ real numbers $a_{j,s}^i$ one can find a unique set of $n \cdot m$ constants $b_k^s$ ($s = 1, \ldots, m; k = 1, \ldots, n$) such that the differences $a_{j,k}^i - \sum_s C_{j,s}^i a_k^s$ are symmetric in $i, j$. Then the complexification of $\mathfrak{g}$ is conjugate to $\mathfrak{so}_n(\mathbb{C})$.

In fact, Cartan solved a somewhat more general problem and showed that already the existence of the $b_k^s$ forces $\mathfrak{g}$ to belong to a very small list.

For the sake completeness, I should point out that Weyl had approached this question in 1919, in his comments to Riemann's Habilitationsschrift [**Ri**], but rather briefly, in a differential geometric way. It was noticed later that the problem stated there was in fact a variant of the one of [**WI**], but that Cartan's solution to the latter also gave a solution to the former. See [**F**] for a discussion, references, and another variation on the problem, and [**K**] for an hermitian analogue.

[4] (31) The main point is to find the automorphisms of $\mathbf{SL}_n(\mathbb{R})$. This was essentially reduced to a Lie algebra problem: to show that all derivations of the Lie algebra $\mathfrak{sl}_n(\mathbb{R})$ are inner. Cartan had already proved this fact in his thesis [**C5**, p. 137], for all semisimple Lie algebras (over $\mathbb{C}$ really, his standing assumption there, but this anyhow implies the result over $\mathbb{R}$). Prompted by Weinstein's result, he determined the structure of the group of automorphisms of all complex simple Lie algebras in [**C82**].

[5] (31) He did not have to go far to find a topic, and soon studied a problem in two-dimensoinal hydrodynamics suggested to him, together with a method of attack, by Weyl (Math. Zeitschr. **19** (1924), 265–275), a problem on which Weyl subsequently also wrote a paper [**W76**].

[6] (33) This was called "characteristic" at the time. Weyl introduced "character" from about 1927 on, first as an alternate possibility, and then later he shifted to it. In [**WV**] he suggested it in order to avoid using characteristic, which has another meaning pertaining to eigenvalues or eigenspaces of linear transformations, but reversed himself in Chap. V, using characteristic for finite groups and characters for continuous groups. In a lecture in French [**W74**] he had used "caractère".

[7] (34) These results are to be found in many textbooks, e.g., [**Bo2**], [**J**]. I shall recall them later in an equivalent global formulation, favored by Weyl. For the sake of completeness I give here a capsule review of some salient features in Cartan's framework.

Let $\mathfrak{g}$ be a complex semisimple Lie algebra, $\mathfrak{h}$ a Cartan subalgebra. A linear form $\lambda$ on $\mathfrak{h}$ is a *weight* of a representation $\pi \colon \mathfrak{g} \to \mathfrak{gl}(V)$ if

$$V_\lambda = \{v \in V | \pi(h)v = \lambda(h) \ (h \in \mathfrak{h})\} \ne 0.$$

The space $V$ is always the direct sum of the $V_\lambda$. The weights of all finite dimensional representations generate a lattice $P$ in the smallest $\mathbb{Q}$-subspace $\mathfrak{h}_\mathbb{Q}^*$ of $\mathfrak{h}^*$ spanned by them, which is a $\mathbb{Q}$-form of $\mathfrak{h}^*$. A nonzero weight of the adjoint representation is a *root*. The roots generate a sublattice $Q$ of $P$. Let $\mathfrak{h}_\mathbb{R}^*$ be the real span of $P$. For each root $\alpha$ there is a unique automorphism $s_\alpha$ of order 2 of $\mathfrak{h}_\mathbb{R}^*$ leaving the set of roots stable, transforming $\alpha$ to $-\alpha$ and having a fixed point set of codimension one.

Fix a set $\Delta$ of "simple roots", i.e. $l = \dim \mathfrak{h}$ linearly independent roots such that any other root is a integral combination of the $\alpha \in \Delta$ with coefficients of the same sign, and call positive those with positive coefficients. Introduce a partial ordering among the weights by saying that $\lambda \geq \mu$ if $\lambda - \mu$ is a positive linear combination of simple roots. Then $\lambda \in P$ is said to be dominant if $\lambda \geq s_\alpha \lambda$ for all simple $\alpha$'s. Let $P^+$ be the set of dominant weights. [The group $W$ of automorphisms of $\mathfrak{h}_\mathbb{R}^*$ generated by the $s_\alpha$ is one realization of the group $(S)$ introduced by Weyl in [**W68**] and later called the Weyl group. A weight can also be defined as dominant if it belongs to a suitable fundamental domain $C$ of $W$, namely the intersection of the half-spaces $E_\alpha$ $(\alpha \in \Delta)$, where $E_\alpha$ is the half-space bounded by the fixed point set of $s_\alpha$ and containing $\alpha$.]

Let $\pi$ be irreducible. Cartan showed first that it has a unique highest weight $\lambda_\pi$, i.e. a weight which is greater than any other weight. This weight has multiplicity one. The main result of [**C37**] is that any $\lambda \in P^+$ is the highest weight of one and only one (up to equivalence) irreducible representation. The dominant weights are linear combinations with positive integral coefficients of the so-called fundamental weights $\omega_i$ $(1 \leq i \leq l)$, where $l = \dim \mathfrak{h}$ is the rank of $\mathfrak{g}$. Cartan's strategy for proving this was first to exhibit a representation $V_i$ for each fundamental dominant weight $\omega_i$, and then to locate an irreducible representation having a given dominant weight $\lambda = \sum c_i \cdot \omega_i$ as its highest weight as the smallest $\mathfrak{g}$-submodule in the tensor product $V_1^{c_1} \otimes \cdots \otimes V_l^{c_l}$ of $c_1$ copies of $V_1, \ldots, c_l$ copies of $V_l$ containing $V_{\omega_1}^{c_1} \otimes \cdots \otimes V_{\omega_1}^{\omega_1}$. It is this last point that Weyl criticized (see below).

[8](34) Cartan agreed there was a gap, but in fact he could have easily bypassed full reducibility to realize his goal at the point, namely to show the existence of an irreducible representation with a given dominant weight as its highest weight. For this, a slight refinement of his argument would have sufficed, as is customarily done nowadays in the algebraic proof found later, or to show the existence of an irreducible highest weight module in situations where full reducibility is not true or not known, such as linear reductive groups in positive characteristic or Kac-Moody Lie algebras: One takes the smallest $\mathfrak{g}$-submodule containing the given highest weight vector, shows that it has a greatest proper invariant subspace, and then divides out by the latter.

Since E. Study was somewhat of a villain in the 1923 incident related earlier, let me add as a counterpart that he was well aware of this problem around 1890 and had brought the first contribution to it. In fact, Lie reports in [**L**, 785-8] that Study had proven full reducibility (phrased differently however, in terms of projective representations) for $\mathfrak{sl}_2$ in an unpublished manuscript and that it quite certainly would be true more generally for $\mathfrak{sl}_n$. In a letter to Lie (December 31, 1890), referred to in [**Hw**], Study even went so far as to conjecture that it should hold for all simple or semisimple Lie algebras. To both of them, this was an important problem. The manuscript was not published, apparently because the proof appeared too complicated and simplifications were hoped for (see Chapter V, §1).

In [**C5**, p. 134] Cartan referred to Lie's summary of Study's results and commented that it would be easy to prove all those pertaining to $\mathfrak{sl}_2$ using some results proved before. In [**Hw**], the author surmises that Cartan must have had full reducibility in mind, and not just the determination of the irreducible representations of $\mathfrak{sl}_2$, which was due to Lie anyway. Prompted by this remark, which T. Hawkins reminded me of in a useful correspondence on a first draft of this chapter, for which I am glad to thank him, I checked that this is indeed the case: Cartan's arguments from line 6, p. 100 to line 9, p. 102 do provide all the ingredients for an algebraic proof, but this consequence is not explicitly stated since Cartan's goal at that point was to supply a complete proof to a theorem announced earlier by F. Engel, namely that every non-solvable Lie algebra contains a subalgebra isomorphic to $\mathfrak{sl}_2$. (As Cartan pointed out on p. 103, the original argument by Engel was not complete, but Engel also published another proof in 1893.) However, Cartan casually uses full reducibility later (p. 116, lines 13–23), referring to the previous chapter, showing that he felt he had indeed proved it there. See Chapter II for more details.

Cartan's next reference in print to that problem that I am aware of, explicit this time, occurred in a footnote to a 1925 paper [**C80**, p. 30], added shortly before publication, after Cartan had seen Weyl's announcements [**W61, W62**]. Part of this paper was devoted to the decomposition into irreducible tensors of the curvature and torsion tensors of an affine connection. Cartan stated in the footnote that at the time he was writing the paper (December 1922) he viewed the full reducibility of tensor representations of semisimple Lie algebras as likely. At about the same time he wrote in [**C81**] that this full reducibility had seemed to be extremely probable.

I do not know of any other statement by Cartan (expect for a similar one in a letter to Weyl) referring to that problem; this indicates that he may not have thought about it between his thesis and Weyl's work. See [**Hw**] for a detailed survey of the early stages of the representation theory of Lie algebras.

[9] (34) Cartan had checked this case by case in [**C38**], but, as he wrote later to Weyl (3.28.1925), without looking for it, without realizing its importance, while determining all real forms of complex simple Lie algebras.

[10] (34) Weyl takes this for granted. However, as Cartan later pointed out [**C116**, p. 7], one has to know also that $G_u$ is a closed subgroup of the orthogonal group of $B_u$. But this follows immediately from the fact that $G_u$ is the identity component of the group of automorphisms of $\mathfrak{g}_u$ (see [4] above).

[11] (34) In view of its importance and novelty, I'll try to sketch Weyl's argument, using some concepts introduced in the next 10 lines of the main text, with $K = G_u^0$. Let us say that $t \in T$ is regular if $T$ is the identity component of its centralizer, singular otherwise. Using the fact that the roots come in pairs $\pm \alpha$, Weyl shows that the set of singular elements has codimension three, hence that any loop based on the identity is homotopic to one containing only regular elements, except for the base point. Any regular element is conjugate to one and only one element contained in a fundamental domain $C_0$ of $W$ in $T$. From this it can be shown that the loop is homotopic to one contained in $C_0$. Under the exponential map $C_0$ is the image of a fundamental domain $C_1$ of the group of affine transformations of $t$ generated by $W$ and the translations by the kernel of the exponential map. This is a simplex if $G_u^0$ is simple. Then a loop in $C_0$ is the image under the exponential map of a path in $C_1$ going from the origin to one vertex mapping to the identity, and its homotopy class is determined by that vertex, whence the sought-for finiteness.

[12] (35) There is a quibbling point which might bother the fastidious reader: $2\rho$ is a well defined character of $T$, but not necessarily $\rho$ itself; hence $\lambda + \rho$ in (8) might not be a character. However, it is if $K$ is simply connected. We could make that assumption to be safe, but it is not necessarily because the ambiguities of numerator and denominator cancel out in (8) and $|A(\rho)|^2$ is anyhow well defined.

[13] (35) The differential of $\lambda_\pi$, multiplied by $\sqrt{-1}$, is then the highest weight of the representation of the complexification $\mathfrak{g} = \mathfrak{k}_{\mathbb{C}}$ of $\mathfrak{k}$ defined by the differential $d\pi$ of $\pi$, in the sense of Cartan. To see this and relate the concepts named similarly

here and in [7], note that $\mathfrak{t}_C = \mathfrak{h}$ is a Cartan subalgebra of $\mathfrak{g}$ an that the map which assigns to $\lambda \in X(T)$ the linear form $i \cdot d\lambda$ on $\mathfrak{h}$ extends to an isomorphism of $X(T)_{\mathbb{R}}$ onto $\mathfrak{h}^*_{\mathbb{R}} = i \cdot \mathfrak{t}^*$ which commutes with the Weyl group and maps roots to roots and $X(T)$ to a lattice sitting between $Q$ and $P$ (equal to $Q$ if $K$ is the adjoint group, to $P$ if $K$ is simply connected).

[14] (35) In his answer to the letter where Cartan outlines the argument sketched in [16] below, Weyl remarks that the starting point and general framework of both approaches are quite similar, and adds:

> Insbesondere entnehme ich aus Ihren Mitteilungen auch, daß es offenbar einigermaßen aussichtslos ist, den Satz von der vollen Reduzibilität zu beweisen, ohne den infinitesimalen Ansatz zu verlassen,

echoed by Cartan in his next letter (3.18.25):

> La difficulté, je n'ose dire l'impossibilité, de trouver une démonstration directe ne sortant pas du domaine strictement infinitésimal montre bien la nécessité de ne sacrifier aucun des deux points de vu. . . .

[15] (35) Formal definitions were provided soon after by O. Schreier, Abh. Math. Sem. Hamburg **4** (1926), 15–32, **5** (1927), 233–244, and then by Cartan in [**C128**], the first exposition of Lie groups and homogeneous spaces from the global point of view (see Chapter IV).

[16] (36) "sans être obligé de se livrer à des considérations d'analysis situs, toujours délicates," as he wrote to Weyl (3.1.25). First the existence of a compact form $\mathfrak{g}_u$ had been checked case by case in his determination of all the real forms [**C38**], as recalled in [9]. Second, in order to apply the integration method, it was not really needed to know that the universal covering of $G^0_u$ was compact, but only that any covering $G_{\pi,u}$ provided by a linear representation $\pi$ had a finite center. For this he gave an argument based on his discussion of the weights of irreducible representations [**C37**]. Let $\mathfrak{t}$ be a Cartan subalgebra of $\mathfrak{g}_u$ and $T_{\pi,u}$ the integral group associated to it in the given representation $\pi$. Cartan's argument is that the weights of $\pi$ in his sense are rational linear combinations of roots (because $Q$ has finite index in $P$, in the notation of [7]), and therefore, if, say, $k$ is a common denominator, then at most $k$ elements of $T_{\pi,u}$ may lie over a given element of the image of $T_{\pi,u}$ in the adjoint group. As I understand it, for this argument to be valid one would have to know that every element of $G_{\pi,u}$ is conjugate to an element of $T_{\pi,u}$. This is indeed true and contained in [**W68**], but its proof uses compactness and I do not see that it was already available to Cartan at that point.

Cartan would come back to that question in [**C116**]. See 2.5.1.

[17] (37) He also speaks of promising "Ansätze" of Cartan, which would bring noteworthy simplifications, known to him through letters, but I did not find any in the correspondence I have seen.

[18] (37) Weyl did not hesitate to claim it was the better point of view [**W74**]:

> Et même dans ce cas particulier, notre méthode est supérieure aux méthodes amciennes et classiques de la théorie des séries de Fourier, car elle permet, comme je le crois, de se rendre compte pour la première fois des véritables raisons de la validité de la formule de Parseval. J'en vois aussi une confirmation dans le fait qu'elle put être appliquée aussi sans modification au cas traité dernièrement par H. Bohr des fonctions *presque périodiques*. . .

[19] (38) In [**HC**, IV], it appears in the relation between the function $\mu_\omega(\nu)$, which gives the Plancherel measure for a continuous family of unitarily induced principal series representations and the $c$-functions, which govern the asymptotic behavior of Eisenstein integrals (see the corollary on p. 144).

[20] (39) In fact, [**Cs**] is just mentioned in [**WIII**, II, p. 33] with no attempt to give an idea of the proof, even in the case of $\mathfrak{sl}_2$, which should have pleased Weyl in view of its origin, although [**WIII**, I] describes the Casimir operator (p. 52) and its use to prove the Peter-Weyl theorem (pp. 100–102).

Soon, R. Brauer provided another algebraic proof [**Br2**], also using the Casimir operator, but in a quite different and simpler way, to which the argument of [**Bo1**] is very closely related, followed a few years later by J. H. C. Whitehead, whose argument, cohomological in spirit, was a boost to the development of the cohomology theory of Lie algebras (see [**J**]).

The Casimir operator was the first example of an operator in the center $\mathcal{Z}(\mathfrak{g})$ of the universal enveloping algebra $\mathcal{U}(\mathfrak{g})$ of a complex semisimple Lie algebra $\mathfrak{g}$. The search for others was taken up again by a physicist, G. Racah, whom I mention here because he wrote two letters to Weyl about it in 1947 and 1949. He determined in fact the degrees of the generators of $\mathcal{Z}(\mathfrak{g})$, see [**Ra**]. He was hoping that the eigenvalues of these generators would separate the irreducible representations (which is true). This would then allow one to adapt directly to the general case the full reducibility proof for $\mathfrak{sl}_2$, avoiding the technical complications of van der Waerden's argument.

In view of the other proofs mentioned above, this was not a goal of interest to mathematicians, but, independently, and for other reasons, Chevalley and Harish-Chandra also determined the structure of $\mathcal{Z}(\mathfrak{g})$ [**HC**, I, 292–360]. Soon $\mathcal{U}(\mathfrak{g})$ and $\mathcal{Z}(\mathfrak{g})$ became a basic tool for the study of finite or infinite dimensional representations of semisimple Lie groups, the foundational paper in that respect being the one of Harish-Chandra's just quoted. One of the first uses of $\mathcal{U}(\mathfrak{g})$ made there is an algebraic, classification-free, proof of the existence of an irreducible finite dimensional $\mathfrak{g}$ module with a given highest weight. The final step towards the algebraic treatment of all the main results of [**W68**] was taken by H. Freudenthal, who supplied an algebraic proof of the character formula, and also one giving the multiplicity of a given weight in an irreducible representation (see [**Bo2**], [**J**]).

[21] (40) Notwithstanding the foreword to [**St2**], Study had very early advocated a broadening of invariant theory in this direction, following Lie, as can be seen from [**St1**], a book which incidentally also describes all irreducible representations of $\mathbf{SL}_3$ (see Chapter V, §§1, 2).

[22] (42) Many mathematicians not steeped in classical invariant theory would concur with [**ABP**] in viewing the role of Capelli's identity as somewhat "mysterious". It becomes maybe less so, once it is pointed out that we are dealing with an element in the center of a universal enveloping algebra, as was made very explicit in [**Ho2**], on which I shall comment later (see [25]). It is remarkable that without the terminology, this was quite clearly seen by Capelli himself: In [**Ca3**] he writes the commutation relations for the $D_{ij}$'s (which show that they form a Lie algebra isomorphic to $\mathfrak{gl}_p(\mathbb{C})$: If we view the $p$ vectors as the $p$ columns of an $n \times p$ matrix, then the action of $\mathbf{GL}_n(\mathbb{C})$ is given by left multiplication on $\mathbf{M}_{n,p}(\mathbb{C})$ and the $D_{ij}$'s span the Lie algebra of $\mathbf{GL}_p(\mathbb{C})$, acting by right multiplication). Capelli then considered its enveloping algebra $U$, i.e. the associative subalgebra of differential operators on $\mathbf{M}_{n,p}(\mathbb{C})$ it generates, and proved a theorem about its structure, which

amounts to saying, in present day terminology, that it is isomorphic to $\mathcal{U}(\mathfrak{gl}_p)$ and satisfies the Poincaré-Birkhoff-Witt theorem. More precisely, he defined a notion of "irreducible degree" (the usual filtration of $\mathcal{U}(\mathfrak{gl}_p)$) and showed that the associated graded algebra is a polynomial algebra. He added that $H$ belongs to the center of this algebra, as proved in [**Ca1**], and that he had described $n-1$ other operators commuting with the $D_{ij}$'s in another paper [**Ca2**]. Their expressions in terms of the polarization operators provide other "Capelli identities". Later on he showed that these operators are algebraically independent and generate the center of $U$ (Rend. Reale Accad. Sci. Napoli (2) VII (1893), 29–38, 155–162). Altogether, he had determined the structure of the center of $\mathcal{U}(\mathfrak{gl}_p)$. I thank C. Procesi, who told me Capelli had done that, which led me to look for it. See also [**P2**]. Of these papers, Weyl quotes only the first one, as far as I could see.

[23] (44) It is also an illustration of some of the comments of Chevalley and Weil [**CWi**] on Weyl's mathematical personality:

> Plutôt que de saisir l'idée brutalement au risque de la meurtrir, il aimait bien mieux la guetter dans la pénombre, l'accompagner dans ses évolutions, la décrire sous ses multiples aspects, dans sa vivante complexité. Etait-ce de sa faute si ses lecteurs, moins agiles que lui, éprouvaient parfois quelque peine à le suivre?

[24] (44) For the convenience of the reader, I recall some definitions pertaining to linear algebraic groups, and refer to either my book (Benjamin 1979; 2nd enlarged ed., Grad. Texts Math. **126**, Springer, 1991) or the one by J. E. Humphreys (Grad. Texts Math. **21**, Springer, 1975) for more details.

Let $K$ be an algebraically closed ground field. A linear algebraic group over $K$ is a subgroup $G$ of $\mathbf{GL}_n(K)$ whose elements are all the invertible matrices whose coefficients annihilate a given family of polynomials in $n^2$ indeterminates, with coefficients in $K$. It may be viewed as an affine variety with coordinate ring $K[G]$ generated by the matrix entries and the inverse of the determinant. A linear representation $\sigma\colon G \to \mathbf{GL}_m(K)$ is rational if the coefficients of $\sigma(g)$ belong to $K[G]$. Let $G$ be irreducible as an algebraic variety. It is semisimple if it has no infinite normal commutative subgroup, reductive if its center consists of semisimple elements. The flag varieties of $G$ are the homogeneous spaces of $G$ which are projective varieties. The isotropy groups of points in those are the parabolic subgroups of $G$. If $G = \mathbf{SL}_n(K)$, these are the stability groups of the flags, i.e., strictly increasing sequences of subspaces of $K^n$ whose dimensions form a given sequence of integers, of which Weyl has studied the invariant theory (in characteristic zero) in [**W60**], as already pointed out.

Now, let $K = \mathbb{C}$ and let $G$ be semisimple. $G$ is also a complex Lie group, but the representation theories of $G$ viewed as a complex Lie group or as an algebraic group are essentially the same since any holomorphic finite dimensional representation is automatically rational. Moreover, any irreducible representation occurs in a tensor product of copies of the defining representation of $G$ and of its contragredient. This had been shown by Schur [**Sc2**] for $\mathbf{SO}_n(\mathbb{C})$ and by Weyl [68] for $\mathbf{SL}_n(\mathbb{C})$ and $\mathbf{Sp}_{2n}(\mathbb{C})$, and now supplies the natural framework for an algebraic theory over more general ground fields. However, this does not hold for $\mathbf{GL}_n(\mathbb{C})$, i.e. for reductive non-semisimple groups, as was already stressed by Weyl in [**W68**, I, §8]: the rational representations of $\mathbf{GL}_n(\mathbb{C})$ are fully reducible [**Sc1**], but not all of the holomorphic ones are.

[25] (45) Let $(H, H')$ be a reductive pair in $\mathbf{GL}_N(\mathbb{C})$ and assume $H, H'$ to be connected. Instead of the enveloping algebras of $H$ and $H'$, consider the enveloping algebras $B$ and $B'$ of their Lie algebras $\mathfrak{h}$ and $\mathfrak{h}'$. They are the centralizers of one another in $\mathbf{M}_N(\mathbb{C})$ and are fully reducible. Their intersection is the center of each. Now $B$ and $B'$ are quotients of $\mathcal{U}(\mathfrak{h})$ and $\mathcal{U}(\mathfrak{h}')$ and, by full reducibility, the center of $B$ (resp. $B'$) is the image of $\mathcal{Z}(\mathfrak{h})$ (resp. $\mathcal{Z}(\mathfrak{h}')$). This remark is made in [**Ho3**] and in particular applied to the Capelli identities, in the case of $p$ vectors in $n$-space. There we get back to the situation described in [19]: $N = n \cdot p$, $H$ is $\mathbf{GL}_n(\mathbb{C})$ operating on $\mathbf{M}_{n,p}(\mathbb{C})$ by left translations, and $H'$ is $\mathbf{GL}_p(\mathbb{C})$ operating by right translations. $B'$ is isomorphic to $\mathcal{U}(\mathfrak{gl}_p)$ and $B \cap B'$ is a polynomial algebra in the Capelli elements.

# References for Chapter III

[A]    M. Artin, *On Azumaya algebras and finite dimensional representations of rings*, J. Algebra **11** (1969), 532–563

[ABP]   M. Atiyah, R. Bott, V. Patodi, *On the heat equation and the index theorem*, Invent. Math. **19** (1973), 279–330

[Bo1]   N. Bourbaki, Groupes et Algèbres de Lie, Chap. I, Hermann, Paris, 1971

[Bo2]   _____ , Groupes et Algèbres de Lie, Chaps. 7, 8, Hermann, Paris, 1975

[Br1]    R. Brauer, *Eine Bedingung für vollständige Reduktion von Darstellungen gewöhnlicher und infinitesimaler Gruppen*, Math. Zeitschr. **41** (1936), 330–339; Collected Papers III, 462–471

[Br2]    _____ , *On algebras which are connected with the semisimple continuous groups*, Annals of Math. **38** (1937), 857–872; Collected Papers III, 446–461

[Ca1]   A. Capelli, *Über die Zurückführung der Cayley'schen Operation $\Omega$ auf gewöhnliche Polar-Operationen*, Math. Annalen **29** (1887), 331–338

[Ca2]   _____ , *Ricerca delle operazioni invariantive fra più serie di variabili permutabili con ogni altra operazione invariantiva fra le stesse serie*, Soc. Reale Napoli: Atti Reale Accad. Sci. Fis. Mat. (2) 1 (1888), 1–17

[Ca3]   _____ , *Sur les opérations dans la théorie des formes algébriques*, Math. Annalen **37** (1890), 1–37.

$$- \quad - \quad - \quad -$$

## Papers by Élie Cartan

The numbering follows his "Oeuvres Complètes" (O.C.), 6 vols., Gauthier-Villars, Paris, 1952.

[C5]    É. Cartan, Sur la structure des groupes de transformations finis et continus, Thèse, Nony, Paris, 1894; 2$^e$ édition, Vuibert, Paris, 1933; O.C. I$_1$, 137–253

[C37]   _____ , *Les groupes projectifs qui ne laissent invariante aucune multiplicité plane*, Bull. Soc. Math. France **41** (1913), 53–96; O.C. I$_1$, 355–398

[C38]   _____ , *Les groupes réels simples, finis et continus*, Ann. Sci. École Norm. Sup. (3) **31** (1914), 263–355; O.C. I$_1$, 399–491

[C65]   _____ , *Sur un théorème fondamental de M. H. Weyl*, J. Math. Pures Appl. (9) **2** (1923), 167–192; O.C. III$_1$, 633–648

[C80]   _____ , *Sur les variétés à connexion affine et la théorie de la relativité généralisée*, Ann. Sci. École Norm. Sup. (3) **42** (1925), 17–88; O.C. III$_2$, 921–995

[C81]   _____ , *Les tenseurs irréducibles et les groupes linéarires simples et semisimples*, Bull. Sci. Math. **49** (1925), 130–152; O.C. I$_1$, 531–553

[C82]   _____ , *Le principe de dualité en la théorie des groupes simples et semi-simples*, Bull. Sci. Math. **49** (1925), 361–374; O.C. I$_1$, 555–568

[C93]   _____ , *Sur une classe remarquable d'espaces de Riemann*, Bull. Soc. Math. France **54** (1926), 214–264; O.C. I$_2$, 587–637

[C94]   _____ , *Sur une classe remarquable d'espaces de Riemann*, Bull. Soc. Math. France **55** (1927), 114–134; O.C. I$_2$, 639–659

[C103]   _____ , *La géométrie des groupes simples*, Annali di Mat. Pura Appl. (4) **4** (1927), 209–256; O.C. I$_2$, 793–840

[C105]   _____ , *La théorie des groupes et la géométrie*, Enseignement Math. **26** (1927), 200–225; O.C. I$_2$, 841–866

[C107]   _____ , *Sur certaines formes riemanniennes remarquables des géométries à groupe fondamental simple*, Ann. Sci. École Norm. Sup. (3) **44** (1927), 345–467; O.C. I$_2$, 867–989

[C111]   _____ , *Sur les nombres de Betti des espaces de groupes clos*, C.R. Acad. Sci. Paris **187** (1928), 196–198; O.C. I$_2$, 999–1001

[C113]   _____ , *Complément au mémoire "Sur la géométrie des groupes simples"*, Annali di Mat. Pura Appl. (4) **5** (1928), 253–260; O.C. I$_2$, 1003-1010

[C116]   _____ , *Groupes simples clos et ouverts et géométrie riemannienne*, J. Math. Pures Appl. (9) **8** (1929), 1–33; O.C. I$_2$, 1011–1043

[C117]   _____ , *Sur la détermination d'un système orthogonal complet dans un espace de Riemann symétrique clos*, Rend. Circ. Mat. Palermo **53** (1929), 217–252; O.C. I$_2$, 1045–1080

[C118] ———, *Sur les invariants intégraux de certains espaces homogènes clos et les propriétés topologiques de ces espaces*, Ann. Soc. Polon. Math. **8** (1929), 181–225; O.C. I$_2$, 1081–1125

[C128] ———, *La théorie des groupes finis et continus et l'analysis situs*, Mém. Sci. Math. XLII, Gauthier-Villars, Paris, 1930; O.C. I$_2$, 1165–1224

– – – – –

[Cs] H. Casimir, *Über die Konstruktion einer zu den irreduziblen Darstellungen halbeinfacher kontinuierlicher Gruppen gehörigen Differentialgleichung*, Koninkl. Akad. Wetensch. Amsterdam Proc. Section Sci. **34** (1931), 844–846

[CW] H. L. Casimir, B. L. van der Waerden, *Algebraischer Beweis der vollständigen Reduzibilität der Darstellungen halbeinfacher Liescher Gruppen*, Math. Annalen **111** (1935), 1–12

[CWi] C. Chevalley, A. Weil, *Hermann Weyl (1885–1955)*, Enseignement Math. (2) **3** (1957), 157–187; reprinted in H. Weyl, Gesammelte Abhandlungen IV, 655–685

[CP] C. De Concini, C. Procesi, *A characteristic free approach to invariant theory*, Adv. in Math. **21** (1976), 300–354

[DRS] P. Doubilet, G. C. Rota, J. Stein, *On the foundations of combinatorial theory. IX: Combinatorial methods in invariant theory*, Studies in Applied Math. **53** (1974), 185–216

[DHS] W. Dwyer, W. C. Hsiang, R. Staffeldt, *Pseudo isotopy and invariant topology*: I, Topology **19** (1980), 367–385; II, Topology Symposium Siegen 1979, Lecture Notes in Math. **788**, Springer, 1980, 418–441

[F] H. Freudenthal, *Zu den Weyl-Cartanschen Raumproblemen*, Archiv der Math. **11** (1960), 107–115

[Ha] W. Haboush, *Reductive groups are geometrically reductive*, Annals of Math. **102** (1975), 67–84

[HC] Harish-Chandra, Collected Papers, 4 vols., Springer, 1984

[Ho1] R. Howe, *Invariant theory and duality for classcal groups over finite fields with applications to their singular representation theory*, preprint

[Ho2] ———, *Remarks on classical invariant theory*, Trans. Amer. Math. Soc. **313** (1989), 539–570; erratum, ibid. **318** (1990), 823

[Ho3] ———, *θ-series and invariant theory*, in Automorphic Forms, Representations and L-functions, Proc. Sympos. Pure Math. **33**, Part I, Amer. Math. Soc., 1979, 275–285

[Ho4] ———, *On the role of the Heisenberg group in harmonic analysis*, Bull. (N.S.) Amer. Math. Soc. **3** (1980), 821–843

[Ho5] ———, *Transcending classical invariant theory*, J. Amer. Math. Soc. **2** (1989), 535–552

[Hu] A. Hurwitz, *Über die Erzeugung der Invarianten durch Integration*, Nachr. Kön. Ges. Wiss. Göttingen, Math.-Phys. Kl. **1897**, 71–90; Mathematische Werke II, 546–564

[Hw] T. Hawkins, Emergence of the theory of Lie groups. An essay in the history of mathematics 1869–1926, Springer, 2000

[J1] N. Jacobson, *Rational methods in the theory of Lie algebras*, Annals of Math. **36** (1935), 875–881

[J] ———, Lie Algebras, Interscience, New York, 1962

[K] W. Klingenberg, *Eine Kennzeichnung der Riemannschen sowie der Hermiteschen Mannigfaltigkeiten*, Math. Zeitschr. **70** (1959), 300–309

[L] S. Lie, Fr. Engel, Theorie der Transformationsgruppen III, Teubner, Leipzig, 1893

[M] G. Mackey, *Hermann Weyl and the applicatons of group theory to quantum mechanics*, in Exakte Wissenschaften und ihre Philosophische Grundlegung: Vorträge des Internationalen Hermann Weyl Kongresses (Kiel, 1985), P. Lang, Berlin, 1986

[Na] M. Nagata, *Invariants of a group in an affine ring*, J. Math. Kyoto Univ. **3** (1963/64), 369–377

[Ne] J. von Neumann, *Die Eindeutigkeit der Schrödingerschen Operatoren*, Math. Annalen **104** (1931), 570–578; Collected Works II, 221–229

[P1] C. Procesi, *The invariant theory of n × n matrices*, Adv. in Math. **19** (1976), 306–381

[P2] ———, *Sulla formula di Gordan Capelli*, in La Matematica nella Cina Antica e di Oggi, Università di Ferrara 1979

[Ra] G. Racah, *Sulla caractterizzazione delle rapresentazioni irreducibili dei gruppi semisemplici di Lie*, Atti Accad. Naz. Lincei Rend. Cl. Sci. Fis. Mat. Nat. (8) **8** (1950), 108–112

[Ri]    B Riemann, Über die Hypothesen, welche der Geometrie zu Grunde liegen, edited and commented by H. Weyl, Springer, Berlin, 1919

[Sc1]   I. Schur, Über eine Klasse von Matrizen, die sich einer gegebenen Matrix zuordnen lassen, Dissertation, Berlin, 1901; Ges. Abh. I, 1–72

[Sc2]   _____, Neue Anwendungen der Integralrechnung auf Probleme der Invariantentheorie. I, II, Sitzungsber. Preuss. Akad. Wiss. Berlin Phys.-Math. Kl. **1924**, 189–208, 297–321; Ges. Abh. II, 440–484

[Se]    C. S. Seshadri, Standard monomial theory and the work of Demazure, in Algebraic Varieties and Analytic Varieties (Tokyo, 1981), Advanced Studies in Pure Math. **1**, Kinokuniya, Tokyo, and North-Holland, Amsterdam, 1983, 355–384

[So]    M. H. Stone, Linear transformation in Hilbert space. III: Operational methods and group theory, Proc. Nat. Acad. Sci. USA **16** (1930), 172–175

[St1]   E. Study, Methoden zur Theorie der ternären Formen, Teubner, Leipzig, 1889

[St2]   _____, Einleitung in die Theorie der Invarianten linearer Transformationen auf Grund der Vektorrechnung, Vieweg, Braunschweig, 1923

[Wi1]   A. Weil, L'intégration dans les groupes topologiques et ses applications, Hermann, Paris, 1940

[Wi2]   _____, Sur certains groupes d'opérateurs unitaires, Acta Math. **111** (1964), 143–221, and Sur la formule de Siegel dans la théorie des groupes classiques, ibid. **113** (1965), 1–87; Oeuvres Scientifiques III, 1–69, 71–156.

[Wn]    A. Weinstein, Fundamentalsatz der Tensorrechnung, Math. Zeitschr. **16** (1923), 78–91; Selecta, 1–15.

− − − − −

## Books and Papers by Hermann Weyl

The numbering of the papers follows his Gesammelte Abhandlungen (G.A.), 4 vols., Springer, Berlin, 1968. Also, SHW stands for Selecta Hermann Weyl, Birkhäuser, Basel, 1956.

[WI]    H. Weyl, Raum, Zeit, Materie: Vorlesungen über allgemeine Relativitätstheorie, $4^{th}$ ed., Springer, Berlin, 1921; $7^{th}$ (variorum) ed., 1988; English transl., Methuen, London, and Dover, New York, 1922

[WII]   _____, Mathematische Analyse des Raumproblems, Springer, Berlin 1923; Wissenschaftliche Buchgesellschaft, Darmstadt, 1977

[WIII]  _____, The structure and representations of continuous groups, I (Notes by N. Jacobson), II (Notes by R. Brauer), The Institue for Advanced Study, 1934–1935

[WIV]   _____, I, Elementary Theory of Invariants (Notes by Weyl and L. M. Blumenthal); II, Invariant Theorey (Outline by A. H. Clifford), The Institute for Advanced Study, 1935–1936

[WV]    _____, Gruppentheorie und Quantenmechanik, Hirzel, Leipzig 1928; $2^{nd}$ ed., 1931; English transl. (by H. P. Robertson), Dover, New York, 1950

[WVI]   _____, The Classical Groups, Their Invariants and Representations, Princeton Univesity Press, 1939; $2^{nd}$ edition, 1946

[W8]    _____, Über gewöhnliche Differentialgleichungen mit Singularitäten und die zugehörigen Entwicklungen willkürlicher Funktionen, Math. Annalen **68** (1910), 220–269; G.A. I, 248–297; SHW, 1–58

[W49]   _____, Die Einzigartigkeit der Pythagoreischen Maßbestimmung, Math. Zeitschr. **12** (1922), 114–146; G.A. II, 263–295

[W60]   _____, Randbemerkungen zu Hauptproblemen der Mathematik, Math. Zeitschr. **20** (1924), 131–150; G.A. II, 433–452

[W61]   _____, Zur Theorie der Darstellung der einfachen kontinuierlichen Gruppen (Aus einem Schreiben an Herrn I. Schur), Sitzungsber. Preuss. Akad. Wiss. Berlin Phys.-Math. Kl. **1924**, 338–345; G.A. II, 453–460

[W62]   _____, Das gruppentheoretische Fundament der Tensorrechnung, Nachr. Ges. Wiss. Götteningen, Math.-Phys. Kl. **1924**, 218–224; G.A. II, 461–467

[W63]   _____, Über die Symmetrie der Tensoren und die Tragweite der symbolischen Methode in der Invariantentheorie, Rend. Circ. Mat. Palermo **48** (1924), 29–36; G.A. II, 468–475

[W68]  ———— , *Theorie der Darstellung kontinuierlicher halbeinfacher Gruppen durch lineare Transformatinen.* I, II, III und Nachtrag, Math. Zeitschr. **23** (1925), 271–309; **24** (1926), 328–376, 377–395, 789–791; G.A. II, 543–647; SHW, 262–366

[W69]  ———— , *Zur Darstellungstheorie und Invariantenabzählung der projektiven, der Komplex und der Drehungsgruppe,* Acta Mathematica **48** (1926), 255–278; G.A. III, 1–24

[W70]  ———— , *Elementare Sätze über die Komplex- und die Drehungsgruppe,* Nachr. Ges. Wiss. Göttingen, Math.-Phys. Kl. **1926**, 235–243; G.A. III, 25–33

[W71]  ———— , *Beweis des Fundamentalsatzes in der Theorie der fastperiodischen Funktionen,* Sitzungsber. Preuss. Akad. Wiss. Berlin Phys.-Math. Kl. **1926**, 211–214; G.A. III, 34–37

[W72]  ———— , *Integralgleichungen und fastperiodische Funktionen,* Math. Annalen **97** (1927), 338–356; G.A. III, 38–57; SHW, 367–386

[W73]  ———— , *Die Vollständigkeit der primitiven Darstellungen einer geschlossenen kontinuier-lichen Gruppe* (gem. mit F. Peter), Math. Annalen **97** (1927), 737–755; G.A. III, 58–75; SHW, 387–404

[W74]  ———— , *Sur la représentation des groupes continus,* Enseignement Math. **26** (1927), 226–239; G.A. III, 76–89

[W75]  ———— , *Quantenmechanik und Gruppentheorie,* Zeitschr. Physik **46** (1927), 1–46; G.A. III, 90–135

[W76]  ———— , *Strahlbildung nach der Kontinuitätsmethode behandelt,* Nachr. Ges. Wiss. Göttingen, Math.-Phys. Kl. **1927**, 227–237; G.A. III, 136–146

[W79]  ———— , *Der Zusammenhang zwischen der symmetrischen und der linearen Gruppe,* Annals of Math. **30** (1929), 499–516; G.A. III, 171–188

[W80]  ———— , *Kontinuierliche Gruppen und ihre Darstellungen durch lineare Transformatio-nen,* Atti Congr. Internat. Math. (Bologna 1928), Vol. I, Zanichelli, Nologna, 1929, 233–246; G.A. III, 189–202

[W96]  ———— , *Über Algebren, die mit der Komplexgruppe in Zusammenhang stehen, und ihre Darstellungen,* Math. Zeitschr. **35** (1932), 300–320; G.A. III, 359–379

[W98]  ———— , *Harmonics on homogeneous manifolds,* Annals of Math. **35** (1934), 486–499; G.A. III, 386–399

[W105]  ———— , *Spinors in n dimensions* (with R. Brauer), Amer. J. Math. **57** (1935), 425–449; G.A. III, 493–516; SHW, 431–454

[W107]  ———— , *Generalized Riemann matrices and factor sets,* Annals of Math. **37** (1936), 709–745; G.A. III. 534–570

[W117]  ———— , *Invariants,* Duke Math. J. **5** (1939), 489–502; G.A. III, 670–683

[W122]  ———— , *On the use of indeterminates in the theory of the orthogonal and symplectic groups,* Amer. J. Math. **63** (1941), 777–784; G.A. III, 670–683

[W147]  ———— , *Relativity theory as a stimulus in mathematical research,* Proc. Amer. Philos. Soc. **93** (1949), 535–541; G.A. IV, 394–400

# CHAPTER IV

# Élie Cartan, Symmetric Spaces
# and Lie Groups

In the years 1925-30, Élie Cartan single-handedly developed a theory of semi-simple groups and of a category of homogeneous Riemannian manifolds (symmetric spaces) which enormously increased the scope and applications of Lie groups. Some highlights were briefly enumerated in Chapter III. We give here a more detailed treatment.

Around 1925, some joint work with J.A. Schouten ([C91], 1926) drew Cartan's attention to a class of Riemannian manifolds, those whose curvature tensor is invariant under parallel transport or, equivalently, has zero covariant derivative, called initially spaces $\mathcal{E}$ ("espaces $\mathcal{E}$"), and, later (from [C117], 1929, on), symmetric spaces.[0] The study of these spaces, a problem in differential geometry (at first, of rather limited interest as Cartan said later [C138], 1932), turned out to have deep connections with semisimple groups, which Cartan explored and developed in a number of papers mainly between 1926 and 1935. They are all contained in Vol. II, Part 1 of his Oeuvres Complètes [C0]. For the sake of this discussion, and without wanting to propose a rigid classification, I would like to group them into two parts, A: The building of the theory and B: Further developments.

In Part A the starting point (§1) is essentially local in character, as pointed out on occasion by Cartan himself, even though there is often some ambiguity in the formulation of the concepts and results. The next part (§2), strongly marked in its initial stages by the influence of Hermann Weyl, is decidedly global, and culminates in a beautiful theory which combines differential geometry and semisimple Lie groups in a striking manner. This quickly led Cartan to outline the foundations of a global theory of Lie groups and homogeneous spaces, in particular symmetric spaces, in his monograph [C128], 1930 (see §3).

Part B is devoted to further explorations in three different directions: §4 is concerned with [C117], 1929, where Cartan extends the Peter-Weyl theory to compact symmetric spaces, and §5 deals with algebraic topology. Cartan proposes the use of "integral invariants" (exterior differential forms) to compute the Betti numbers of compact Lie groups and their homogeneous spaces, an idea which greatly influenced the course of algebraic topology. §6 deals with a later paper [C145], 1935, to which Cartan was led by the work of his son Henri Cartan on automorphisms of bounded domains. It singles out the subclass of irreducible symmetric spaces which

---

[0] The terminology "espace $\mathcal{E}$" has completely disappeared nowadays. Still, I shall keep it in the first two sections, to be faithful to Cartan and also to remind us that Cartan was more likely to view such a space as a totally geodesic submanifold of the group manifold than as a quotient space.

Élie Cartan

admit an invariant complex structure, later called hermitian symmetric spaces, the non-compact representatives of which are the bounded symmetric domains.

This still leaves out one contribution of Cartan to the theory of (compact) Lie groups. It will be included, however, in Chapter VI, §6.

*If $M$ is a smooth manifold and $x \in M$, we let $TM$ be the tangent bundle of $M$ and $TM_x$ the tangent space at $x$.*

## A. Building Up the Theory

### §1. The spaces $\mathcal{E}$. Local theory

**1.1.** Let $M$ be a Riemannian manifold (connected—this will always be understood). In [C93], 1926, and subsequent papers, Cartan calls it a space $\mathcal{E}$ if its curvature tensor is invariant under parallel transport or, equivalently, has covariant derivative zero. It is noted in §14 of [C93] that $M$ is in that category if and only if the local symmetry around each point is isometric. Here, by a local symmetry at $x$ is meant the diffeomorphism of a geodesic neighborhood of $x$ which maps a geodesic segment with origin $x$ and initial tangent vector $t$ isometrically onto the geodesic segment with origin $x$ and initial tangent vector $-t$. However, it is only from [C117], 1929, on that Cartan put that property to the foreground and called $M$ a symmetric space. More precisely, what has been defined so far is a Riemannian locally symmetric space. $M$ is symmetric (globally) if the local symmetry at each point is induced by a global isometry, necessarily of order two. Although Cartan does not make at that stage a sharp terminological distinction between locally and globally symmetric, it is very much on his mind and he distinguishes between the two in his results, see e.g. footnote 1, p. 263 in [C93].

**1.2.** Let $M$ be a space $\mathcal{E}$. Given a geodesic arc $\overline{xy}$ with end points $x, y \in M$ which is small, i.e. contained in a geodesic neighborhood of its middle point $m$, say, the local symmetry $s_m$ exchanges $x$ and $y$. Since any two points $x, y \in M$ may be joined by a path made up of small geodesic arcs, there always exists a local isometry of $M$ mapping some neighborhood of $x$ onto one of $y$, so $M$ is Riemannian locally homogeneous. If $M$ is (globally) symmetric, this argument shows the existence of a global isometry mapping $x$ onto $y$, and $M$ is Riemannian homogeneous.

Let $\overline{xy}$ again be a geodesic arc, small enough to be contained in geodesic neighborhoods of its end points. Then the product $s_y.s_x$ of the local symmetries at $x$ and $y$ maps $x$ to $x' = s_y(x)$, and its differential induces an isomorphism of $TM_x$ onto $TM_{x'}$. The map $s_y.s_x$ is isometric, is shown to define the parallel transport from $TM_x$ to $TM_{x'}$ along $\overline{xx'}$, and is called a *transvection* along that geodesic. Therefore the holonomy group $\Gamma_x$ at $x$ is a group of *isometries* of $TM_x$ (with respect to the given metric). [Recall that, by definition, $\Gamma_x$ is the subgroup of $GL(TM_x)$ generated by parallel transport along loops based at $x$, which may be assumed to be broken geodesic polygons.] $\Gamma_x$ is also the greatest connected group of local isometries leaving $x$ fixed ([C93], n° 43). Up to isomorphism, it is independent of $x$.

**1.3.** The spaces with constant curvature are obviously among the spaces $\mathcal{E}$, but Cartan soon realized there were many others. The main goal of [C93] and [C94] is to give a classification, up to local isomorphisms. The problem is easily reduced to the search for "irreducible" ones, i.e. those which cannot be written locally as isometric products of two Riemannian manifolds. This is equivalent to $\Gamma_x$ being an irreducible group of automorphisms of $TM_x$. Cartan indicates two methods to classify them, based respectively on the consideration of $\Gamma_x$ and of the identity component $G$ of the group of isometries of $M$. Underlying the first one is a classification of subgroups of the orthogonal groups which leave the curvature tensor invariant. It is the subject matter of Chapter II in [C93]. Cartan does not fully carry out all the computations which would yield a complete list, but proves a number of important results, notably:

To a given $\Gamma_x$ there correspond two spaces $\mathcal{E}$, one positively and one negatively curved. If $\Gamma_x$ is not absolutely irreducible, then it is the product of a one-dimensional central group by one or two simple non-abelian groups. If it is absolutely irreducible, it is the product of one or two simple non-abelian groups. In the latter case Cartan determines $\Gamma_x$ when it has no 3-dimensional factor. The case of a complex simple Lie algebra, viewed as a real simple Lie algebra, and the corresponding case of positive curvature are also considered. He also points out that, since $\Gamma_x$ is irreducible, any $\Gamma_x$ invariant quadratic form on $TM_x$ is a multiple of the metric, and hence the "contracted Riemann tensor" (the Ricci form) is a multiple of the metric tensor.

**1.4.** [C94], 1927, is devoted to the second method, of which he says, at the beginning of [C93]:

> Cette dernière méthode est celle qui conduit aux problèmes les plus inattendus; en particulier, la recherche des espaces $\mathcal{E}$ irréductibles revient à celle de toutes les structures simples *réelles* qui correspondent à un même type de structure simple *complexe*. C'est un problème que j'ai résolu dans un mémoire d'avant guerre [C38] et c'est en partant des résultats que j'ai alors obtenus que la détermination des espaces $\mathcal{E}$ irréductibles se présente de la manière la plus simple.[1]

Cartan first discusses "the form $\varphi(e)$", giving the squares of the eigenvalues of the characteristic equation, i.e. of $\operatorname{ad} x$ ($x \in \mathfrak{g}$), which I shall call the Killing form, even though Cartan never did so (see I, 2.1 for that terminology). He first arrives at the now familiar *Cartan decomposition* of the Lie algebra $\mathfrak{g}$ of $G$, all expressed in terms of basis elements. I shall allow myself to get closer to present day notation by using subspaces. Let $\mathfrak{k}$ be the Lie algebra of $\Gamma_x$. The restriction of $\varphi(e)$ to it is negative non-degenerate. The orthogonal complement $\mathfrak{p}$ to it may be identified to $TM_x$ and the restriction of $\varphi(e)$ to it is also definite, being invariant under $\Gamma_x$, positive or negative. Then we have $\mathfrak{g} = \mathfrak{k} \oplus \mathfrak{p}$ and

$$(1) \qquad\qquad [\mathfrak{k}, \mathfrak{k}] \subset \mathfrak{k}, \quad [\mathfrak{k}, \mathfrak{p}] \subset \mathfrak{p}, \quad [\mathfrak{p}, \mathfrak{p}] \subset \mathfrak{k}.$$

See (12), p. 115, which in fact goes back to (12), p. 225, in [C93].

Assume $\mathfrak{g}$ is simple unitary, or, equivalently, that the restriction of $\varphi$ to $\mathfrak{p}$ is also negative non-degenerate. Then Cartan associates to the given space $\mathcal{E}$ another one, with negative curvature. Its connected group of isometries $G'$ has as Lie algebra

$$(2) \qquad\qquad \mathfrak{g}' = \mathfrak{k} \oplus \mathfrak{p}', \quad \text{where } \mathfrak{p}' = i\mathfrak{p} \ (i = \sqrt{-1}),$$

viewed as a real subalgebra of the complexification $\mathfrak{g}_c = \mathfrak{g} \otimes_{\mathbb{R}} \mathbb{C}$ of $\mathfrak{g}$. [It has to be checked that $\mathfrak{g}'$ is a real Lie algebra, but this follows from (1)]. Both $\mathfrak{g}$ and $\mathfrak{g}'$ are real forms of $\mathfrak{g}_c$. He calls $\mathfrak{g}$ and $\mathfrak{g}'$ two simple real structures, belonging to the same complex simple type, between which there is what he calls a "normal isomorphism", namely the map

$$x + y \mapsto x + i\,y \qquad (x \in \mathfrak{k}, y \in \mathfrak{p}).$$

[This is of course not an isomorphism of real Lie algebras, but it extends to an automorphism of $\mathfrak{g}_c$.] He asks whether this is a general phenomenon by raising two questions on p. 123 (again stated here in a different language):

1. Given a simple non-unitary real structure belonging to a given complex type $\mathfrak{g}_c$, does there exist a normal isomorphism onto a compact form of $\mathfrak{g}_c$?

2. Given two such structures, do the normal isomorphisms between them always define the same holonomy group?

---

[1] "This last method is the one which leads to the most unexpected problems; in particular the search for the irreducible spaces $\mathcal{E}$ amounts to a search for all the *real* simple structures which correspond to the same type of *complex* simple structure. This is a problem which I have solved in a paper written before the war [C38] and it is by starting from the results then obtained that the determination of the irreducible spaces $\mathcal{E}$ presents itself in the simplest manner".

He then states that the answer is affirmative, but he can only check it type by type, leaning on [C38]. He will later find a conceptual answer to his questions, so I shall not give more details here. In these questions, $\mathfrak{g}$ is supposed to be simple. If not, he shows that $\mathfrak{g}$ is a product of two isomorphic unitary simple algebras $\mathfrak{g}$, and $\mathfrak{g}'$ is a complex form of $\mathfrak{g}$, viewed as a real Lie algebra.

To arrive at his questions, Cartan also uses a characterization of the space $\mathcal{E}$ as a totally geodesic submanifold of the group, using the fact that $\mathfrak{p}$ satisfies the condition $[\mathfrak{p}, [\mathfrak{p}, \mathfrak{p}]] \subset \mathfrak{p}$ (i.e. is a Lie triple system) in view of (1). The construction of totally geodesic submanifolds tangent to Lie triple systems is discussed at length, in a more general framework, in [C101], 1927, which appeared after [C93], but before [C94], to be discussed below.

**Remark.** The space $\mathcal{E}$ is at first a homogeneous space, a coset space. If so, the operations of $G$ are simply induced by left translations. However, very often, Cartan prefers to view the space $\mathcal{E}$ as a totally geodesic submanifold $P = \exp \mathfrak{p}$ of $G$. We shall later describe the operations of $G$ in that setting (see 2.4.4).

**1.5.** Cartan's paper [C101] presents a systematic exposition of the theory sketched in the joint paper with J.A. Schouten [C91]. It is in the more general context of affine connections and, logically, should precede the previous paragraphs. However, Cartan quickly focused on the Riemannian case, and the added generality of [C101] does not seem to have played much role in his subsequent work. From the point of view of Riemannian symmetric spaces, it is more of a digression (which can be skipped in part), and I have preferred to relegate it to the end of §1.

It is not entirely a digression, however: on two points at least this paper is the only reference in Cartan's work for proofs of the formula (1) below for the curvature form and of the relation between Lie triple systems and totally geodesic manifolds.

The main objects of study in [C101] are three canonical affine connections on a group manifold in which the geodesics through the identity are one-parameter subgroups. Two of them, discussed in Chapter II, have zero curvature, but non-trivial torsion, and an absolute parallelism given either by left or right translations. Their torsion is described by the exterior derivatives of the Maurer-Cartan forms. Chapter III and IV are devoted to a third connection, without torsion. A first definition ($\mathrm{n}^{\mathrm{os}}$ 50-56) displays its parallel transport along a small geodesic as a geometric mean between the parallel transports of the first two connections. Another definition ($\mathrm{n}^{\mathrm{os}}$ 58-59) is given via local symmetries. Note first that in an affine connection, the geodesics are canonically endowed with a structure of an affine line, or of a segment of one, so that, in particular, the middle point of a geodesic arc has an intrinsic meaning, which allows one to define as in 1.2 the local symmetry $s_x$ with center a given point $x$. Then let $\overline{xy}$ be a geodesic arc, with middle point $m$. The product $s_m.s_x$ maps $x$ onto $y$ and defines an isomorphism of $TG_x$ onto $TG_y$. It is the parallel transport from $x$ to $y$ along $\overline{xy}$ in this third connection, again as in 1.2. Given $X, Y \in \mathfrak{g}$, the endomorphism $R(X, Y)$ of $\mathfrak{g}$ given by the curvature tensor is

$$(1) \qquad\qquad R(X; Y) = c.\,\mathrm{ad}[X, Y],$$

where $c$ is a constant depending on some normalizations (n° 60), and it is deduced in n° 62 that the curvature tensor has covariant derivative zero. Furthermore, the Killing form is also invariant under the connection. If $G$ is semisimple, it provides an invariant pseudo-riemannian metric ("pseudo" because not necessarily definite), and the connection under consideration is indeed the associated Riemannian connection (the connection which keeps the metric tensor invariant and has no torsion, usually called the Levi-Cività connection). Next (n$^{os}$ 67, 68), Cartan shows that if $\mathfrak{p} \subset \mathfrak{g}$ is such that $[\mathfrak{p}, [\mathfrak{p}, \mathfrak{p}]] \subset \mathfrak{p}$, then the geodesics tangent to $\mathfrak{p}$ span a totally geodesic submanifold $P$ of $G$. By restriction, the connection induces on $P$ again a torsion free connection with curvature tensor invariant under parallel transport. Cartan's next goal in the paper is to prove a converse. For this we start with a manifold $M$ endowed with a torsion-free connection whose curvature tensor has zero covariant derivative. As in 1.2, one sees, using local symmetries, that it is homogeneous under the greatest group of automorphisms of the connection. Let $H$ be the identity component of the latter and $K$ the isotropy group of a point $x$. Then (p. 88, n° 82) we again have the decomposition $\mathfrak{g} = \mathfrak{k} \oplus \mathfrak{p}$ with the familiar relations 1.4(1) (expressed in terms of a basis). In particular, $\mathfrak{p}$ is a Lie triple system, hence tangent to a totally geodesic submanifold $P$, which is isomorphic to $M$ (n° 82). Assume $H$ to be semisimple. Then the restriction of the Killing form to $\mathfrak{p}$ defines on $M$ an invariant pseudo-riemannian metric. It is in particular invariant under ad $\mathfrak{k}$. If it is definite, then $M$ is a Riemannian space $\mathcal{E}$ and we are back to the situation of the previous sections. If not, then we are dealing with what goes nowadays under the name of affine or semisimple or pseudo-riemannian symmetric spaces. This generalization of Cartan's Riemannian symmetric spaces has been systematically investigated in these last twenty years or so, and the Cartan structure theory (to be recalled in §2) has been extended. These spaces have proved to be highly interesting for harmonic analysis and many natural extensions of eigenvalue problems on Riemannian symmetric spaces. Except here, Cartan limited himself to the Riemannian case. Still, it is worth noting that these later generalizations were already included in the framework of [C101].

## §2. Spaces $\mathcal{E}$ and semisimple groups. Global theory

The theory developed so far was essentially local, even though it had some global aspects. In a way, this bears some analogy with the early phase of Lie theory: the difference between local and global was not always spelled out, sometimes blurred, but when it came to actual precise theorems, one would fall back on the local. From [C103], 1927, on, however, Cartan combined his techniques with results and points of view of Weyl in [W68], 1925-6, and embarked on a global theory, pursued mainly in [C103], [C107], [C116], [C145], and given a systematic exposition in [C128], the first monograph on Lie groups from the global point of view.

**2.1.** Cartan remarks in [C116], 1929, p.2, that Weyl was the first to show the great importance of the distinction between compact and non-compact groups.[2] ("C'est M. H. Weyl qui a montré le premier la grande importance de la distinction entre *groupes clos* et *groupes ouverts*.")

As we saw in Chapter III, Weyl had shown that a compact connected semisimple Lie group has a finite fundamental group, and is covered by its maximal tori, which are all conjugate under inner automorphisms. Given a complex simple Lie algebra, he had also shown the existence of a compact form, and from the construction could deduce some main structural properties of compact groups. With this as a

---

[2]I shall use compact and non-compact, but Cartan never did so. He spoke first of simple unitary groups or Lie algebras, characterized by having a (negative) definite Killing form, and later of "groupes clos" and "groupes ouverts" (literally, closed groups and open groups) for compact and non-compact groups respectively.

starting point, Cartan begins [C103], 1927, with a systematic discussion of compact semisimple groups:

> Un premier chapitre introductif est consacré à la topologie des groupes simples unitaires; il a son point de départ dans les recherches de H. Weyl relatives à la théorie des groupes semisimples ([W68]), la question y est reprise en entier; les résultats de H. Weyl sont complétés et les questions qui se posent sont résolues jusqu'au bout en utilisant un de mes mémoires récents ([C81]).[3]

This paper starts with a discussion of compact semisimple groups. Apart from the language, it is already in substance almost definitive: it and its complement [C113], 1928, contain many of the nowadays standard features of any textbook treatment.

**2.2.** Let $G$ be a compact connected simple group, $T$ a maximal torus, $n$ and $\ell$ their dimensions $\mathfrak{g}, \mathfrak{t}$ their Lie algebras, $G_{\mathrm{ad}}$ its adjoint group (Cartan denotes it $\Gamma$). The roots of $\mathfrak{t}$ in $\mathfrak{g}$ are purely imaginary and denoted by $2\pi i \varphi_\alpha$. The linear forms $\varphi_\alpha$ are called the angular parameters. If $Y \in \mathfrak{t}$, then $\exp \operatorname{ad} y$ has the eigenvalues 1 in $\mathfrak{t}$ and $\exp 2\pi i \varphi_\alpha(x)$ outside $\mathfrak{t}$. From the Killing-Cartan theory, we know that one can choose $\ell$ such roots, to be called simple, such that any other root is a linear combination of the simple roots with integral coefficients of the same sign, and that there is one *dominant root*, in which these coefficients have the maximal possible value.

Let $(R)$ be the lattice in $\mathfrak{t}$ on which all angular parameters take integral values. Then the exponential map yields an isomorphism of $\mathfrak{t}/(R)$ onto the maximal torus $T_{\mathrm{ad}}$ generated by $\mathfrak{t}$ in $G_{\mathrm{ad}}$.

Following Weyl, Cartan considered the group of automorphisms of $\mathfrak{t}$ induced by inner automorphisms of $G$ or $G_{\mathrm{ad}}$ leaving $\mathfrak{t}$ stable, denoted $(S)$ by Weyl, $\mathcal{G}'$ by Cartan, and nowadays called the Weyl group of $\mathfrak{g}$ with respect to $\mathfrak{t}$ (or of $G$ with respect to $T$), denoted $W$ or $W(\mathfrak{t}, \mathfrak{g})$, or $W(T, G)$. He knows from [W68] that the fundamental group of $G$ or $G_{\mathrm{ad}}$ is finite. Here, he desires not only to reprove it but also to determine explicitly the fundamental group of $G_{\mathrm{ad}}$. To this effect, he introduces the now familiar Cartan polyhedron or Cartan alcove, given by

$$(1) \qquad \varphi_\alpha \geq 0 \quad (\alpha \text{ simple}), \qquad \varphi_\alpha \leq 1 \quad (\alpha \text{ dominant}).$$

His argument is in essence the same as Weyl's. The singular elements in $G$ (i.e. those whose centralizers have dimension $> \ell$) form a subvariety of codimension three, and any loop based at 1 can be deformed to one consisting of regular elements (except for 1). Such a loop can then be continuously deformed to one not only in $T$, but in the exponential of $(P)$. The non-zero elements of $\pi_1(G_{\mathrm{ad}})$ are then shown to be represented (bijectively) by the exponentials of the edges of $(P)$ joining the origin to a vertex contained in $(R)$. Cartan then describes $\pi_1(G_{\mathrm{ad}})$, the group of connection of $G_{\mathrm{ad}}$ in his terminology, for all simple compact adjoint groups, as well as $(R)$ and $(P)$ in $\mathfrak{t}$.

Each conjugacy class in $G_{\mathrm{ad}}$ is represented by the exponential of $h$ elements of $(P)$, where $h$ is the order of $\pi_1(G_{\mathrm{ad}})$. Cartan now proceeds to define "abstractly"

---

[3] "A first introductory chapter is devoted to the topology of simple unitary groups; it has its starting point in the research of H. Weyl pertaining to the theory of semisimple groups ([W68]); the problems are completely treated; the results of H. Weyl are completed and the questions which occur are solved completely, using one of my recent memoirs [C81]."

the universal covering, call it here $G_{sc}$, of $G_{ad}$. For this group, $\exp(P)$ is an exact set of representatives of conjugacy classes.[4] Cartan also points out that every group $G_{sc}$ has a faithful linear representation (not necessarily irreducible). He checks this case by case, using his determination of irreducible representations of simple Lie algebras, and finds that one irreducible representation suffices whenever $G$ has a cyclic center (a necessary condition). This includes all cases except for the two-fold covering of $\mathbf{SO}_{4n}$, for which one can use the full spinor representation, direct sum of the two half-spinor representations. He remarks that Weyl had shown this existence only for the four infinite classes, but, in fact, it also follows from the Peter-Weyl theorem [W73], which was being published at that time. Cartan also describes the lattice $(\bar{R})$, contained in $(R)$, which is the kernel of the exponential map from $\mathfrak{t}$ to $G_{sc}$. Then the center of $G_{ad}$ is isomorphic to $(R)/(\bar{R})$, and $(P)$ is the fundamental domain for the group generated by $(\bar{R})$ and $W$ (the lattice $(\bar{R})$ is the coroot lattice, in present day terminology.)

Sections 26 to 29 in Chapter II describe in substance, for each simple unitary type, all the locally isomorphic groups with the given Lie algebra, but Cartan sees this differently, from the point of view of the spaces $\mathcal{E}$. He identifies $G$ to "a representative of the associated space $\mathcal{E}$"; otherwise said, he views $G$ as the homogeneous space of $G \times G$, acting by left and right translations. I shall call it here $\mathcal{E}_{G \times G}$. The element $(u, v) \in G \times G$ acts on $\mathcal{E}_{G \times G}$ by $(u, v).x = u.x.v^{-1}$. The isotropy group of the identity is the diagonal $K = \{(u, u)\}$; the symmetry $s_1$ at the identity is induced by the permutation $s$ of the two factors. On $\mathcal{E}_{G \times G}$ it is the inversion $x \mapsto x^{-1}$. This space can also be viewed as a totally geodesic submanifold of $G \times G$, namely, the "antidiagonal" $\{(u, u^{-1})\}$ $(u \in G)$.

The "mixed isotropy group" (groupe adjoint d'isotropie mixte) is the group of all automorphisms of $\mathcal{E}_{G \times G}$ leaving 1 fixed. It is the same as the group of all linear invertible transformations of the tangent space to $\mathcal{E}_{G \times G}$ at 1 leaving the curvature tensor invariant. It contains all automorphisms of $G$ and also the symmetry $s_1$, and it is generated by those.

The following sections of Chapter II discuss geodesics, antipodal varieties, and the types of families of geodesics which join the identity to a given point.

Chapter III is devoted to complex simple Lie groups, but approached by means of the theory of symmetric spaces.

The space $\mathcal{E}_{G \times G}$ has positive curvature. By the general theory (1.3, 1.4), there corresponds to it a space $\mathcal{E}$ with the same isotropy group, but negative curvature, and the passage from one to the other is labeled a normal isomorphism. If $\mathfrak{g} = \mathfrak{k} \oplus \mathfrak{p}$ is the Cartan decomposition of one, the other is $\mathfrak{g}' = \mathfrak{k} \oplus i\mathfrak{p}$ (see 1.4). In the present case we start from a unitary group product of two isomorphic simple groups; hence, in the adjoint group, $\mathfrak{g}, \mathfrak{k}, \mathfrak{p}$ are given by skew-hermitian matrices, and therefore $i\mathfrak{p} = \mathfrak{p}'$ is represented by hermitian matrices. It is easily checked that $\mathfrak{g}'$ is the complexification of our initial simple unitary Lie algebra, but viewed as a real Lie algebra of twice the dimension. Let $G'$ be the adjoint group of $\mathfrak{g}'$. The space $\mathcal{E}$ of $\mathfrak{g}'$, call it $\mathcal{E}_{G'}$, may be identified to the totally geodesic submanifold of $G'$ tangent to $\mathfrak{p}'$. But $\mathfrak{p}'$ consists of hermitian $r \times r$ matrices ($r = \dim \mathfrak{p}$) and the

---

[4] It seems to me that Cartan implicitly takes for granted that two elements in $T$ which are conjugate in $G$ are already conjugate under $W$. This is indeed true, by an easy, now standard, argument, but I did not find it in Cartan. Later on, he mentions a more general fact of this type as one he could check *a posteriori* (see the end of 2.4.2).

exponential induces a diffeomorphism of the space of hermitian $r \times r$ matrices onto the space of positive non-degenerate hermitian forms. The space $\mathcal{E}_{G'}$ associated to $G'$ is therefore diffeomorphic to euclidean space, and it is shown that the Cartan decomposition of $\mathfrak{g}'$ induces at the group level an isomorphism of manifolds of $G'$ onto $G \times P'$.

If $G' = \mathbf{SL}_n(\mathbb{C})$, and $K = \mathbf{SU}(n)$, then $\mathfrak{p}'$ is the space of $n \times n$ non-degenerate hermitian forms and $G' = K.P'$ is the polar decomposition of a complex invertible $n \times n$ matrix. The global Cartan decomposition just arrived at is a generalization of the polar decomposition of invertible complex matrices.

Thus $G'$ is homeomorphic to the product of a euclidean space by our initial compact group $G$. In particular they have the same fundamental group. Cartan points out that this extends to any covering of the adjoint group of $G$, and that, consequently, every connected simple complex group has a faithful linear representation. He also remarks that on $P'$ any two points are on a unique geodesic, and wonders (n° 49) whether this does not already follow from the fact that $P'$ is simply connected with negative curvature (a question he will soon answer positively under a further assumption of completeness). He also notes that $P'$ has no "Klein form", i.e. no quotient by a discrete group which is still homogeneous, a question I shall come back to in the next section.

**2.3.** In the complement [C113], 1928, Cartan proves geometrically, from properties of reflection groups, a number of facts partly checked case by case in [C103]. In particular: $W$ has a fundamental domain consisting of a cone defined by $\ell$ faces $\varphi_\alpha = 0$, and there exists a unique dominant root so that the polyhedron $(P)$ is defined by (1) above. It also follows a priori that the highest weights are linear combinations with positive integral coefficients of $\ell$ fundamental ones, a fact which Cartan had had to check case by case in [C39], 1914.

**2.4.** The next paper [C107], 1927, is devoted to the spaces $\mathcal{E}$ attached to absolutely simple groups (i.e., to groups whose real Lie algebra remains simple after extension of the ground field to the complex numbers). In this paper and also in [C116], 1929, they are always identified to totally geodesic submanifolds of the group endowed with the metric (and the canonical connection) defined by the Killing form (hence positive if $G$ is compact, indefinite otherwise), and I shall do so, too. Two main themes in that series of papers, global properties and interactions between Lie groups and differential geometry of the space $\mathcal{E}$, are restated in the introduction (p. 346):

> L'étude faite ici est celle des propriétés *intégrales* et non plus *locales* de l'espace. Les problèmes qui se posent sont du reste différents suivant que la courbure est positive ou négative. Leur résolution, basée sur des méthodes générales applicables à toutes les classes, s'appuie sur la théorie des groupes; inversement, l'existence des espaces en question permet de résoudre des problèmes importants de la théorie des groupes.[5]

---

[5] "The study carried out here is that of the global, not local, properties of the spaces. The problems which arise are different according to whether the curvature is positive or negative. Their solution, based on general methods applicable to all the classes, leans on the theory of groups; conversely, the existence of these spaces allows one to solve important problems in group theory".

**2.4.1.** We now consider the situation parallel to 2.2, but where the underlying groups of isometries are simple, and are real forms of a given simple complex type. Chapter I is devoted to the structure theory, developed for both spaces. In particular, it attaches to a space a root system and a Weyl group, later often called the restricted root system and Weyl group. Important properties of non-compact simple groups are derived. These data and further global properties are described for the various types, of positive curvature in Chapter II, of negative curvature in Chapter III. Models for the classical symmetric spaces are given explicitly. They include a number of spaces which were classically known in various contexts, many properties of which are now given a uniform treatment in Cartan's theory.

We first summarize Cartan's starting point, and also fix some notation. At the Lie algebra level, we have the decompositions

$$(1) \qquad \mathfrak{g}_u = \mathfrak{k} \oplus \mathfrak{p}_u \qquad \mathfrak{g} = \mathfrak{k} \oplus \mathfrak{p}, \text{ where } \mathfrak{p} = i.\mathfrak{p}_u,$$

$\mathfrak{g}_u$ being a compact real form and $\mathfrak{g}$ a non-compact real form of their common complexification $\mathfrak{g}_c$. We let $s_u$ and $s$ be the involutive automorphisms of $\mathfrak{g}_u$ and $\mathfrak{g}$ with fixed point set $\mathfrak{k}$ and $(-1)$-eigenspaces, $\mathfrak{p}_u$ and $\mathfrak{p}$. We fix groups $G_u$ and $G$ with Lie algebras $\mathfrak{g}_u$ and $\mathfrak{g}$, and let $\mathcal{E}_{G,u}$ and $\mathcal{E}_G$ be the corresponding spaces $\mathcal{E}$, to be identified with the totally geodesic submanifolds

$$(2) \qquad P_u = \exp \mathfrak{p}_u, \qquad P = \exp .\mathfrak{p}.$$

[The notation is by and large not that of Cartan, who expresses everything in terms of bases. We have however kept $G$ and $G_u$. We also note that Cartan views $\mathcal{E}_{G,u}$ and $\mathcal{E}_G$ as the two spaces attached to $G$; indeed, $\mathcal{E}_{G,u}$ is not determined entirely by $G_u$.] Recall that we have

$$(3) \qquad [\mathfrak{k}, \mathfrak{k}] \subset \mathfrak{k}, \quad [\mathfrak{k}, \mathfrak{p}_u] \subset \mathfrak{p}_u, \quad [\mathfrak{k}, \mathfrak{p}] \subset \mathfrak{p}, \quad [\mathfrak{p}_u, \mathfrak{p}_u], [\mathfrak{p}, \mathfrak{p}] \subset \mathfrak{k}.$$

The metrics on $\mathfrak{k}, \mathfrak{p}, \mathfrak{p}_u$ are defined by the restriction of the Killing form (up to sign). In the formula for the curvature form in [C101], n° 60 (see 1.5), the constant $c$ is then $\pm 1$ and the endomorphism of $\mathfrak{p}_u$ or $\mathfrak{p}$ is defined by the curvature

$$(4) \qquad R(X, Y) = -\mathrm{ad}\,[X, Y] \qquad (X, Y \in \mathfrak{p} \text{ or } X, Y \in \mathfrak{p}_u).$$

As a consequence, if $X, Y$ are orthonormal, then the Riemannian curvature of the facet spanned by $X$ and $Y$ is

$$-c\varphi([X, Y], [X, Y]), \quad \text{where } c = 1 \text{ if } X, Y \in \mathfrak{p}_u \text{ and } c = -1 \text{ if } X, Y \in \mathfrak{p};$$

hence it is $\geq 0$ on $\mathcal{E}_{G,u}$ and $\leq 0$ on $\mathcal{E}_G$. Again, Cartan states this in terms of a suitable basis and finds that the curvature is $\pm$ a sum of squares of structural constants ([C107], 1927, p. 383). In fact, it is stated there for $\mathcal{E}_G$. But, since $\mathfrak{p}_u = i\mathfrak{p}$, it follows also in the other case, though I do not see that Cartan states it explicitly.

**2.4.2.** In view of (3), a subspace $\mathfrak{a}$ of $\mathfrak{p}_u$ (or $\mathfrak{p}$) is a subalgebra if and only if it is commutative. By (4), the totally geodesic submanifold $A = \exp \mathfrak{a}$ is flat, hence covered by euclidean space. Following usual custom, I shall call $A$ or any transform of it under an isometry a *flat*, and a *maximal flat* if $\mathfrak{a}$ is maximal commutative in $\mathfrak{p}_u$ or $\mathfrak{p}$. Cartan developed a theory in which the maximal flats play a role similar to that of maximal tori in the theory of compact groups. In fact, it reduces to the latter in the special case considered in [C103], 1927, of a compact group, if the compact group is viewed as a space $\mathcal{E}$ for $G \times G$ (see 2.2).

Since $K$ acts in the same way on $\mathfrak{p}_u$ and $\mathfrak{p}$, the arguments provide a similar theory for spaces of negative curvature. However, there is *a priori* no analogy with the standard theory of roots, except for the so-called split real simple Lie algebras, where roots and Weyl groups are again familiar objects. But it turned out later that there is a strong connection with the theory of simple algebraic groups (see Chapter VI, §7).

If $\mathfrak{b}$ is a subspace of $\mathfrak{p}$, it is stable under $s$; hence so is its centralizer $\mathfrak{z}(\mathfrak{b})$, which is therefore the direct sum of its intersections $\mathfrak{z}(\mathfrak{b})_\mathfrak{k}$ and $\mathfrak{z}(\mathfrak{b})_{\mathfrak{p}_u}$ with $\mathfrak{k}$ and $\mathfrak{p}_u$ respectively. Fix a maximal commutative subalgebra $\mathfrak{a}$ of $\mathfrak{p}$, a maximal torus $T$ containing $A$, and let $\lambda = \dim \mathfrak{a}$. Let $X \in \mathfrak{a}$. Then $\mathfrak{z}(X)_{\mathfrak{p}_u} \supset \mathfrak{a}$. By definition $\mathfrak{z}(\mathfrak{a})_{\mathfrak{p}_u} = \mathfrak{a}$. The roots of $\mathfrak{g}$ with respect to $\mathfrak{a}$ are zero with multiplicity the dimension of $\mathfrak{z}(\mathfrak{a})$ and non-zero linear forms, to be written as before $2\pi i\varphi_\alpha$, where the $\varphi_\alpha$ are real valued on $\mathfrak{a}$ and are again called the angular parameters. $\varphi_\alpha$ and $-\varphi_\alpha$ occur with the same multiplicity. An element $X \in \mathfrak{a}$ is regular if no $\varphi_\alpha$ is zero on $X$. If $X$ is regular, then $\mathfrak{z}(X) = \mathfrak{z}(\mathfrak{a})$, in particular $\mathfrak{z}(X)_{\mathfrak{p}_u} = \mathfrak{a}$. Those elements form an open dense set in $\mathfrak{a}$. If $X$ is singular (some $\varphi$ is zero on $X$), then $\dim \mathfrak{z}(X) \geq \lambda + 2$ (since the roots occur in complex conjugate pairs), from which it follows that the set of singular elements in $\mathfrak{p}_u$ has codimension at least two in $\mathfrak{p}_u$. By an argument borrowed from Weyl, already used in [C103], see 2.2, Cartan deduces from it that any two maximal $\mathfrak{a}$'s are conjugate under $K$.[6] Their common dimension $\lambda$ is, by definition, the *rank* of the space $\mathcal{E}$. These maximal $\mathfrak{a}$ are the groups or infinitesimal groups $\gamma_\lambda$ in [C103], $n^{os}$ 5, 6, and are nowadays often called Cartan subalgebras of the symmetric pair $(\mathfrak{g}_u, \mathfrak{k})$; similarly $i\mathfrak{a}$ is a Cartan subalgebra of the symmetric pair $(\mathfrak{g}, \mathfrak{k})$.

**2.4.3.** The next goal, expressed in today's language, is to show that the angular parameters form an irreducible root system, with Weyl group induced by elements of $K$ normalizing $\mathfrak{a}$. The angular parameters span $\mathfrak{a}^*$ (otherwise the center of $\mathfrak{g}_u$ would have strictly positive dimension) and are invariant under any automorphism induced by an element of $K$ normalizing $\mathfrak{a}$. These elements moreover leave a lattice in $\mathfrak{a}$ invariant, the kernel of the exponential map, hence also the dual lattice in $\mathfrak{a}^*$. So the main point is to show that the reflections in the hyperplanes $\varphi_\alpha = 0$ are induced by elements of $K$ normalizing $\mathfrak{a}$. Let $\varphi_\alpha$ be an angular parameter and $X$ an element of $\mathfrak{a}$ annihilated by $\varphi_\alpha$ but by no angular parameter which is not a multiple of $\varphi_\alpha$. Then $\mathfrak{z}(X)$ consists of $\mathfrak{z}(\mathfrak{a})$ and of the intersection of $\mathfrak{g}$ with the sum of the two eigenspaces corresponding to the roots $2\pi i\varphi_\alpha$ and $-2\pi i\varphi_\alpha$ in $\mathfrak{g}_c$. It is then not difficult to see that it contains a copy $\mathfrak{s}_\alpha$ of $\mathfrak{su}_2$, so that the Weyl group element in the group $S_\alpha$ with Lie algebra $\mathfrak{s}_\alpha$ (which is either $\mathbf{SU}_2$ or $\mathbf{PSU}_2$) provides the reflection of $\mathfrak{a}^*$ with respect to $\varphi_\alpha = 0$. The geometric discussion of [C113] then applies and yields a fundamental domain $(D)$ for the group $(S)$ generated by these reflections ($n^{os}$ 7 to 10). As usual, I shall denote $(S)$ by $W(\mathfrak{a}, \mathfrak{g}_u)$.

Cartan remarks that any regular element $X \in \mathfrak{a}$ is conjugate to exactly one element in the interior of $(D)$, "at any rate if any two elements of $\mathfrak{a}$ conjugate under $K$ are already conjugate under the group $(S)$", a *proviso* of which he says in a footnote, p. 360, that it "is true, as can be checked *a posteriori*". This is indeed an important fact, which can now be proved directly. This shows that, although

---

[6] G. Hunt gave a very simple proof of the conjugacy of maximal tori in a compact group (Proc. A.M.S. **7** (1956), 307-8), which was immediately seen to extend to the more general case under consideration here, and this has been the standard argument since then.

Cartan did not have an *a priori* proof, he was aware of it and of its importance. It implies in particular that $W$ may be identified to the quotient of the normalizer of $\mathfrak{a}$ in $K$ by the centralizer of $\mathfrak{a}$ in $K$.[7]

**2.4.4.** We now assume that $G$ and $G_u$ are the adjoint groups (denoted $\Gamma$ and $\Gamma_u$ by Cartan) and we identify $\mathfrak{g}$ and $\mathfrak{g}_u$ to their images under the adjoint representation. In n$^{\text{os}}$ 17 and 22, Cartan studies the structure of $G$ and in particular obtains a generalization of the polar decomposition of an invertible $m \times m$ matrix ($m = \dim \mathfrak{g}$), now called the Cartan decomposition of $G$.

Fix a basis $(e_i)$ of $\mathfrak{g}$ consisting of orthonormal bases of $\mathfrak{p}$ (resp. $\mathfrak{k}$) with respect to $\varphi$ (resp. $-\varphi$). Let $x_i$ be the corresponding coordinates and $\psi$ the unit quadratic form with respect to these coordinates.[8] Using the invariance of $\varphi$ under $s$ and under $\mathfrak{g}$, it is readily checked that $x$ is symmetric (resp. antisymmetric) if $x \in \mathfrak{p}$ (resp. $x \in \mathfrak{k}$). Since $\mathfrak{p}$ consists of symmetric matrices, the exponential $\mathfrak{p} \to P$ is an isomorphism of $\mathfrak{p}$ onto $P$; this is well-known and elementary. The space $P$ consists of positive non-degenerate symmetric matrices. Moreover, if $p = e^X$ ($X \in \mathfrak{p}$), then the one-parameter group $e^{\mathbb{R}.X}$ is the unique geodesic containing 1 and $p$, the element $p$ has a unique square root $p^{1/2} = e^{X/2}$, and any two points in $P$ are contained in a unique geodesic.

The automorphism $s$ of $\mathfrak{g}$ extends to an automorphism of $G$, also denoted $s$. Let $K$ be its fixed point set. The Lie algebra of $K$ is $\mathfrak{k}$. If $k \in K$, then $\operatorname{Ad} k$ leaves $\mathfrak{k}$ and $\mathfrak{p}$ stable, hence also $\psi$, and so $K$ belongs to the orthogonal group of $\psi$. The automorphism $s$ leaves $P$ invariant and is, on $P$, the inversion $p \mapsto p^{-1}$. N$^{\text{os}}$ 17 to 22 define the operation of $G$ on $P$. The main point is to show that if $p \in P$, then $p.P.p \subset P$. The proof is simple, but relies essentially on the point of view of [C101] (see 1.5), namely, that the connection on $\mathcal{E}_G$ is the restriction of the third canonical connection on $G$, for which the symmetry $s_1$ at the origin is the inversion $g \mapsto g^{-1}$. Let $p \in P$. The transformation $g \mapsto p^{-1/2}.g.p^{-1/2}$ is an isomorphism of the connection on $G$ bringing $p$ to 1; therefore the symmetry of $G$ at $p$ is the map

$$(1) \qquad g \mapsto p^{1/2}.s_1(p^{-1/2}.g.p^{-1/2}).p^{1/2},$$

so that the product of the symmetries $s_p.s_1$ is the map

$$(2) \qquad g \mapsto p^{1/2}.s_1(p^{-1/2}.s_1(g).p^{-1/2}).p^{1/2} = p.g.p.$$

But $s_p$ and $s_1$ leave $P$ stable, since the latter is totally geodesic; hence so does their product, and therefore $pgp \in P$ if $g \in P$. This also shows that the map $q \mapsto p.q.p$ ($q \in P$) is the transvection (see 1.3) of $\mathcal{E}_G$ along the geodesic joining 1 to $p$.

This argument is basically the argument of Cartan (n° 18) but, in the context of [C101], he views it as valid only locally, and so he completes it in n° 19 by a simple limit argument, which does not really seem necessary to me.

---

[7] I spoke of a "root system" for the sake of brevity, although such a notion did not exist at the time, and Cartan does not check explicitly all the conditions we include in the present definition. He has the angular parameters, the reflections and knows that the group $(S)$ leaves the set of roots stable. He does not worry about the integrality condition, but since $(S)$ leaves a lattice invariant (it is crystallographic), it follows anyhow. Later in the paper, he enumerates all possible sets of roots.

[8] More invariantly, $\psi$ can be defined by the relation $\psi(X, Y) = -\varphi(X, s(Y))$   ($X, Y \in \mathfrak{g}$).

Given $g \in G$, define $g^* = s(g^{-1})$. We have[9]

(3)
$$g^{**} = g, \qquad (g.h)^* = h^*.g^*,$$
$$g^* = g^{-1} \text{ if } g \in K \quad \text{and} \quad g^* = g \text{ if } g \in P.$$

For $p \in P$, define $\alpha_g(p) = g.p.g^*$. Then, by (3),

(4)
$$\alpha_{g.h}(p) = \alpha_g\big(\alpha_h(p)\big) \qquad (g, h \in G, \ p \in P).$$

If $g$ belongs to $K$ or $P$, then $\alpha_g$ leaves $P$ stable, as we saw. Around the identity, every element of $G$ is a product of one element in $K$ and one in $P$; hence there is a neighborhood of the identity $U$ such that $\alpha_g(P) = P$ for $g \in U$. Since $U$ generates $G$, this is then true for all $g$, and $g \mapsto \alpha_g$ defines the action of $G$ on $P$ by isometries.

Let $g \in G$ and $p = g.g^*$. Let $q = p^{-1/2}$. Then $q.g.g^*.q = 1$, i.e. $q.g.(q.g)^* = 1$, hence $q.g = s(q.g)$, and therefore $q.g = k \in K$. Thus we have $G = K.P$. This decomposition is unique: if $g = k.p$ ($k \in K, p \in P$), then $g^*.g = p^2$, and hence $p = (g^*.g)^{1/2}$. Since $p \mapsto p^{1/2}$ is a smooth map on $P$, this also shows that $k$ and $p$ depend smoothly on $g$, so the map $(k, p) \mapsto k.p$ defines an isomorphism of manifolds of $K \times P$ onto $G$. This is the *Cartan decomposition* of $G$. It also implies that $K$ is connected.

Ad $K$ has no fixed point in $\mathfrak{p}$, except for the origin (remember that the isotropy group is irreducible); hence $K$ has only the origin as fixed point on $P$. Since $G = K.P$ any subgroup of $G$ conjugate to $K$ is conjugate to $K$ by an element $p \in P$, and hence has $p^2$ as its only fixed point on $P$; therefore (p. 371) *"We can view the space $\mathcal{E}$ as the space of all conjugate subgroups of $K$"*, a remark which, in retrospect, is Cartan's first step toward a theory of maximal compact subgroups.

**Remark.** In the current presentation of the theory, no use is made of the third canonical connection on $G$. For comparison, let me sketch the usual proof of the relation $p.P.p \subset P$. It relies on the fact that $G$ is the identity component of a real algebraic group, namely the group Aut $\mathfrak{g}$ of automorphisms of $\mathfrak{g}$, i.e., there is a family $J$ of polynomials on the space of linear transformations of $\mathfrak{g}$ into itself, such that Aut $\mathfrak{g}$ is the set of invertible matrices which are the common zeroes of the polynomials in $J$. Let $q \in P$ and $r = p.q.p$; then $r$ is a positive non-degenerate symmetric form. There exists $X$ in the Lie algebra of $SL(\mathfrak{g})$ such that $r = e^X$. Now let $F$ be in $J$. Then $F$ is annihilated by $r$, hence by all powers $e^{mX}$ ($m \in \mathbb{Z}$) of $r$. But this implies that it is annihilated by all transformations $e^{tX}$ ($t \in \mathbb{R}$) by lemma 2.4 in [M] (also reproduced in [B], VII, 3.6). Hence $X \in \mathfrak{g}$. But we have

$$s(r) = p^{-1}.q^{-1}.p^{-1} = r^{-1};$$

hence $s(X) = -X$. Consequently $X \in \mathfrak{p}$ and $r \in P$.

On the other hand, the argument in the text proving the existence of the Cartan decomposition, essentially Cartan's, seems complete to me.

**2.4.5.** Any element in $K$ or $P$ is an exponential. Therefore, as remarked by Cartan in the introduction (p. 347, bottom), any $g \in G$ is the product of two exponentials, a best result since $G$ may have elements which are not exponentials.

$G$ being the product of a euclidean space by $K$, its "groupe de connexion" (i.e. fundamental group $\pi_1(G)$) is the same as that of $K$. The group $K$ is either semisimple, in which case $\pi_1(G)$ is finite, or locally the product of a semisimple

---

[9] In this paper, Cartan does not use a notation for $g^*$. However, he does so in [C117], n° 19, where it appears as $\bar{g}^{-1}$ (since he writes $\bar{g}$ for our $s(g)$).

group by a circle group, and then $\pi_1(G)$ is the product of a finite group by an infinite cyclic group. To each subgroup of the fundamental group is associated a covering $G'$ of $G$. The space $P$ is simply connected; hence the exponential in $G'$ is an isomorphism of $\mathfrak{p}$ onto $P$, so all these groups have the same space $\mathcal{E}_G$. If $\pi : G' \to G$ is the covering map, then $G' = \pi^{-1}(K).P$ (see n° 34, formulated differently), and $\pi^{-1}(K)$ is connected. It is compact if $\ker \pi$ is finite, and otherwise it is the product of a compact semisimple group by a line.

So far, we have used the adjoint group. It is the "continuous group of isometries of $\mathcal{E}_G$", i.e. the identity component of the full group of isometries, say $I(\mathcal{E}_G)$, of $\mathcal{E}_G$. The transvections are already transitive on $\mathcal{E}_G$; hence any isometry is the product of a transvection by an isometry leaving 1 fixed. There is therefore a natural bijection between the connected components of $I(\mathcal{E}_G)$ and of the mixed isotropy group $\tilde{K}$ (all isometries leaving 1 fixed) (n° 26). As to the latter, Cartan points out that an invertible linear transformation of $\mathfrak{p}$ defines an isometry of $P$ if and only if it leaves $K$ and the metric invariant (n° 27). Cartan then proceeds to describe $\tilde{K}$ in a number of cases, depending on the structure of $K$ (n°$^{\text{os}}$ 28, 29, 30).

Cartan also discusses the relation between $I(\mathcal{E}_G)$ and the group Aut $\mathfrak{g}$ of automorphisms of $\mathfrak{g}$ ("le groupe adjoint mixte"). The former is obviously of finite index in the latter. More precisely, the transforms of $K$ under Aut $\mathfrak{g}$ form finitely many conjugacy classes under $G$. This number is the index of $I(\mathcal{E}_G)$ in Aut $\mathfrak{g}$. If there is only one such class, then Aut $\mathfrak{g} = I(\mathcal{E}_G)$. Cartan adds that he has checked this for the spaces $\mathcal{E}$ attached to the four big classes of simple groups. (He soon found a general argument; cf. 2.5.2.)

In the final paragraph of this chapter (n° 37), Cartan comes back to the maximal flats, the "euclidean manifolds $E_\lambda$" in his terminology. Let $(D)$ be the exponential of a cone in $\mathfrak{a}$ which is a fundamental domain for $W$. Cartan points out that every geodesic ray starting at 1 in $P$ is the transform under $K$ of one in $(D)$, and hence $P$ is the union of the transforms $k.(D).k^{-1}$ of $(D)$ by $K$. Nowadays, $(D)$ is usually denoted $A^+$ (the positive Weyl chamber defined by some ordering on the roots), and this relation is written $G = K.A^+.K$, also called a Cartan decomposition.

**2.4.6.** Chapter II of [C107] is devoted to examples of spaces $\mathcal{E}$ with negative curvature. For each type, Cartan gives the rank $\lambda$, the Weyl group, the roots and their multiplicities, a set of simple roots, and a fundamental domain $(D)$. The root systems are those he knows from the Killing-Cartan theory of semisimple complex Lie algebras, except for one new type, in which the double of a root may be a root (called non-reduced nowadays). Later on (n° 98, p. 430) he shows that this is the only possibility: if $\varphi_\beta = m\varphi_\alpha$ $(m > 1)$, then $m = 2$.

In the classical cases, Cartan also describes the "mixed isotropy group" and gives explicit realizations of the spaces $\mathcal{E}$.

**2.4.7.** Chapter III of [C107] is concerned with compact spaces, so we now come back to the case already studied in 2.4.1 to 2.4.3, (n°$^{\text{os}}$ 6, 7, 8 of [C107]). The discussion there also showed that $A$ is compact, hence so is $P_u$, which is the union of the conjugates $k.A.k^{-1}$ $(k \in K)$. However, it is not *a priori* clear that $P_u$ is a submanifold. If it is known that $p.P.p \subset P$ $(p \in P)$, then the last point follows, because $P_u$ is a manifold in the neighborhood of the identity and $u \mapsto p.u.p$ will map this neighborhood isomorphically onto a neighborhood of $p^2$ in $P_u$. However, as long as $P_u$ cannot be assumed to be a manifold, the earlier argument for proving this inclusion is not clearly valid.

To establish this, and for other reasons, Cartan extends to the present case the construction of the polyhedra he had introduced for compact groups (see 2.2, where they were called the Cartan polyhedra). Recall that in the non-compact case (2.4.2), an element $x \in \mathfrak{a}$ is called regular if it is not annihilated by a root. Now Cartan considers the stronger condition: $\alpha(X) \notin \mathbb{Z}$ for all angular parameters. If $X$ is in none of the hyperplanes $\alpha = m$ ($m \in \mathbb{Z}, \alpha$ an angular parameter), then the centralizer of $a = \exp X$ in $G_u$ has the same dimension as the centralizer of $\mathfrak{a}$. An element of $G_u$ is called regular if it has that property, singular otherwise, I shall say that $X$ is strongly regular if no $\alpha$ is integral on $X$. The closure of a connected component of the set of strongly regular elements is a polyhedron (a simplex if $G_u$ is simple) called a "polyhedron $(P)$" by Cartan ($\text{n}^{\text{os}}$ 95 to 98). These polyhedra are permuted transitively by the group generated by the Weyl group $W(\mathfrak{a}, \mathfrak{g}_u)$ and the group $T$ of translations by elements of the lattice $(R)$ of points on which all angular parameters are integral. In fact ($\text{n}^\circ$ 97) each polyhedron is a fundamental domain for the group generated by the symmetries to its "hyperfaces" (faces of codimension one), a group which is contained in the group generated by $W(\mathfrak{a}, \mathfrak{g}_u)$ and $T$, but may be strictly smaller (of finite index).

The closures of the connected components are therefore congruent polyhedra, and any element of $P_u$ is conjugate to one in such a polyhedron, to one in the interior of that polyhedron if it is regular. Because the roots occur in complex conjugate pairs, the centralizer of a singular element has dimension at least $\ell + 2$, and as before it follows that the singular elements form a closed subset of codimension $\geq 2$.

Fix $p \in P_u$. The proof of the inclusion $p.P_u.p \subset P_u$ is contained in $\text{n}^\circ$ 92, p. 426. First assume $q \in P_u$ to be regular. Then it can be joined to the identity by a curve consisting only of regular elements (except for 1). Cartan asserts that, if $r$ varies continuously from 1 to $q$ in the variety of regular elements, "it is always possible to find $r' \in P_u$ such that $r' = p.r.p$." If $q$ is singular, it is a limit of regular elements, whence $p.q.p \in P_u$. [I have to say that I do not find the argument for regular $q$ utterly convincing. Cartan himself came back to this point later in [C117] (see §4). Moreover, at the end of this subsection, I shall indicate a simple argument which would have been available at the time.]

It follows, as before, that $G_u$ operates on $P_u$ transitively by means of the transformations $\alpha_g : p \mapsto g.p.g^*$, so that the isotropy group of 1 is the fixed point set $K$ of $s$, and we get $G_u = K.P_u$, but this decomposition is not always unique, since $K \cap P_u$ may contain several elements (of order $\leq 2$, clearly).

We drop the assumption that $G_u$ is adjoint. Let $T$ be a maximal torus of $G_u$ containing $A$, and $\mathfrak{t}$ its Lie algebra. The group $G_u$ operates on $P_u$ transitively, and the isotropy group of 1 is the subgroup $K$ of fixed points of $s$. The space $P_u$ is the exponential in $G_u$ of $\mathfrak{p}_u$. Contrary to what happens in the negatively curved case, it may depend on the choice of $G_u$, among the (at most) finitely many groups with the given Lie algebra. Cartan investigates their dependence $G_u$, in particular their fundamental groups.

Let $(R)$ and $(R_u)$ be the kernels of the exponentials on $\mathfrak{t}$ and $\mathfrak{a}$. Of course $(R_u) = \mathfrak{a} \cap (R)$. In [C103], Cartan constructs a polyhedron $(D)$ in $\mathfrak{t}$ with one vertex at the origin having, among others, the following property: the exponentials of the edges joining the origin to one vertex in $(R)$ represent the elements of $\pi_1(G_u)$ (see 2.2). Here ($\text{n}^\circ$ 99), Cartan constructs a similar polyhedron $(D_u)$ with the same property for $P_u$. These polyhedra are defined by inequalities involving angular parameters. Those pertaining to $P_u$ are the non-zero restrictions of those of $G_u$. It can be so arranged that $(D_u) = \mathfrak{a} \cap (D)$. Therefore $\pi_1(P_u)$ is a subgroup of $\pi_1(G_u)$. In particular,

$(*)$                     *If $G_u$ is simply connected, then so is $P_u$.*

(cf. n° 100). The fibration $G_u/K = P_u$ then shows that $K$ *is connected*. Therefore the fixed point set of an involutive automorphism of a compact simply connected group is connected (a result which I extended much later to any automorphism, Tôhoku Math. J. (2) **13** (1961), 216-40, 3.4). This need not be so if $G_u$ is not simply connected.

Now let $\tilde{G}_u, G_u$ and $G'_u$ be the simply connected, the adjoint group and a group with the given Lie algebra $\mathfrak{g}_u$, and $\tilde{P}_u, P_u, P'_u$ the corresponding spaces. The canonical projections $\tilde{G}_u \to G'_u \to G_u$ induce covering maps $\tilde{P}_u \to P'_u \to P_u$. In particular, $\tilde{P}_u$ is the universal covering of $P'_u$. The various $P'_u$ are the *Klein forms of $\tilde{P}_u$* in Cartan's terminology (n° 102).

All these data are then described explicitly for the spaces $\mathcal{E}$ associated to the classical groups (n°s 108 to 150). The special properties of the type $\mathbf{D}_4$, due to the existence of triality, are discussed in n° 150. Some indications on the Klein forms of the exceptional types are also given in n° 150. Finally, in n° 151, Cartan points out that the elliptic space is the only one in which any two points are joined by a *unique* geodesic. He singles out the cases of rank 1, where two points are "in general" on a unique geodesic.

**Remark.** As a counterpart to the remark at the end of 2.4.4, let us indicate a simple direct argument to prove the relation $p.P.p \subset P$ $(p \in P)$.

Let $Q$ be the fixed point set of the diffeomorphism $g \mapsto g^* = s(g^{-1})$. It is a compact submanifold, which contains $P_u$ and is obviously stable under the maps $\alpha_q : x \mapsto p.x.p$ $(p \in Q)$. To establish our assertion, or in fact that $p.P_u.p = P_u$, it suffices to show that $P_u$ is equal to the connected component $Q^o$ of the identity in $Q$. First we show that they coincide around the identity. Fix a neighborhood $U$ of the identity in $G_u$ which is the isomorphic image under the exponential of a neighborhood $V$ of the origin in $\mathfrak{g}_u$. We may assume that $V$ is the product of its intersections with $\mathfrak{k}$ and $\mathfrak{p}$, hence $U = (U \cap K).(U \cap P_u)$. Let $x \in U$ and assume that $x^* = x$. Write $x = k.p$ $(k \in K \cap U, p \in P_u \cap U)$. We have

$$k.p = x = x^* = s(x^{-1}) = p.k^{-1};$$

hence $k = p.k^{-1}.p^{-1}$. There are unique elements $X \in \mathfrak{k}$ and $Y \in \mathfrak{p}$ such that $k = e^X$ and $p = e^Y$. We therefore get Ad $p(-X) = X$; hence $[X, Y] = X$. But $[X, Y] \subset [\mathfrak{k}, \mathfrak{p}] \subset \mathfrak{p}$, which forces $X = 0$. Therefore $P_u \cap U = Q^o \cap U$. Let $p \in P_u$. It follows that $p.P_u.p$ and $Q^o$ coincide around $p^2$. Since every element of $P_u$ has at least one square root, we see that $P_u$ is an open and compact submanifold of $Q^o$, hence is equal to $Q^o$.

**2.5.** We now come to [C116], the culmination of this series of papers, which contains *a priori* proofs of a number of basic facts Cartan had had to check case by case. After recalling the subject matter of the previous papers, Cartan writes in the introduction (p. 2):

> Dans les problèmes, de nature assez variée, qui se sont ainsi présen-
> tés, il s'est trouvé quelques théorèmes importants que j'ai vérifiés
> pour chaque structure simple particulière, mais dont je n'avais pu

réussir à mettre en évidence la raison profonde. Je me propose dans ce Mémoire de revenir sur ces théorèmes fondamentaux et d'en donner une démonstration générale. Il résultera de là non seulement que la théorie des espaces $\mathcal{E}$ reposera sur une base logique beaucoup plus satisfaisante, mais encore que les calculs souvent pénibles que j'ai été obligé de faire peuvent être réduits énormément; bien plus, le long Mémoire dans lequel j'ai déterminé toutes les formes réelles des groupes simples ([C38]) pourrait maintenant être réduit de 90 pages à une vingtaine.[10]

**2.5.1.** The first chapter is devoted to some properties of compact (connected) Lie groups. Let $G$ be one, and $\Gamma$ its adjoint group. For the latter to be compact, it is necessary that the (Killing) form $\varphi(e)$ be negative. If $G$ is compact, so is $\Gamma$. Assume $\varphi$ to be negative non-degenerate. Cartan then proves that $\Gamma$ is compact, pointing out that this fact, which is Weyl's starting point, was assumed without proof by him. To show this he remarks first that, in suitable coordinates, $\Gamma$ is an orthogonal group. It remains to prove that $\Gamma$ is closed, and for this Cartan notes that $\Gamma$ is the identity component of an algebraic variety, namely, the group of automorphisms of the Lie algebra $\mathfrak{g}$ of $G$. Moreover, $\Gamma$ has a finite fundamental group by a fundamental theorem of Weyl; hence $G$ is compact, too. Cartan also shows that any compact connected Lie group is, locally, the product of a semisimple group by some circle groups. (Section 8. In the italicized statement ending that section, the word "simple" should clearly be replaced by "semi-simple".)

In section 9, Cartan points out that the compactness of $G$ can be proved by a simple elementary argument if it is known that $G$ has a faithful linear irreducible (or fully reducible) representation. It relies on a theorem proved in [C28], 1909, Thm. XVIII, p. 148, which we would derive now from Schur's lemma (I do not know whether Cartan was ever aware of it) and Lie's theorem on solvable linear groups:

(∗) *Let $M$ be a connected irreducible linear Lie group. Then $M$ is either semisimple or the direct product of a semisimple group by the group of one-dimensional homothetic transformations.*

This implies in particular that if our previous group $G$ is linear and irreducible, then its center consists of multiples $\rho I$ of the identity matrix. On the other hand, since $G$ is semisimple, it is equal to its derived group, hence consists of matrices of determinant 1. Therefore $\rho$ is an $n$-th root of one, $n$ being the dimension of the representation space. Hence $\mathcal{C}G$ is finite, and $G$ is compact. Cartan also remarks that this argument uses only the fact that $G$ is its own commutator subgroup, and that it is also valid if the given representation of $G$ is fully reducible. In section 11, he comments that it does not seem easy to prove in this way the full reducibility of any representation of $G$ and that, in fact, Weyl went the other way around, showing directly that the universal covering of $\Gamma$ is compact. At this point, therefore, he

---

[10] (free translation) "In the various problems which presented themselves, there were some important theorems I had checked for each simple structure, but for which I could not give a deeper reason. In this Memoir, I intend to come back to these fundamental theorems and to give a general proof. As a result, not only will the theory of the spaces $\mathcal{E}$ rest on a much more satisfying logical basis, but also the often painful computations I had had to carry out can be reduced enormously; even more, the long Memoir in which I determined all real forms of the simple groups ([C38]) could now be reduced from 90 pages to about twenty."

does not see how to bypass Weyl's argument, contrary to what he had attempted in [C81] (see Note [16] to Chapter III).

**2.5.2.** Chapters II and III deal with non-compact simple groups. First let $\mathfrak{g}$ be an (absolutely) simple real Lie algebra. Recall that the starting point of [C107] (see 2.4) is a "Cartan decomposition" $\mathfrak{g} = \mathfrak{k} \oplus \mathfrak{p}$ satisfying 2.4.1(1), with in particular the following properties: $s : x + y \mapsto x - y$ $(x \in \mathfrak{k}, y \in \mathfrak{p})$ is an automorphism, $G = K.P$, where $K$ is compact with Lie algebra $\mathfrak{k}$ and $P = \exp \mathfrak{p}$ is the space $\mathcal{E}_G$ of $G$, diffeomorphic to euclidean space, with negative curvature. Cartan's concern in Chapter II is to show the uniqueness of this structure, up to inner automorphisms. He gives two proofs. The first one (n° 14) uses properties of the space $\mathcal{E}_G$ and Lie algebra techniques. I shall not try to summarize it, since the second one is much more conceptual and far reaching, being based on his famous fixed point theorem.

Interestingly, this theorem, which is now a cornerstone of the theory, entered in Cartan's work by the back door, so to say, first in his book on Riemannian geometry [C114], 1926. There, he calls a Riemannian manifold normal (we say complete) if every bounded set (with respect to the metric) is relatively compact. In the last part of that book, Note III, Cartan investigates the normal manifolds $M$ with negative curvature and proves a number of now familiar properties, established by J. Hadamard in the two-dimensional case, and which led one to call the simply connected ones Cartan-Hadamard manifolds: the exponential from the tangent space $TM_x$ at any point is a diffeomorphism onto $M$ which increases distances, any two points are contained in a unique geodesic,[11] and in a geodesic triangle with vertices $a, b, c$ we have the triangular inequality:

$$(1) \qquad |\overline{bc}|^2 \geq |\overline{ac}|^2 + |\overline{bc}|^2 - 2|\overline{ac}| \; |\overline{bc}| \cos \alpha$$

where $|\overline{xy}|$ is the length of a geodesic arc $\overline{xy}$ with end points $x, y$, and $\alpha$ is the angle between $\overline{ac}$ and $\overline{bc}$ (p. 261).

Still assuming $M$ simply connected, Cartan considers in the very last section a proper quotient $M'$ of $M$ by a discrete covering group of isometries $\mathcal{G}'$ acting freely, and wants to prove that $\mathcal{G}'$ is infinite. To establish this, he shows, using (1), that, given a finite set of points $C = \{x_1, \ldots, x_n\}$ of $M$, there is a unique point $x_0$ which realizes the minimum of the sum of the square distances $|\overline{x_i x_0}|^2$, hence any isometry leaving the set $\{x_1, \ldots, x_n\}$ stable leaves $x_0$ fixed, so cannot act freely (hence $\mathcal{G}'$ has no element of finite order, except for the identity).

Let us now come back to [C116]. There (n° 16), Cartan asserts (without further elaboration) that the uniqueness of the point $x_0$, call it a center of gravity, still holds if the finite set is replaced by a compact subset.

If $\gamma$ is a compact subgroup of $G$, the $\gamma$-orbit of a point in $\mathcal{E}_G$ is a compact submanifold, stable under $\gamma$; therefore its center of gravity is fixed under $\gamma$. In the framework of [C114], this argument shows in fact that any compact Lie group of isometries of a Cartan-Hadamard manifold has a fixed point.[12]

---

[11] In the second edition of his book on Riemannian geometry [C183], Note IV, Cartan shows more generally that in a normal connected Riemannian manifold, any two points are the end points of a geodesic segment whose length is the distance between the two points. This was proved first (with a converse) by H. Hopf and W. Rinow (Comment. Math. Helv. **3** (1931), 209-225), but Cartan does not seem to have been aware of it.

[12] As far as I know, Cartan never gave more details, but the argument has been described in detail in several places. Of course, some differentiability is used. Later, F. Bruhat and J. Tits gave a proof based on a different inequality, which does not require any differentiability and is

As Cartan had pointed out in [C117] (see 2.4), this result, which he had checked for the space $\mathcal{E}_G$ attached to the classical cases, implies that Aut $\mathfrak{g}$ is the group of isometries $I(\mathcal{E}_G)$ of $\mathcal{E}_G$.

In n° 18, Cartan remarks that if $M$ is a Riemannian homogeneous space of $G$, then dim $M \geq \dim \mathcal{E}_G$ and there is equality if and only if $M$ is isomorphic to $\mathcal{E}_G$ (because the isotropy group of a point in $M$ is compact).

In n° 17, Cartan gives some information on Betti numbers. Let $b_i(X)$ denote the $i$-th Betti number of a space $X$. Since $G$ is the product of $K$ by a euclidean space, $G$ and $K$ have the same Betti numbers. If $K$ is semisimple, Weyl's theorem implies that $b_1(K) = 0$. Cartan points out that Weyl's argument, which relies on the fact that the singular elements in $K$ form a subset of codimension 3, also implies that any two-dimensional compact surface can be deformed to a point, and hence $b_2(K) = 0$. [In fact, it shows that the second homotopy group $\pi_2(K)$ of $K$, hence of $G$, is zero.] If now $K$ is not semisimple, it is locally the product of a semisimple group by a circle; hence $b_1(K) = 1$ and $b_2(K) = 0$.

**2.5.3.** The uniqueness of the Cartan decomposition gives a positive answer to the second question raised by Cartan in [C94], n° 54 (see 1.4). In Chapter III of [C116] he also settles the first question affirmatively, by establishing the existence of a Cartan decomposition. In fact, the proof leads more precisely to a new way to classify real forms.

Recall that, given the absolutely simple non-compact Lie algebra $\mathfrak{g}$, the problem is to find a compact form $\mathfrak{g}_u$ of the complexification $\mathfrak{g}_c$ of $\mathfrak{g}$ admitting an involution $s$ such that

$$(1) \qquad\qquad \mathfrak{g}_u = \mathfrak{k} \oplus \mathfrak{p}, \qquad \mathfrak{g} = \mathfrak{k} \oplus i\mathfrak{p},$$

where $\mathfrak{k}$ is the fixed point set of $s$ and $\mathfrak{p}$ its $(-1)$-eigenspace. ($\mathfrak{g}_u$ and $\mathfrak{g}$ are then related by a "normal isomorphism" in the terminology of [C94]; see 1.4.)

I shall give here first a simple proof, not that of Cartan on which I shall comment below. It uses only the tools he had developed, and bears some relation to his argument. Somehow, he did not see the most direct way to achieve his goal.

Viewed as a real Lie algebra, $\mathfrak{g}_c$ is the direct sum of $\mathfrak{g}$ and $i\mathfrak{g}$. The complex conjugation $\sigma$ of $\mathfrak{g}_c$ with respect to $\mathfrak{g}$ is an automorphism, and $\mathfrak{g}$ (resp. $i\mathfrak{g}$) is its fixed point set (resp. $(-1)$-eigenspace). We let $G_c$ be the adjoint group of $\mathfrak{g}_c$. The map $\sigma$ extends to an automorphism of $G_c$, viewed as a real Lie group, also denoted by $\sigma$, which fixes $G$ pointwise.

We now know that the space $\mathcal{E}$ of $G_c$ may be identified to the space of maximal compact subgroups of $G_c$, on which $G_c$ acts by conjugation. More generally, any automorphism of $G_c$ defines an isometry of $\mathcal{E}$. This is in particular the case for $\sigma$, which has a fixed point, say $x_o$, on $P$. [For the existence of $x_o$, one does not need to invoke the fixed point theorem: if $\sigma$ permutes two points $x$ and $y$, then it leaves stable the unique geodesic segment joining them, hence also its middle point.] Let $G_u$ be the stability group of $x_o$. It is invariant under $\sigma$, which induces on it an involutive automorphism, and we have $\mathfrak{g}_u = \mathfrak{k} \oplus i\mathfrak{p}$, where $\mathfrak{k} = \mathfrak{g}_u \cap \mathfrak{g}$ and $\mathfrak{p} = \mathfrak{g}_u \cap i\mathfrak{g}$, so $\mathfrak{g}_u$ has the desired relationship with $\mathfrak{g}$.

---

also valid in a different context, that of the Bruhat-Tits buildings (Inst. Hautes Études Sci. Publ. Math. **41** (1972), 5-252, n° 32), for which, in fact, it was established.

In n° 23, Cartan points out that a (non-trivial) involution on $\mathfrak{g}_u$ yields a non-compact real form in the obvious way, and concludes that the search for non-compact real forms amounts to the determination of the involutions of a compact form. He also adds that the non-compact forms not conjugate in the adjoint group correspond bijectively to the conjugacy classes within the adjoint group of the elements of order $\leq 2$ in a maximal compact subgroup. (This is indeed true, but it seems to me that Cartan overstates slightly what he had actually proved.)

In n° 24, he comments about the effective determination of real forms. If Aut $\mathfrak{g}_u$ is reduced to Ad $\mathfrak{g}_u$, this is easy because the conjugacy classes of elements of order 2 in $G_u$ are explicitly known (the elements of order 2 of a maximal torus, modulo the Weyl group). He asserts that the only difficult case in which $\sigma$ is not inner occurs for $\mathbf{E}_6$. In the next section, as an application, he treats the "four general classes", duly taking into account the triality automorphisms of the type $\mathbf{D}_4$.[13]

**2.5.4.** Cartan's proof for the existence of $\mathfrak{g}_u$ satisfying (1) above is contained in n° 21. He expresses everything in terms of bases, a point on which I shall not follow him, but I'll use his notation for the maps. Choose a compact form $\mathfrak{g}_u$. There is then a complex automorphism $S$ of $\mathfrak{g}_c$ which brings $\mathfrak{g}$ onto $\mathfrak{g}_u$. Let $^-$ be the conjugation with respect to $\mathfrak{g}$ (the $\sigma$ of the previous subsection). Then $\bar{S}.\mathfrak{g} = \mathfrak{g}'_u$ is another compact form of $\mathfrak{g}_c$. Let $G_u$ and $G'_u$ be the corresponding groups and $A, A'$ their fixed points on the space $\mathcal{E}$ of $G_c$. There is a unique geodesic segment joining them. Let $O$ its middle point and $G''_u$ the isotropy group of $O$. Then Cartan proves that it satisfies the condition (1). [Indeed, the complex conjugation exchanges $A$ and $A'$, hence leaves $O$ fixed.] He uses the transvection $\theta$ which brings $O$ to $A$. Then $\theta^{-1}$ maps $O$ to $A'$. He claims that $\theta^{-1} = \bar{\theta}$. This is true because complex conjugation acts on $\mathcal{E}$, permuting $A$ and $A'$, leaving $O$ fixed, but his argument does not seem so convincing to me, using, it seems to me, the complex conjugation with respect to $G''_u$. Let $\Sigma = \theta^{-1}.S$. Then $\Sigma$ and $\bar{\Sigma}$ bring $\mathfrak{g}$ onto $\mathfrak{g}''_u$. Cartan shows that $\bar{\Sigma} = R.\Sigma$, where $R$ is an automorphism of $\mathfrak{g}''_u$, which he has already proved is of order two. Then (1) follows, with $\mathfrak{g}_u$ replaced by $\mathfrak{g}''_u$.

**2.5.5.** All this relies first of all on the existence of a compact form, on which Cartan comments in n° 19. He had checked it case by case, without realizing its importance, as he wrote to Weyl (see Chapter III), and Weyl had given a general proof. It relied on the full strength of the Killing-Cartan structure theory of $\mathfrak{g}_c$, and Cartan wondered whether it would not be possible to give a more direct proof, using only the most elementary part of Lie theory. In his opinion, this would allow a considerable simplification in the exposition of the theory of simple Lie groups. He then describes an unsuccessful attempt of his to prove it. This idea was taken up much later, and pushed through to a proof by R. Richardson [R].

## §3. An exposition of Lie group theory from the global point of view

As far as Lie groups are concerned, all the above mainly dealt with semisimple ones. Going beyond those, Cartan felt it was time to outline a theory of Lie groups stressing global aspects, which he did in his monograph [C128] of 1930.

---

[13] The program outlined by Cartan was implemented by P. Lardy (Comment. Math. Helv. **8** (1936), 189-234). Since then, several versions of the classification of involutions of compact simple Lie algebras have appeared (e.g. F. Gantmacher, Mat. Sbornik **5** (1939), 217-248; S. Murakami, J. Math. Soc. Japan **4** (1952), 103-133, **5** (1953), 105-112; S.Araki, J. Math. Osaka City Univ. **13** (1962), 1-34). This classification may also be proved in the framework of Galois cohomology; see Borel and Serre, Comment. Math. Helv. **39** (1964), 111-164, section 6.8.

**3.1.** The first chapter deals with general definitions and concepts. It defines (topological) manifolds. A topological group is a *finite and continuous group of order* $r$ if its underlying space is a manifold of dimension $r$. No differentiability is required. General properties are discussed, in particular homogeneous spaces, spaces of left or right cosets with respect to a subgroup, covering groups, and universal covering group.

Recall that, for Lie, a "finite and continuous group" is a local analytic transformation group (see Chapter I). Here, that terminology is reserved for the topological case, and besides for an abstract group, not necessarily a transformation group, though it is pointed out that a group can always be viewed as a transformation group, by letting it act on itself by means of left or right translations (the parameter groups in the sense of Lie).

**3.2.** In Chapter II a finite and continuous group is *a Lie group* if coordinates can be introduced so that product and inverse are analytic functions of their arguments. It is pointed out that, according to F. Schur, it suffices to postulate the existence of continuous first and second derivatives. The Maurer-Cartan equations (see V, 4,7) are recalled.

Whether a "finite and continuous group" is a Lie group was part of Hilbert's fifth problem, but this had not been seriously investigated until then. Cartan himself does not seem to have been aware of Hilbert's problem and says simply that this question appears never to have been considered: "*le problème de savoir s'il existe des groupes finis et continus d'ordre* [i.e. dimension] $r > 1$ *qui ne soient pas des groupes de Lie n'a en somme jamais été abordé*" (n° 19).

**3.3.** Cartan then reviews the "fundamental theorems" of classical Lie theory. He stresses the fact that the existing proofs of a converse to the third theorem (namely, a Lie algebra is always the Lie algebra of a Lie group) prove only the existence of a local group with the given Lie algebra, except for the first proof of Lie, when it applies, i.e. when the center of the Lie algebra is reduced to zero (in which case one can use the adjoint representation). In n° 22, he sketches a general proof, starting with solvable groups.

A first step beyond Schur had been carried out by von Neumann [N], who had shown that a finite and continuous subgroup of $\mathbf{GL}_n(\mathbb{R})$ is a Lie group. Moreover, any representation bounded around the identity is continuous. In n$^{os}$ 26, 27, Cartan proves more generally that any finite and continuous subgroup, or any closed subgroup, of a Lie group is a Lie group. This implies that any continuous homomorphism of Lie groups is analytic in suitable coordinates. He then passes to some general remarks about homogeneous spaces $G/H$ of a connected Lie group $G$ (in particular, if $G$ is simply connected, then $G/H$ is simply connected if and only if $H$ is connected), and deduces some topological properties of $\mathbf{SL}_2(\mathbb{R})$. He concludes this section by listing all two-dimensional homogeneous spaces.[14]

**3.4.** Chapter IV is devoted to the work of Weyl and Cartan on compact connected Lie groups, and Chapter V to "symmetric spaces". It is here (and, slightly earlier, in [C117]) that Cartan introduces that notion. A Riemannian manifold is symmetric if for every point $x$, the local symmetry around $x$ is the restriction of a global isometry. From then on, he adhered to that terminology. Chapter V gives an extensive summary of the work discussed in the previous sections of this chapter.

---

[14] It was pointed out later by G.D. Mostow (Annals of Math. **52** (1950), 606-636) that the list was not quite complete.

Cartan ends up with some remarks about Betti numbers. In particular, he shows that the third Betti number of a compact simple group is not zero, and ends by pointing out: "*Il y a là un sujet très important de recherches qu'on peut dire à peu près inexploré*". (There is there a very important topic for research which may be said to be practically unexplored). It was not to stay unexplored for long (see §5).

# B. Further Developments

## §4. Complete orthogonal systems on homogeneous spaces of compact Lie groups

The paper [C117], 1929, to which this section is devoted, "*was inspired by the beautiful Memoir where H. Weyl shows that the different irreducible linear representations of a continuous closed group supply a complete orthogonal system in the space of the group* [W73]" (a paper due in fact to F. Peter and H. Weyl). Cartan's goal is to extend the Peter-Weyl theorem to homogeneous spaces of compact connected Lie groups (his "continuous closed groups"), in particular to compact symmetric spaces.

**4.1.** Let $E$ be a compact space on which a compact connected Lie group $G$ acts continuously and transitively. It is a smooth manifold and, as we know, it carries an invariant Riemannian metric. [This does not seem evident to Cartan, who promises a proof in another paper. I take it for granted.] Given $O \in E$, Cartan also feels the need to prove that the subgroup leaving $O$ fixed is compact, with finitely many connected components. As before, he calls it the *isotropy group* of $O$, and usually denotes it $g$, writing $E = G/g$, a point on which I shall not follow him, preferring to use, as usual, some capital letter, say $U$.

The first four sections of the paper discuss the general situation. Cartan considers finite dimensional subspaces of functions on $E$ invariant under $G$ and calls a basis of such a subspace a *fundamental sequence*. He argues mostly in terms of fundamental sequences. Here too, I shall not do so, and shall usually speak of invariant subspaces or $G$-subspaces. There is a positive non-degenerate hermitian product between two such functions, defined by integration of the hermitian scalar product of values at points of $E$ with respect to the Riemannian volume element $d\tau_M$, and so all $G$-subspaces are unitary. Cartan points out first that an irreducible $G$-subspace contains exactly one line pointwise fixed under $U$, and proposes to call the elements of the latter *zonal functions* (n° 3). Conversely, any finite dimensional irreducible representation of $G$ containing a line pointwise fixed under $U$ has a realization as a $G$-subspace on $E$ (n° 4). Two distinct irreducible $G$-submodules intersect only at zero. Two irreducible non-equivalent $G$-submodules are orthogonal (n° 8). A finite dimensional $G$-invariant subspace which is a sum of equivalent irreducible representations can be written as an orthogonal direct sum of irreducible $G$-subspaces (n° 7). So the space spanned by the finite dimensional $G$-subspaces may be written as an orthogonal direct sum of irreducible ones. To prove that this sum is *complete* (i.e. dense in $L^2(E, d\tau_m)$), he uses, following [W73], integral equations with kernels $K(M, P)$ which are continuous real functions of the distance between the two arguments. In particular, they are symmetric and invariant under $G$.

To each such kernel and each $\lambda \in \mathbb{C}$ is associated an integral equation of the first kind

$$(1) \qquad\qquad \varphi(M) = \lambda \int K(M, P).\varphi(P)d\tau_p.$$

The space of solutions of (1) is finite dimensional and invariant under $G$. Using a known result on integral equations, Cartan shows that any real continuous function orthogonal to all those subspaces, for variable $K(M, P)$, is zero, whence the completeness (n° 12).

**4.2.** In section IV, Cartan first summarizes the main results of [C37] on irreducible representations of a complex semisimple Lie algebra $\mathfrak{g}_c$. Each such representation is characterized by a dominant weight. There are $\ell = \text{rank } \mathfrak{g}_c$ such representations, with dominant weights $\Pi_1, \ldots, \Pi_\ell$, such that the dominant weights are all the positive integral linear combinations of the $\Pi_i$. In fact, Cartan presents this as a statement on the irreducible representations of $G$, which is a compact connected group, not necessarily semisimple. So some minor adjustment is in order. Strictly speaking, this would apply to $G$ if $G$ is semisimple and simply connected.

Also, if $G$ is the circle group, identified to the group of complex numbers of modulus one, say $\{\exp 2\pi i\varphi\}(\varphi \in \mathbb{R}$, modulo 1), then the irreducible representations are of degree one, of the form $z \mapsto z^m$ $(m \in \mathbb{Z})$, so the weights are the integral multiples of $\varphi$, not just the positive multiples of $\varphi$.

The group $G$ always has a finite covering $\tilde{G}$ which is a product of a simply connected compact semisimple group by circle groups, and so this description is valid for $\tilde{G}$, except that the coefficients of the weights associated to the circle factors are arbitrary integral multiples of a given one.

However, this is not an issue for what Cartan has in mind. Later on, $G$ will always be semisimple. In section IV his main goal is to prove that if $\omega$ and $\omega'$ are the highest weights of two irreducible representations occurring on $E$, then the irreducible representation with highest weight $\omega + \omega'$ also occurs in $E$. Equivalently, if two irreducible representations have non-zero $U$-fixed elements, then so does their "Cartan product" (which is how this is proved in n° 14). But, those highest weights are not necessarily all the integral linear combinations of some linearly independent ones. To get more precise results, he goes over to symmetric spaces, from section V on.

**4.3.** Cartan starts section V by giving in n° 16 the definition of a *symmetric (Riemannian) space* $E$: the "symmetry" at each point is isometric.[15] He recalls that, given a geodesic, there is a transvection which leaves the geodesic invariant and induces the parallel transport along that geodesic.

He then assumes $E$ to be compact, and more precisely takes $G$ to be the greatest connected group of isometries of $E$. He first shows that if $V$ is an irreducible $G$-subspace, then the symmetry $s_0$ with respect to any point $0 \in E$ changes it into another $G$-subspace. More precisely, he shows that the symmetry $s_0$ changes "a fundamental sequence into the complex conjugate one", i.e., in invariant terms, it changes a $G$-subspace $V$ into one isomorphic to the contragredient representation. Then, in n° 17, he concludes that a given irreducible representation of $G$ has at most one realization as a $G$-subspace on $E$.

---

[15] Here a symmetry is, by definition, a global isometry which is locally around $x$ the symmetry as defined earlier. It could also be defined as an isometry fixing $x$, with differential at $x$ equal to $-\text{Id}$.

Cartan next discusses in n$^{os}$ 18, 19 some foundational material on symmetric spaces. As we saw, for him the symmetric space has usually been identified with the subvariety $P = \exp \mathfrak{p}$ of $G$, consisting of the transvections. On the other hand, one may consider the quotient $E' = G/K$ of left $K$-cosets. If $q \in p.K$ $(q, p \in P)$, then

$$q^* = k^{-1}.p, \text{ hence } q^2 = p.k.k^{-1}p = p^2.$$

Assume $G$ simply connected. Then so is $P$ (see 2.4.7) and $K$ is connected. Hence $G/K$ is also simply connected, and therefore $(pK) \mapsto p^2$ defines a map of $E'$ onto $E$. Then Cartan states that since $E$ is simply connected, $E'$ must coincide with $E$ ("comme l'espace $E$ est simplement connexe, cela exige que l'espace $E'$ se confonde avec $E$"). However, this map is not a covering map, it may have fibres of strictly positive dimension, so I do not see how to justify this remark. On the other hand, the conclusion he draws, namely that if $p^2 = q^2$ $(p, q \in P)$, then $p \in qK$, is correct. [To define the map $E' \to E$, let $\alpha : g \mapsto g.g^*$. Since $G = K.P = P.K$, it maps $G$ onto $P$ and is constant along the cosets $pK$, hence defines a map $E' \to E$, the fibres of which are precisely the cosets $pK$.]

Next, Cartan comes back to the proof of the relation $p.q.p \subset P$ $(p, q \in P)$, see 2.4.7, finding that "it will not be useless to give another proof, which will be useful in other respects" (p. 18). In fact, although he does not quite express it in this way, he computes the differential of the exponential mapping $\mathfrak{p} \to P$ and its determinant.

Let $x \in \mathfrak{p}$ and consider $\mathrm{ad}\, x$. The relation $[\mathfrak{p}, [\mathfrak{p}, \mathfrak{p}]] \subset \mathfrak{p}$ implies that $(\mathrm{ad}\, x)^m$ leaves $\mathfrak{p}$ (and $\mathfrak{k}$) invariant for all even $m$'s, so that the (convergent) power series

$$(1) \qquad 2\frac{\sinh \mathrm{ad}\, x}{\mathrm{ad}\, x} = \frac{e^{\mathrm{ad}\, x} - e^{-\mathrm{ad}\, x}}{\mathrm{ad}\, x} = \sum_{m \geq 0} \frac{(\mathrm{ad}\, x)^{2m}}{(2m + 1)!} := \mu(x)$$

is an endomorphism of $\mathfrak{p}$. Let $p = e^x$ and $r = e^{-x/2}$. Thus $r^2.p = 1$. Recall that $\alpha_p$ is the map $q \mapsto p.q.p$. Now $\alpha_r \circ (\delta e)_x$ is an endomorphism of $\mathfrak{p}$. Cartan shows on p. 19 that it is equal to the endomorphism $\mu(x)$ in (1). We can also write that

$$(2) \qquad (\delta e)_x = \alpha_q \circ \mu(x) \qquad (q = e^{x/2}),$$

an important computation of functional determinant which can be found in more recent presentations of the theory (see e.g. [H], IV, §4, or [B], IV, 3.6). The eigenvalues of $(\mathrm{ad}\, x)^2$ on $\mathfrak{p}$ are 0 with multiplicity $\lambda = \dim \mathfrak{a}$ and one of $\pm 2\pi i\alpha$ for each pair of complex conjugate roots. The determinant of $\mu(x)$ is, up to a numerical factor independent of $x$, equal to

$$(3) \qquad \det \mu(x) = C_0 \prod \frac{\sin 2\pi a(x)}{2\pi \alpha(x)},$$

where $c$ is a non-zero constant independent of $x$, and the product is over pairs of complex conjugate roots (counted with multiplicities). Now assume $x$ to be regular. Then $\det(\delta e)_x \neq 0$ and the exponential is, around $x$, an isomorphism of $\mathfrak{p}$ onto $P$. If we identify $\mathfrak{p}$ to $P$ around $x$, by that map, then the invariant volume element $dv$ on $P$ around $p$ becomes

$$(4) \qquad dv = c. \prod \frac{\sin 2\pi \alpha}{\alpha} dv_o,$$

where $dv_o$ is the Lebesgue measure on $\mathfrak{p}$ (p. 20). Since the complement of the regular elements has lower dimension, this provides one with the invariant volume element on $P$.

**4.4.** Recall that an irreducible representation of $G$ occurs in $L^2(P)$ if and only its fixed point set under $U$ is $\neq 0$, in which case the latter is one dimensional, and its elements are represented by "zonal functions" (see 4.1). To borrow from a later, quite usual, terminology, say that such representations are of class one. Chapter VI is devoted to a parametrization of representations of class one. We assume that $G$ is simply connected. Cartan first studies them by means of what are often called nowadays restricted weights, namely the restrictions of the weights to $\mathfrak{a}$. Fix a Cartan subalgebra $\mathfrak{t}$ of $\mathfrak{g}$ containing $\mathfrak{a}$, hence belonging to $\mathfrak{z}(\mathfrak{a})$. On $\mathfrak{a}$ we have the root system $\Phi(\mathfrak{a}, \mathfrak{g})$ consisting of the non-zero restrictions of the roots of $\mathfrak{g}_c$ with respect to $\mathfrak{t}_c$; call this system $\Phi = \Phi(\mathfrak{t}, \mathfrak{g})$. We identify $\mathfrak{a}^*$ to a subspace of $\mathfrak{t}^*$ via the Killing form. Orderings on $\Phi$ and $\Phi(\mathfrak{g}, \mathfrak{a})$ may be chosen so as to be compatible, i.e. the restriction of a positive root is positive, and so the set $\Delta(\mathfrak{t}_c, \mathfrak{g}_c)$ of simple roots restricts to the simple roots of $\Phi(\mathfrak{a}, \mathfrak{g})$ (or zero). Let $P(\mathfrak{a}, \mathfrak{g})$ be the weights of the restricted root system $\Phi(\mathfrak{a}, \mathfrak{g})$, in other words, the linear forms $\mu \in \mathfrak{a}^*$ such that $2(\mu, \alpha).(\alpha, \alpha)^{-1} \in \mathbb{Z}$ $\left(\alpha \in \Phi(\mathfrak{a}, \mathfrak{g})\right)$. They are the differential of all the characters of $A$, and the restrictions to $\mathfrak{a}$ of weights of representations of $G$. The restricted dominant weights are positive linear integral combinations of $\lambda = \dim \mathfrak{a}$ fundamental restricted dominant weights, which Cartan denotes

$$2\pi i \omega_1, \dots, 2\pi i \omega_\lambda$$

(n° 24). The *order* of a representation $\Gamma$ of class one is defined by the coefficients of the $\omega_i$ in the highest restricted weight of $\Gamma$. Cartan points out (n° 24) that these coefficients are necessarily *even* (because a representation of class one is trivial on the elements of order $\leq 2$ of $A$, which form $A \cap K$).

Then Cartan argues in n°s 27 and 28 that if we associate to a representation of class one its order, then we establish a bijective correspondence between those representations and the positive linear integral combinations of $2\omega_1, \dots, 2\omega_\lambda$.

Now the dominant weights of the irreducible representations of $G$ are positive linear integral linear combinations of $\ell = \dim \mathfrak{t}$ fundamental ones. Cartan has shown that the subset of the highest weights of representations of class one is a positive sublattice of rank $\lambda$. What he does not do, however, is to try to characterize those highest weights directly. He simply points out that all one has to do is to find those representations with orders

$$(2, 0, \dots, 0), (0, 2, 0, \dots, 0), \dots, (0, \dots, 0, 2).$$

He then adds, as a consequence of the completeness theorem (4.2), that the restrictions of the zonal functions on a fundamental polyhedron form a complete orthogonal system with respect to the volume element $\prod(\sin \pi \varphi_\alpha) d\varphi_1, \dots, d\varphi_\lambda$, where the $\varphi_i$ are the simple restricted roots.

The last two paragraphs are devoted to examples: complex projective spaces and spheres.

**4.5.** As we can see, Cartan had gone a long way towards the description of representations of class one. His arguments are not easy to follow, at any rate not to me, and I would not vouch that he has ironclad proofs. The way he argues in n°s 28, 29 makes me suspect he had realized that, if we write a $U$-fixed vector as a linear combination of weight vectors, then the highest weight vector occurs with a *non-zero* coefficient, indeed an important point in the proofs.

The first new treatment, and a more complete one, was given by S. Helgason (Adv. Math. **5** (1970), 1-154, III n° 3). The problem was later taken up in the more general context of "complex semisimple symmetric spaces": $G$ is a complex 1-connected semisimple group, $K$ the fixed point set of an involution. The highest weights of the spherical representations form a lattice of rank equal to the rank of the symmetric space (see T. Vust, *Opérations des groupes réductifs dans un type de cones presque homogènes*, Bull. Soc. Math. France **102** (1974), 317-333, C. deConcini and C. Procesi, *Complete Symmetric Varieties*, Springer LNM **996**, Nos 1.4-1.7). A characterization of the highest weights of the representations of class one in a still more general framework may be found in T.A. Springer, *Algebraic groups with involutions*, in Advanced Studies in Pure Math. **6** (1985), 525-543 (see Thm. 5.6).

## §5. Differential forms and algebraic topology

**5.0.** As soon as Cartan adopted a global point of view, he became interested in the topology of group manifolds, in particular in their Betti numbers, as we saw. This interest in "analysis situs", as topology was often called then, grew quickly. André Weil was rather fond of telling that once, around 1927, while he was walking with Élie Cartan, the latter said to him: *"J'apprends l'analysis situs, je crois que je pourrai en tirer quelque chose."* ("I am learning analysis situs, I believe I shall be able to draw something from it").[16]

**5.1.** In his book on integral invariants [C64], 1922, Cartan had essentially laid down the foundation of exterior differential calculus (even though there was no clear cut definition of exterior differential forms). He had introduced the exterior differential, known before (since Frobenius) for one-forms under the name bilinear covariant, and had also proved the Poincaré lemma (without attribution here). Guided by the work of Riemann on algebraic curves and by some rather obscure remarks of Poincaré, he stated in a C.R. note ([C111], 1928) two hypothetical Theorems, A and B. Given a compact connected manifold $M$, they amounted to asking whether, given $p$, the number of linearly independent closed differential $p$-forms modulo the differentials of $(p-1)$-forms was equal to the $p$-Betti number $b_p(M)$ of $M$. [For convenience, I'll denote this number by $c_p(M)$: Cartan does not have a notation.] If so, the determination of the Betti numbers of compact group manifolds would be a purely algebraic, quite accessible, problem. As an example, he pointed out that the Betti number of $\mathbf{SU}_n$ would likely be the coefficients of the polynomial

$$(1) \qquad (t^3 + 1).(t^5 + 1)\cdots(t^{2n-1} + 1).$$

**5.2.** Without waiting for further information on his Theorems A and B, Cartan wrote a full-fledged paper [C118], 1929, devoted to the computation of the $c_p(M)$, when $M$ is acted upon transitively by a compact connected Lie group.[17]

---

[16]Weil included this reminiscence in a letter to Henri Cartan that was published in "Élie Cartan et les mathématiques d'aujourd'hui", Astérisque, numéro hors série, 1985, pp. 5–7.

[17] However, a footnote added on p. 182 points out that G. de Rham had just announced proofs of these theorems (C.R. **188** (1929), 1651-2), known since as the de Rham theorems.

Georges de Rham, lecturing on his thesis in 1931

He proposes the name "Poincaré polynomial of $M$" for the polynomial $\Phi(t)$ of degree $n = \dim M$ defined by

$$\Phi(t) = \sum_{p \geq 0} c_p(M).t^p.$$

By a simple averaging process, he shows that $c_p(M)$ is also equal to the number of $G$-invariant closed $p$-forms modulo the differentials of $G$-invariant $(p-1)$-forms.

Fix $x \in M$ and let $U$ be its isotropy group (Cartan uses $\gamma$), so $M = G/U$. A $G$-invariant form is completely determined by its value at $TM_x$, which is an element of the $p$-th exterior power $\Lambda^p T(M)_x^*$ of the cotangent space, invariant under the action of the linear isotropy group (the natural action of $U$ on $T(M)_x$ and its tensor powers), and conversely. So the determination of the number of linearly independent invariant $G$-invariant $p$ forms is a purely algebraic problem. In the general case it still remains to find the dimension of the space of "coboundaries", i.e. of the spaces of differentials of $G$-invariant $(p-1)$-forms. This, too, is in principle an algebraic problem, but Cartan does not attack it.[18] He really did not have to, because for group manifolds, and more generally for symmetric spaces, his main case of interest, a $G$-invariant $p$-form is automatically closed (and even coclosed, hence harmonic), so that, in that case, $c_p(M)$ is just the dimension of

---

[18] This was carried out later by C. Chevalley and S. Eilenberg, and gave rise to the cohomology theory of Lie algebras (see Chapter VII).

the fixed point set of $U$ in $\Lambda^p T(M)^*_x$. Choose a basis of $T(M)_x$ and write $\big(a_{ij}(u)\big)$ for the image of $u \in U$ in the linear isotropy representation. Then the Poincaré polynomial $\Phi(t)$ is given by the integral over $U$, with respect to the invariant volume element, normalized to give $U$ the volume 1, of

$$(2) \qquad\qquad \Delta_R(t) = \det(a_{ij}(u) + \delta_{ij}t).$$

Using the known results of Weyl on invariant volume elements, he deduces in section VIII various properties of the Poincaré polynomial of a simple compact group $G$: it is divisible by $(1+t)^\ell$, where $\ell$ is the rank of the group, by $(t^3 + 1)$, and the sum of Betti numbers is $2^\ell$. He also conjectures, besides (1) above, that the Poincaré polynomial of $\mathbf{SO}_{2n+1}$ is

$$(3) \qquad\qquad (t^3 + 1).(t^7 + 1) \cdots (t^{4n-1} + 1).$$

He also considers various spaces, notably the "*espace projectif réglé*", i.e. the space of lines in $\mathbb{P}_3(\mathbb{C})$.

Richard Brauer

**5.3.** Cartan came back to these questions in the conference [C150], 1936: He returns to the converse of the third Lie fundamental theorem by showing more precisely that a given Lie algebra is always the Lie algebra of a simply connected Lie group which is, topologically, the product of a euclidean space by a compact subgroup. (He believes, however, that this decomposition is not valid for a non-simply-connected group.)

A last section is devoted to the Betti numbers of compact simple groups. The conjectures (1), (3) above were proved by R. Brauer, who moreover gave the Poincaré polynomials of the other two series of simple groups (C.R. **201** (1955), 419-21). A little before, L.S. Pontrjagin had determined them by a completely different, geometric, method (C.R. Acad. Sci. Paris **200** (1935), 1277-8).

Cartan points out that the Poincaré polynomial of the exceptional group $G_2$ is $(1 + t^3)(1 + t^{11})$, as already follows from [C118], 1929. He hopes however that, even if the Poincaré polynomials of the exceptional groups can be found case by case to have the same shape, it will be possible to find a general reason why these polynomials are products of factors $(t^{2m+1} + 1)$. This challenge was brilliantly met a few years later by Heinz Hopf, who deduced it from the existence of a continuous product with identity (Annals of Math. **42** (1941), 22-52).

**5.4.** In 1937, Cartan gave several lectures in the framework of an "International Conference on Tensor Differential Geometry" in Moscow, one [C154] on the Betti numbers of homogeneous spaces of compact Lie groups. Most of it summarizes work done earlier, but he also draws attention to one special class of irreducible compact symmetric spaces, those admitting an invariant complex analytic structure. They are the compact counterparts of the bounded symmetric domains (see §6). Apart from two exceptions, they form four infinite classes, all consisting of well known projective manifolds. As to their topology, he states:

> Dans tous ces espaces, les nombres de Betti d'ordre impair sont nuls; ceux d'ordre pair sont tous positifs et liés au nombre de représentations linéaires irréductibles du groupe $\gamma$ [the linear isotropy group]. On peut trouver facilement des bases d'homologie pour chaque ordre; elles sont formées de variétés algébriques. Les résultats obtenus se rattachent, tout en étant beaucoup plus généraux, à de nombreux travaux ressortissant à la géométrie énumérative de H. Schubert.[19]

He then refers to a recent thesis of C. Ehresmann (Annals of Math. **35** (1934), 396-443), where the author establishes these results by a different, topological, method, which allows him to get information on torsion coefficients, and is also applicable to certain non-symmetric homogeneous spaces and to some real forms of these spaces.

Even though these spaces are not mentioned in [C145] (see §6), Cartan was well aware of their remarkable properties.

## §6. Bounded symmetric domains

**6.1.** The incentive for Élie Cartan's paper [C145], 1935, on bounded homogeneous spaces in $\mathbb{C}^n$ appears to have been the work of his son Henri Cartan on automorphisms of bounded domains. Let $D$ be one. In particular, H. Cartan had shown that: a) the group Aut $D$ of automorphisms of $D$ is a real Lie group (not necessarily connected); b) elements in the Lie algebra of Aut $D$ which are linearly

---

[19] In all these spaces, the Betti numbers of odd order are zero; those of even order are all positive and related to the number of irreducible linear representations of $\gamma$ [the linear isotropy group]. One can easily find bases for the homology for each order; they consist of algebraic varieties. The results so obtained are related, while being much more general, to many works pertaining to the enumerative geometry of H. Schubert.

independent over $\mathbb{R}$ remain so over $\mathbb{C}$; and c) the isotropy group of a point $o \in D$ is compact. Its elements are completely determined by their differentials at $o$.

Point c) became obvious once Bergmann showed the existence of a canonical complete hermitian metric on $D$, invariant under any automorphism, which I shall take for granted.

In view of a), if $D$ is homogeneous, then it is a quotient $D = G/K$, where $G = (\operatorname{Aut} D)^o$ is a real Lie group, the identity component $(\operatorname{Aut} D)^o$ of $\operatorname{Aut} D$, and the compact subgroup fixing some point $o \in D$. It was natural for him to try to find information in the homogeneous case. He determined all bounded homogeneous domains for $n \leq 3$ (see below). The general problem becoming untractable in higher dimensions, he turned to a special case which brought him back to his earlier work on symmetric spaces.

**6.2.** Say that $D$ is *symmetric* if every point is the isolated fixed point of an involutory automorphism of $D$. Viewed first as a real $2n$-dimensional Riemannian manifold (with respect to the Bergmann metric), $D$ is Riemannian symmetric, so these spaces had to be among those Cartan had classified earlier. He quickly established some general properties: $G$ is semisimple, and $D$ is a product of irreducible domains. Each of those is a quotient $G/K$, where $G$ is simple non-compact with center reduced to $\{1\}$ and $K$ a maximal compact subgroup. Write $D_r$ for $D$, viewed as a *real* symmetric space. Let $\mathfrak{g} = \mathfrak{k} \oplus \mathfrak{p}$ be the usual Cartan decomposition of the Lie algebra $\mathfrak{g}$ of $G$. Then $\mathfrak{p}$ is identified to $(TD_r)_o$ and the isotropy representation of $K$ in $(TD_r)_o$ is irreducible, but it is not absolutely irreducible: since $D$ is a complex manifold, $\mathfrak{p}$ is in fact the complex tangent space $TD_o$, viewed as a real vector space, and its complexifications $\mathfrak{p}_c$ is a direct sum $\mathfrak{p}_c = \mathfrak{p}^+ \oplus \mathfrak{p}^-$, where $\mathfrak{p}^+ = (TD)_o^{1,0}$ is the complex tangent space, spanned by the derivatives $\partial/\partial z_1, \ldots, \partial/\partial z_n$ of the local coordinates $z_i$, and $\mathfrak{p}^- = (TD)_o^{0,1}$ is the antiholomorphic tangent space, spanned by the $\partial/\partial \bar{z}_i$. Both spaces are invariant under $K$. The representations of $K$ in them are irreducible and complex conjugate. Moreover, Cartan showed that $\mathfrak{p}^+$ and $\mathfrak{p}^-$ are commutative subalgebras of $\mathfrak{g}_c$.

**6.3.** Thus we are led to the subclass of irreducible symmetric spaces of negative curvature in which the isotropy representation is not absolutely irreducible. It turns out that this property implies others, apparently much stronger. Consider the following conditions, with $G$ and $K$ as before.

(i) The isotropy representation is not absolutely irreducible.

(ii) $K$ is not semisimple, hence has a one-dimensional center.

(iii) $G/K$ is a bounded symmetric domain.

Of course (ii) $\Rightarrow$ (i). Cartan's classification shows by inspection that (i) $\Rightarrow$ (ii).[20] Clearly, (iii) $\Rightarrow$ (i). The spaces satisfying the first two conditions divide into four infinite classes, where $G$ is a classical group, and two exceptional cases, of complex dimensions 16, 27, where $G$ is a real form of $\mathbf{E}_6$ or $\mathbf{E}_7$. For all four infinite classes Cartan gave a realization of $G/K$ as a bounded symmetric domain. An assertion in n° 45 implies that for him this was also the case for the two exceptional ones, and he takes it for granted in formulating his "conclusion générale", but he did not give an explicit construction.[21]

---

[20] For a classification-free proof, see A. Borel and A. Lichnerowicz, C.R. Acad. Sci. Paris **234** (1952), 2332-2334, or [B], IV, 1.3.

[21] The first general proof that these spaces can be realized as bounded symmetric domains is due to Harish-Chandra, Amer. J. Math. **78** (1956), 564-628; Collected Papers II, 90-154.

**6.4.** The last part of the paper is devoted to the determination of homogeneous bounded domains in $\mathbb{C}^2$. Cartan finds out, a posteriori, that all are symmetric. He also states that he had determined the bounded homogeneous domains in $\mathbb{C}^3$ and had again checked that they are all symmetric. He raises the question of whether this remains true in higher dimensions.[22]

**6.5.** Among the bounded symmetric domains are the Siegel upper half-spaces

$$\{Z \in M_n(\mathbb{C}) \mid {}^t Z = Z, \ \text{Im} \, Z > 0\},$$

which generalize Poincaré's upper half-plane. Indeed, the bounded symmetric domains, or more precisely their quotients by arithmetic groups, have been the framework of the theory of holomorphic or meromorphic automorphic forms in several complex variables, developed first mostly by C.L. Siegel, and later by G. Shimura and many others.

**6.6.** In this paper, Cartan does not mention at all the compact counterparts of the bounded symmetric spaces, in his correspondence between irreducible symmetric spaces of positive and of negative curvature, but he does so in [C154], 1937; see 5.4. Define an hermitian symmetric space to be an hermitian manifold in which every point is an isolated fixed point of an involutive automorphism (of the hermitian manifold). Then the bounded symmetric domains and their counterparts are all the (non-flat) irreducible hermitian symmetric manifolds (see e.g. [B], VI). The correspondence between compact and non-compact ones takes a very concrete form in the sense that in a given pair, the non-compact one admits a canonical embedding in the compact one. The latter can be viewed as $G_u/K$ as usual, but also as $G_c/Q$, where $Q$ is the subgroup of $G_c$ with Lie algebra $\mathfrak{k}_c \oplus \mathfrak{p}^-$. It is a special case of the rational homogeneous spaces to be discussed in VI, 2.3.

---

[22] The first counterexample, in dimension four, was given by I. Piatetski-Shapiro, Dokl. Akad. Nauk SSSR **113** (1957), 980-983. This was the starting point for considerable activity, which led to a full description of bounded homogeneous domains, up to isomorphisms. For a survey of this and related problems on complex homogeneous spaces, see Chapter IV, by D.N. Ahiezer, in Encyclopaedia of Math. Sci. **10** (1986), Springer.

# References for Chapter IV

Except for [C0], the numbers of the references to É. Cartan's work are those of his Oeuvres Complètes (hereafter O.C. for short).

[C0]　　É. Cartan, Oeuvres Complètes, Part I, Gauthier-Villars, Paris, 1952.

[C28]　　_____, *Les groupes de transformations infinis, continus, simples*, Ann. Sci. École Norm. Sup. (3) **26** (1909), 93–161; O.C., Part II₂, 857–925.

[C37]　　_____, *Les groupes projectifs qui ne laissent invariante aucune multiplicité plane*, Bull. Soc. Math. France **41** (1913), 53–96; [C0], 355–398.

[C38]　　_____, *Les groupes réels simples finis et continus*, Ann. Sci. École Norm. Sup. (3) **31** (1914), 263–355; [C0], 399–492.

[C64]　　_____, Leçons sur les invariants intégraux, Hermann, Paris, 1922.

[C81]　　_____, *Les tenseurs irréductibles et les groupes linéaires simples et semi-simples*, Bull. Sci. Math. **49** (1925), 130–152; [C0], 531–553.

[C91]　　É. Cartan and J.A. Schouten, *On the geometry of the group-manifold of simple and semisimple groups*, Koninkl. Akad. Wetensch. Amsterdam Proc. **29** (1926), 803–15; [C0], 573–585.

[C93]　　É. Cartan, *Sur une classe remarquables d'espaces de Riemann*, Bull. Soc. Math. France **54** (1926), 214–264; [C0], 587–637.

[C94]　　_____, *Sur une classe remarquable d'espaces de Riemann*, Bull. Soc. Math. France **55** (1927), 114–134; [C0], 639–659.

[C101]　　_____, *La géométrie des groupes de transformations*, J. Math. Pures Appl. (9) **6** (1927), 1–119; [C0], 673–791.

[C103]　　_____, *La géométrie des groupes simples*, Ann. Mat. Pura Appl. (4) **4** (1927), 209–256; [C0], 793–840.

[C107]　　_____, *Sur certaines formes riemanniennes remarquables des géométries à groupe fondamental simple*, Ann. Sci. École Norm. Sup. (3) **44** (1927), 345–467; [C0], 867–988.

[C111]　　_____, *Sur les nombres de Betti des espaces de groupes clos*, C.R. Acad. Sci. Paris **187** (1928), 196–198; [C0], 999–1001.

[C113]　　_____, *Complément au Mémoire "Sur la géométrie des groupes simples"*, Ann. Mat. Pura Appl. (4) **5** (1928), 253–260; [C0], 1003–1010.

[C114]　　_____, Leçons sur la géométrie des espaces de Riemann, Gauthier-Villars, Paris, 1926.

[C116]　　_____, *Groupes simples clos et ouverts et géométrie riemannienne*, J. Math. Pures Appl. (9) **8** (1929), 1–33; [C0], 1011–1043.

[C117]　　_____, *Sur la détermination d'un système orthogonal complet dans un espace de Riemann symétrique clos*, Rend. Circ. Mat. Palermo **53** (1929), 217–252; [C0], 1045–1080.

[C118]　　_____, *Sur les invariants intégraux de certains espaces homogènes clos et les propriétés topologiques de ces espaces*, Ann. Soc. Polonaise Math. **8** (1929), 181–225; [C0], 1081–1125.

[C128]　　_____, *La théorie des groupes finis et continus et l'Analysis Situs*, Mém. Sci. Math. **XLII**, Gauthier-Villars, Paris, 1930; [C0], 1165–1225.

[C138]　　_____, *Les espaces riemanniens symétriques*, Verh. Internat. Congr. Math. (Zürich, 1932), 152–161; [C0] 1247–1258.

[C145]　　_____, *Sur les domaines bornés homogènes de l'espace de n variables complexes*, Abh. Math. Sem. Hamburg **11** (1935), 116–162; [C0], 1259–1305.

[C150]　　_____, *La topologie des espaces représentatifs des groupes de Lie*, Enseignement Math. **35** (1936), 177–200; Exposés de Géométrie VIII, Hermann, Paris 1936; [C0], 1307–1330.

[C154]　　_____, *La topologie des espaces homogènes clos*, Trudy Sem. Vektor. Tenzor. Anal. **4** (1937), 388–394; [C0], 1331–1337.

[C183]　　_____, Leçons sur la géométrie des espaces de Riemann, 2ème éd., revue et augmentée, Gauthier-Villars, Paris, 1946.

## Further references

[B]　　A. Borel, Semisimple groups and Riemannian symmetric spaces, Texts and Readings in Math., vol. **16**, Hindustan, New Delhi, 1998.

[H]　　S. Helgason, Differential Geometry and Symmetric Spaces,, Pure Applied Math. **12**, Academic Press, 1962.

[M]　　G.D. Mostow, *A new proof of É. Cartan's theorem on the topology of semi-simple Lie groups*, Bull. Amer. Math. Soc. **55** (1949), 69–80.

[N]   J. von Neumann, *Zur Theorie der Darstellung kontinuerlicher Gruppen*, Sitzungsber. Preuss. Akad. Wiss. Berlin Phys.-Math. Kl. **1927**, 76–90; Coll. Wks. I, 134–148.

[R]   R. Richardson, *Compact real forms of a complex semi-simple Lie algebra*, J. Diff. Geometry **2** (1968), 411–419.

[W68]   H. Weyl, *Theorie der Darstellung kontinuerlicher halbeinfacher Gruppen durch lineare Transformationen*. I, II, III und Nachtrag, Math. Zeitschr. **23** (1925), 271–309; **24** (1926), 328–376, 377–395, 789–791; Collected Papers II, 543–647.

[W73]   H. Weyl and F. Peter, *Die Vollständigkeit der primitiven Darstellungen einer geschlossenen kontinuerlichen Gruppe*, Math. Annalen **97** (1927), 737–755; Collected Papers III, 58–75.

# CHAPTER V

# Linear Algebraic Groups in the 19th Century

The various works surveyed in this chapter had various motivations and were carried out to a large extent independently from one another. They are presented here as contributions to a theory of linear algebraic groups a bit by hindsight: they did not necessarily appear so to their authors, who had other goals in mind, except for Maurer, whose aim was indeed to build up such a theory.

## §1. S. Lie, E. Study, and projective representations

**1.1.** The theory of linear representations of Lie algebras was initiated by Killing and later developed by Élie Cartan, as recalled in Chapter I. Lie and Study considered projective representations of a few groups, mainly $\mathbf{PGL}_n(\mathbb{C})$ or (for Study) $\mathbf{SL}_n(\mathbb{C})$ ($n = 2, 3, 4$), from a quite different point of view, which was all but forgotten for over sixty years.

They were concerned with irreducible representations in a projective setting. Let $G = \mathbf{SL}_2(\mathbb{C})$, and let $B$ be its subgroup of upper triangular matrices. Let $\sigma : G \to \mathrm{Aut}(\mathbf{P}_n)$ be a (rational—this is always understood) representation of $G$. By Lie's theorem, $B$ has at least one fixed point, say $z$. Assume it is not fixed under $G$. Then, since $B$ is a maximal proper subgroup, it is the full isotropy group and the orbit $G.z$ is isomorphic to $G/B$, i.e. is a smooth rational curve. If $\sigma$ is irreducible (no proper projective subspace is invariant under $G$), then $Z = G.z$ is not contained in any proper projective subspace. In Lie's terminology, it is "as curved as possible" (möglichst gekrümmt). Assuming this, Lie showed that $Z$ is a normal rational curve of smallest possible *order*, namely $n$.[1] In suitable homogeneous coordinates $(y_o, \dots, y_n)$ in which $z = (1, 0, \dots, 0)$ it is given by

$$(1) \qquad y_i = x^{n-i}.y^i \qquad (i = 0, \dots, n),$$

where $x, y$ are homogeneous coordinates on $\mathbf{P}_1(\mathbb{C})$.

Lie proved it under what appeared to him more general assumptions ([LE], pp. 182-7). Namely, he assumed that $Z$ is a smooth rational curve in $\mathbf{P}_n$ invariant under two non-commuting infinitesimal transformations of $\mathbf{P}_n$, i.e. two non-commuting elements of $\mathrm{Lie}(\mathrm{Aut}\,\mathbf{P}_n)$. They generate a two- or three-dimensional subgroup of $\mathrm{Aut}(Z)$. The latter is of course isomorphic to $\mathbf{PSL}_2(\mathbb{C})$. Lie knew that the two-dimensional subgroups of $\mathbf{SL}_2(\mathbb{C})$ were conjugate to $B$, so that $Z$ is invariant under a group of projectivities $H$ of the ambient space isomorphic to $B$, or rather $B/\{\pm 1\}$. Lie carried out the argument for $n = 3$ (*loc. cit.*, Theorem 1) and said that the proof extends without difficulty to the general case (Theorem 2). Let $n = 3$, and let $p$ be a point of $Z$ fixed under $H$. Then $H$ also leaves invariant a flag $F_1$ at $p$

---

[1] We recall that the *order* of a closed irreducible subvariety of $\mathbf{P}_n(\mathbb{C})$ is the number of intersection points with a general projective subspace of complementary dimension.

Eduard Study

consisting of a projective line $E_1$ through $p$, and a projective plane containing $E_1$. Consider now a "general" point $q \in Z$. Its orbit under $H$ is one-dimensional. Let $H_q$ be the isotropy group of $q$ in $H$, and $Y$ a generator of its Lie algebra. The group $H_q$ also leaves stable a flag $F_2$ at $q$. Using this, Lie showed that $H_q$ fixes the four vertices of a tetrahedron. In other words, $Y$ is diagonalisable. It is then rather easy to show, using the fact that $Z$ is as curved as possible, that in suitable non-homogeneous coordinates $(v, w)$ adapted to the flag $F_2$, the equations of $Z$ are $v = u^2, w = u^3$.

Lie's assumption on $Z$ was not really more general, because if $Z$ is a smooth projective variety having a Picard variety reduced to zero, which is the case for $\mathbb{P}_1$, then any automorphism of $Z$ belonging to the identity component of $\mathrm{Aut}(Z)$ is induced by projective transformations of the ambient space. In particular, any projective embedding of $\mathbf{P}_1$ yields a projective representation of $G$.

Lie determines here the projective embeddings of $\mathbf{P}_1$ which are as curved as possible. Of course, any irreducible projective representation leads to one, but the converse is not *a priori* true. The text on p. 784 of [LE] shows that for Lie and Engel the two problems were equivalent. However, this was surely written after 1890, at a time when full reducibility had been established, as explained below; and, modulo that fact, the two questions are indeed equivalent.

**1.2.** E. Study, in the course of investigations on invariant theory, had been led to consider general projective representations of $G$. Around 1889, he sent to Lie and Engel a manuscript asserting that all such representations are fully reducible, stated in the projective language. Namely, there should exist projective subspaces $E_i$ $(i = 1, \ldots, s)$ of the ambient $\mathbf{P}_n$ stable and irreducible under $G$, hence with empty pairwise intersections, such that $\mathbf{P}_n$ is the join of the $E_i$; in particular,

$$n + 1 = \sum_{1 \leq i \leq s} (\dim E_i + 1).$$

According to ([LE], p.785) the proof had some gaps, but was completed by Engel (as was also acknowledged by Study in [S2]).[2] On pp. 786-7 of [LE] Lie describes a number of results communicated by Study in [S2] and in a later manuscript, providing "almost complete proofs", which seems to be lost.[3] Those pertaining to $\mathbf{SL}_3$ are also stated in [S3]:

Let $G = \mathbf{SL}_3$, and let $B$ be again the group of upper triangular matrices in $G$, i.e. of determinant one. The quotient $G/B$ is the variety of flags in $\mathbf{P}_2$, i.e. a point in $G/B$ consists therefore of a "line element" on $\mathbf{P}_2$: a point of $\mathbf{P}_2$ and a (projective) line containing it. The subgroup $B$ is properly contained in exactly two proper Lie subgroups $P_1, P_2$, one leaving invariant a point, the other a line in $\mathbf{P}_2$. The quotient of $G$ by either is $\mathbf{P}_2$.

Let $G \to \mathrm{Aut}(\mathbf{P}_{N-1})$ be an irreducible projective representation of $G$. Again, $B$ has a fixed point, say $z$, by Lie's theorem. But there are now two main cases: either the orbit $Z$ of $z$ is three-dimensional and the isotropy group $G_z$ is $B$, or it is two-dimensional and $G_z$ is $P_1$ or $P_2$. The representation is characterized by a pair $(m, n)$ of natural integers. They are both non-zero if and only if $Z$ is three-dimensional. In that case, the integer $N$ is given by

(1)                    $$N = (m + 1)(n + 1)(m + n + 2)/2$$

and $Z$ is of order $3mn(m + n)$.

The orbit $Z$ is two-dimensional if and only if one of $m, n$ is zero, depending on whether the isotropy group is $P_1$ or $P_2$. We have accordingly

(2)              $$N = (m + 1)(m + 2)/2 \quad \text{or} \quad N = (n + 1)(n + 2)/2,$$

and the order of $Z$ is $m^2$ or $n^2$.

Study asserts that these are all irreducible representations of $G$. [This is indeed the case. Let $\omega_1$ and $\omega_2$ be the highest weight of the identity representation $\rho$ of $G$ and of its contragredient $\rho^*$ (the two fundamental representations of $\mathbf{SL}_3$). Then the irreducible representation characterized by $(m, n)$ is the one of highest weight $m\omega_1 + n\omega_2$. The degree checks with Weyl's formula. Moreover, the value of the order of $Z$ is confirmed by the general formula in ([BH], 24.10).]

Any invariant (closed) subvariety is a union of closures of orbits. Study calls those "Körper" ([S1], p. 113). He notes that every "Körper" contains $Z$, which is therefore the unique smallest orbit. This amounts to the fact that $z$ is the only fixed point of $B$, or, as we would say now, that the highest weight has multiplicity one.

**1.3.** In [S1], p. 53 and §11, Study describes these irreducible representations in linear terms. Consider the identity representation $(\rho, V)$ of $G$ in $V = \mathbb{C}^3$, with

---

[2]See Chapter II for full reducibility proofs by Cartan, Fano, and others.

[3]According to [W], p. 133, footnote 13, Study's Nachlass contained a copy of it: "*Die allgemeine projective Gruppe* $\mathfrak{g}_8$ *in der Ebene*", 6 pages, which is apparently also lost.

canonical basis $e_1, e_2, e_3$ and coordinates $x_1, x_2, x_3$. Let $(\rho^*, V^*)$ be the contragredient representation. The $m$-th symmetric power $\rho_m$ (resp. $\rho_m^*$) of $\rho$ (resp. $\rho^*$) is the natural representation of $G$ in the space $E^{m,0}$ (resp. $E^{0,n}$) of homogeneous forms of degree $m$ in the $e_i$ (resp. $x_j$). We let $E^{m,n}$ be the space of homogeneous forms in the six variables $x_i, e_j$ of bidegree $(m, n)$, i.e. of degree $m$ (resp. $n$) in the $e_i$ (resp. $x_j$). Thus $E^{m,n} = E^{m,0} \otimes E^{0,n}$ (also as $G$-modules). Then the irreducible representation labeled $(m, n)$ by Study is the restriction of the natural representation in $E^{m,n}$ to the space of so-called "normal forms", to be denoted here $F^{m,n}$. If either $m$ or $n$ is zero, these are all the forms of the given type. If $m.n \neq 0$, the normal forms are the zeroes of the mixed Laplacian

$$(1) \qquad \Delta = \frac{\partial^2}{\partial e_1 \partial x_1} + \frac{\partial^2}{\partial e_2 \partial x_2} + \frac{\partial^2}{\partial e_3 \partial x_3}.$$

[As usual, let $x_i - x_j$ $(i > j)$ be the positive roots of $\mathfrak{g}$ with respect to the Cartan subalgebra of diagonal matrices of trace zero. Then the highest weight line in $V$ is spanned by $e_1$, but in the contragredient representation it is spanned by $x_3$, so that $e_1^m$ (resp. $x_3^n$) is a highest weight vector in $E^{m,0}$ (resp. $E^{0,m}$) and $e_1^m \otimes x_3^n$ is a highest weight vector in $E^{m,n}$. Consequently, $\Delta$ annihilates the highest vector and, since it obviously commutes with $G$, also the orbit of the highest weight vector. Study asserts that it is irreducible, and hence $F^{m,n}$ is spanned by the orbit of the highest weight vector.]

This construction, for $m.n \neq 0$, reminds one of the Cartan product. The representations $\rho$ and $\rho^*$ are the two fundamental representations of $G$. Given $m, n \in \mathbb{N}$, Cartan looks for an irreducible representation of highest weight $m\omega_1 + n\omega_2$ in the tensor product of $m$ copies of $\rho$ and $n$ copies of $\rho^*$. But $E^{m,0}$ (resp. $E^{0,n}$) has highest weight $m\omega_1$ (resp. $n\omega_2$) and happens to be irreducible. Then $F^{m,n}$ is a suitable subspace of $E^{m,n}$, the space of normal forms for Study, the smallest $G$-invariant subspace containing a highest weight vector for Cartan. Study also gives the decomposition of $E^{m,n}$ into irreducible $G$-modules:

$$(2) \qquad E^{m,n} = \bigoplus_{0 \leq j \leq \min(m,n)} F^{m-j,n-j}$$

(with the understanding that $F^{a,b} = 0$ if $a$ or $b$ is $< 0$). We have therefore

$$(3) \qquad E^{m,n} = F^{m,n} \oplus E^{m-1,n-1},$$

which yields (2).

The equality (2) is not stated in this way in [S1]. It is derived from an identity of Gordan in invariant theory. The framework is quite different from the present one, so I prefer to devote a separate section to it (see §2).

**1.4.** In [LE] (*loc. cit.*), Lie and Engel mention some other results of Study, taken from the manuscript mentioned above, about $\mathbf{SL}_4$ or $\mathbf{SO}_4$. In the former case, he states that the orbits of a point $z$ fixed under the upper triangular group $B$ have dimension 3, 4 or 5, in which cases they are respectively isomorphic to $\mathbf{P}_3$, the space of lines on $\mathbf{P}_3$ and the space of flags on $\mathbf{P}_3$. These are indeed the quotients of $G$ by the proper subgroups containing $B$.

The group $\mathbf{PSO}_4(\mathbb{C})$ is the product of two copies of $\mathbf{PSL}_2$. Study gives a description of the irreducible representations of that group, which, translated into modern terminology, exhibits them as tensor products of irreducible representations of the two factors.

In [S2], after having described the irreducible projective representations of **SL**$_3$, Study adds that he has tried to prove full reducibility for **SL**$_3$, but the argument still has a gap. Then he adds:

> and so on. I have moreover many analogous results on the projective group of **P**$_3$, on the conformal group and similar groups; I believe that one can establish such results for simple and semisimple groups.

For **SL**$_3$, what is mainly missing from the material available to us is the proof of the irreducibility of the $F^{m,n}$ and the computation of the order of $Z$ (1.2), but his letters and [S1] make it clear that he had arrived at a remarkable understanding of this topic.[4] Apparently, he never came back to it. His approach was not pursued, being overshadowed by the Cartan theory. But his more geometric and global point of view, the importance attached to $Z$, seem to me closer in spirit to the later approach of Borel and Weil [B1] than to the infinitesimal theory, even though Study did not think in terms of sections of homogeneous line bundles.

## §2. E. Study, Gordan series and linear representations of SL$_3$

**2.1.** The basic tool of Study to arrive at 1.3(2) is a series due to P. Gordan ([G], §§4, 5), established in the framework of the symbolic calculus of A. Clebsch and P. Gordan. To describe it, first I have to introduce some notation.

Fix $p \geq 2$. Let $V = \mathbb{C}^p$, with basis $e_1, \dots, e_p$ and coordinates $x_1, \dots, x_p$. The $x_i$'s are then the dual basis of $V^*$. An element $c_1 e_1 + \cdots + c_p e_p$ $(c_i \in \mathbb{C})$ of $V$ will be denoted $(c, e)$, where $c$ (resp. $e$) stands for the $p$-tuple $\{c_1, \dots, c_p\}$ (resp. $\{e_1, \dots, e_p\}$). Similarly an element $x_1 d_1 + \cdots + x_p d_p$ $(d_i \in \mathbb{C})$ will be denoted $(x, d)$. More generally, given two $p$-tuples $u = \{u_1, \dots, u_p\}$ and $v = \{v_1, \dots, v_p\}$, we let $(u, v)$ denote the sum of the $u_i.v_i$, assuming it makes sense, as in the two previous cases. An important role is played by $(x, e) = x_1 e_1 + \dots + x_p e_p$, the canonical bilinear form putting $V$ and $V^*$ in duality.

$E^{m,0}$ (resp. $E^{0,n}$) is now the space of homogeneous forms of degree $m$ (resp. $n$) on $V$ (resp. $V^*$). A general element of $E^{m,0}$ (resp. $E^{0,n}$) is written $(c, e)^m$ (resp. $(x, d)^n$); hence one of $E^{m,n} = E^{m,0} \otimes E^{0,n}$ is $(c, e)^m.(x, d)^n$. If $m.n \neq 0$, Gordan calls it (following Aronhold) a "Zwischenform", while Study [S1] uses the word "connexe" (for $p = 3$).

$P$ is now the mixed Laplacian 1.3(1) in $2p$ variables. Obviously $P.E^{m,n} \subset E^{m-1,n-1}$ (it being understood that $E^{a,b} = 0$ if $a$ or $b$ is $< 0$). Again, a zero of $P$ is called a normal form.

**Remark.** For the reader who wishes to look at [G], I have deviated from Gordan's notation and terminology in some ways. Gordan would have written $c_e$ and $x_d$ for $(c, e)$ and $(x, d)$ (and Study uses $(ce)$ and $(xd)$). The mixed Laplacian is denoted $\delta$ by Gordan. I am following Study. Gordan has no name for a normal form, only a notation, namely, square brackets [ ] around the form, but Study attributes the terminology "normal form" to Gordan. I shall also use it.

**2.2.** In the sequel, $\mu(m, n)$ denotes the minimum of $m$ and $n$. The result of Gordan needed by Study is the following:

---

[4]Although we do not know for sure, it seems rather likely to me that Study had indeed proved the irreducibility of the $F^{m,n}$. He surely knew that it sufficed to show that the highest weight line was the only one invariant under $B$ in $F^{m,n}$, which may well have been within his means.

(*) A Zwischenform $F \in E^{m,n}$ can be written uniquely as a linear combination

$$(1) \qquad F = \sum_{0 \leq j \leq \mu(m,n)} c_j^{m,n}(x,e)^j . F_j,$$

where $F_j$ is a normal form of bidegree $(m-j, n-j)$ (see (II) in [G], §5, p. 106), and

$$(2) \qquad c_j^{m,n} = \binom{m}{j} \cdot \binom{n}{j} \cdot \binom{m+n+p-j-1}{j}^{-1}.$$

To show this, Gordan first proves ((IV) on p. 103) that, for a Zwischenform $F \in E^{m,n}$,

$$(3) \qquad P\big((x,e)^j . F\big) = (x,e)^j . PF + j(m+n+p+j-1)(x,e)^{j-1}.F$$

by a straightforward computation. From (2) he deduces the "remarkable conse-quence" that there is a unique linear combination [F] of the forms $(x,e)^j.P^jF$ $\big(0 \leq j \leq \mu(m,n)\big)$ which is a normal form, and gets

$$(4) \qquad F = [F] + \sum_{1 \leq j \leq \mu(m,n)} d_j^{m,n}(x,e)^j . P^j F.$$

[For the sake of accuracy, I should add that Gordan writes $F$ as a linear combination of the forms

$$(5) \qquad P^j F . \big(m(m-1)\cdots(m-j+1)n(n-1)\cdots(n-j+1)\big)^{-1} \cdot (x,e)^j$$

with coefficients $c_j^{m,n}$ given by

$$(6) \qquad c_j^{m,n} = (-1)^j \binom{m}{j} \binom{n}{j} \binom{m+n+p-2}{j}^{-1}$$

(pp. 103-4).

The existence of the series (1) then follows by an easy induction. Gordan proves the uniqueness by similar computations (pp. 105-6). [In fact, he could have spared himself these computations, because (3) shows that if $F$ is a normal form, then $(x,e)^j.F$ is an eigenvector of $(x,e)P$ with eigenvalue $j(m+n+p+j-1)$, and those eigenvalues are distinct, as $j$ varies from 0 to $\mu(m,n)$.]

The $F_j$ are the "elementary covariants" of $F$. The uniqueness shows that $F$ is a normal form if and only if it is not divisible by $(x,e)$, and that if $F$ is divisible by $(x,e)^j$, then $F_i = 0$ for $i < j$.

**2.3.** Gordan looks at this series for one $F$ at a time. Now let us go back to [S1]. Study refers to [G] for 2.2(1), but still gives a complete proof. Being interested in representations, he considers the space of such forms. Clearly 2.2(1) implies that

$$(1) \qquad E^{m,n} = \bigoplus_{0 \leq j \leq \mu(m,n)} (x,e)^j . F^{m-j,n-j},$$

where $F^{a,b}$ is the space of normal forms of bidegree $(a,b)$. Since $(x,e)$ is invariant under $\mathbf{SL}_p$ and $P$ commutes with $\mathbf{SL}_p$, this is a decomposition into $\mathbf{SL}_p$-invariant subspaces. In the case $p=3$, to which Study specializes, this is a more precise form of 1.3(2) (which is obviously valid for any $p$).

**2.4.** Let $p = 2$. Then the $E^{m,0}$ are all irreducible representations of $\mathbf{SL}_2$, and they are all self-contragredient. $F^{m,n}$ is the irreducible representation $E^{m+n}$. The equality (1) now implies that

$$(1) \qquad\qquad E^m \otimes E^n = \bigoplus_{0 \leq j \leq \mu(m,n)} E^{m-2j},$$

which is the so-called Clebsch-Gordan series, giving the decomposition of the tensor product of two irreducible representations of $\mathbf{SL}_2$.

For $p = 2$, Clebsch had also introduced a series similar to 2.2(1), independently of, and at the same time as, Gordan, so this became known as the Clebsch-Gordan series. It was noticed by Weyl (in Gruppentheorie und Quantummechanik, 2nd edition, Hirzel, Leipzig 1930, p. 115) that (1) follows immediately from the Clebsch-Gordan series in invariant theory, so I presume this is the reason why this name became attached to the decomposition (1).[5]

Let $p = 3$. Study determined that the $F^{m,n}$ are all the irreducible representations of $\mathbf{SL}_3$, and so we get a decomposition into irreducible $\mathbf{SL}_3$-modules of the $E^{m,n}$, i.e. of the tensor product of certain irreducible representations.

Let $p = 4$. Study realized that the $F^{m,n}$ would not be all irreducible representations. As he puts it ([S1], Note 17, p. 204), there are now three kinds of variables: besides the points and planes in $\mathbf{P}_3$, one has to consider the lines in $\mathbf{P}_3$. (In our language, besides the identity representation $\rho$ and its contragredient $\rho^*$, there is a third fundamental representation, the second exterior power $\Lambda^2\rho$ of $\rho$). He may have also realized that the symmetric powers of $\Lambda^2\rho$ are not irreducible. At any rate, he concluded that the theory of quaternary forms would be much more difficult than that of ternary forms.

**2.5.** Again let $p \geq 2$ be arbitrary. We still have 2.3(1), and hence we again get 1.3(2). Since the dimension of $E^{m,0}$ or $E^{0,m}$ is $\binom{m+p-1}{p-1}$, we get

$$(1) \qquad \dim F^{m,n} = \binom{m+p-1}{p-1}\binom{n+p-1}{p-1} - \binom{m+p-2}{p-2}\binom{n+p-2}{p-2}.$$

Let $\omega_1$ and $\omega_{n-1}$ be the highest weight of $\rho$ and $\rho^*$. Then the highest weight in $F^{m,n}$ is $m\omega_1 + n\omega_{n-1}$, and hence $F^{m,n}$ contains the irreducible representation with that highest weight. But Weyl's degree formula (III, 4(9)) shows readily that the dimension of the latter is the same as that of $F^{m,n}$. Therefore the $F^{m,n}$ are irreducible, and 2.3(1) also yields the decomposition of $E^{m,n}$ into irreducible invariant subspaces.

## §3. Emile Picard

**3.1.** Sophus Lie said very early that one goal of his theory was to establish a Galois theory for systems of differential equations. As pointed out in Chapter I, §1, he had remarked—and this played a role in the genesis of his theory—that some classical methods of integration of ordinary differential equations could be unified by noting that they relied, at least implicitly, on the existence of a one-parameter group of transformations leaving the system invariant. This led to the idea that if a system is invariant under some continuous group, this might be used to simplify the

---

[5] In the notes *Elementary theory of invariants*, Institute for Advanced Study, 1935-36, p. 136, Weyl calls (1) the "Clebsch-Gordan development in the language of representation theory".

Emile Picard

integration of the system. The Galois group would then be the group leaving the system invariant. But the analogy with Galois theory is weak, since most systems are not invariant under a Lie group of positive dimension.

**3.2.** In the early 1880's, Picard set out to establish a Galois theory of a linear homogeneous differential equation

$$(1) \qquad \frac{d^n y}{dx^n} + P_1(x)\frac{d^{n-1}y}{dx^{n-1}} + \cdots + P_{n-1}(x)\frac{dy}{dx} + P_n(x)y = 0,$$

where the $P_i$'s are rational functions of a complex variable $x$, much closer formally to the Galois theory of algebraic equations. We refer to Chapter VIII for a discussion and references. Here we shall concentrate on the algebraic group aspects of his work. Announced first in [P1], it is developed in various papers (see his Collected Papers, Vol. II) and then in the Traité d'Analyse [P2].

A set of solutions $y_1, \ldots, y_n$ of (1) is fundamental if its Wronskian

$$\begin{vmatrix} y_1 & \cdots & y_n \\ \frac{dy_1}{dx} & \cdots & \frac{dy_n}{dx} \\ \vdots & & \\ \frac{d^{n-1}y_n}{dx^{n-1}} & \cdots & \frac{d^{n-1}y_n}{dx^{n-1}} \end{vmatrix}$$

is $\neq 0$. If so, any other solution is a linear combination with constant coefficients of the $y_i$'s. By looking at linear transformations in $n$ variables going from one fundamental system to another without altering algebraic relations between the $y_i$'s and their derivatives, in a suitable sense, Picard was led to a linear group $G$, which he called the *group of transformations of* (1). It has to be distinguished from the "*group of the equation*", i.e., for us, the monodromy group of (1), a discrete group which, as he shows, is contained in $G$. The group $G$ is clearly a Lie group of some dimension $r$. In particular, around the identity (Picard is vague about such points) the transformations depend on $r$ parameters. He shows that in fact, those parameters can be chosen so that the matrix entries are algebraic functions of them. This condition defines for him the notion of algebraic group of linear transformations, or continuous algebraic group.

Although there is no indication of the domain of definition of the parametrization, this concept is basically equivalent to the notion of algebraic group in the present sense (see 3.3), so that his papers are historically the first ones using that terminology. Also, he shows that $G$ contains the product of any two elements, but does not worry as to whether it contains the inverses of its elements. (This was established later by Loewy; see Chapter VIII for a reference.) He then comes to the problem of determining the linear algebraic groups. He first recalls the fundamental mechanism of Lie theory which, via the exponential mapping, reduces problems on groups to questions on infinitesimal substitutions generating the group, i.e. on the Lie algebra of the group, as we would say now.

Promising to come back to the general question later, he limits himself to the determination of two- and three-dimensional solvable Lie algebras (without that terminology: for him it is a type of group "considered incidentally by Lie in an 1888 paper").

He comes back to this question in Chapter XVII, Section 12 of [P2]. He first shows that if $G$ is one-dimensional, then the parameter can be chosen so that the matrix coefficients are rational functions of it. He then sketches the proof of an analogous statement in dimension two and asserts that the argument is general, leading to the conclusion that it may always be arranged that the matrix coefficients are rational functions of the parameters, for any continuous linear algebraic group.

If we grant that $G$ is a linear algebraic group in the present sense and that the parametrization is valid in a "Zariski-open subset", i.e. outside a proper algebraic subset, this would imply that $G$ is a unirational variety.

The whole treatment is rather vague, so there does not seem much point to me in trying to sort out exactly what we would view as rigorously proved. What I would like to retain is that Picard had at least the intuition that a linear algebraic group is a rational variety.

As we shall see in the next section, Picard's results, and more, had been proved a bit earlier by Ludwig Maurer. It is likely that Picard was not aware of it.

**3.3.** Let $G \subset \mathbf{GL}_N(\mathbb{C})$ be a connected Lie group which is algebraic in Picard's sense, and let $n$ be its dimension. By definition there are $n$ one-parameter subgroups associated to a basis of the Lie algebra $\mathfrak{g}$ of $G$ which provide rational maps $\varphi_i : \mathbb{C} \to \mathbf{GL}_N(\mathbb{C})$ $(i = 1, \dots, n)$. The map $\varphi_i$ is defined outside finitely many points. Let $\varphi : \mathbb{C}^n \to \mathbf{GL}_N(\mathbb{C})$ be defined by $(c_1, \dots, c_n) \mapsto \varphi_1(c_1) \dots \varphi_n(c_n)$; it is a rational map defined on a Zariski-open subset $U$ of $\mathbb{C}^n$. Let $V = \varphi(U)$. It is contained in $G$. Let $H$ be the smallest algebraic subgroup of $\mathbf{GL}_N(\mathbb{C})$ (in the usual sense) containing $V$. It is elementary that $H$ is generated by $V$ and $V^{-1}$ (see e.g.

Ludwig Maurer

I, 2.2 in [B2]). $H$ is therefore contained in $G$, but its Lie algebra contains $\mathfrak{g}$, hence is equal to $\mathfrak{g}$ and so $G = H$. Thus Picard's definition implies the usual one. The converse is clear.

## §4. Ludwig Maurer

**4.0.** Maurer's thesis [M0] deals with the normal form of an $n \times n$ complex matrix. Leaning on Weierstrass, who had introduced the elementary divisors (Monatsberichte der Berliner Akademie, 1868), he proves, "in a way slightly different from Weierstrass' " the existence of what we call the Jordan normal form. He then proceeds to find all the linear transformations which transform a given matrix into its normal form, establishing a result announced without proof by Frobenius (J. Reine Angew. Math. **84** (1878)). Contrary to what is stated in [LE], pp. 801-2, there is no discussion of one-parameter groups. Theorems 15 and 16 stated there are proved by Maurer first in [M1], to which we now turn.

**4.1.** Given a rational homogeneous function $f$ on $\mathbb{C}^n$, Maurer considers in [M1] the set of linear transformations leaving it invariant. This condition translates into algebraic relations for the coefficients, so this set is an algebraic subset of $\mathbf{M}_n(\mathbb{C})$, say $M(f)$. As such it is the union of finitely many irreducible components. Maurer shows that only one irreducible component of $M(f)$ contains the identity, and he restricts his attention to the invertible elements in that component. It is a group, to be denoted $G$. To him, it is also the arcwise connected component of $M(f)$

containing the identity. This is indeed easily seen, though I am not sure he proves it explicitly. He is thus dealing with a connected linear algebraic group defined by one invariant. [Later on, in [M3], he will arrive at the general notion of linear algebraic groups, called by him regular groups (see 4.8), and will note that all results proved in the present paper remain valid in general.] The discussion is a bit convoluted in some sections because of the notion of parametrization and a rather awkward notation.

**4.2.** Let $m$ be the dimension of $G$, as an affine algebraic set. Maurer assumes that the coefficients of the matrices in $G$ depend on $m$ independent parameters $p_1, \ldots, p_m$ (meaning, I presume, that they cannot be expressed as functions of a strictly smaller number of variables). He also assumes that "one can give the parameters values such that $m$ coefficients take values prescribed in advance". What this means exactly I am not sure. However, as the subsequent discussion shows, practically the condition seems to be that around a given element of $G$ there are $m$ matrix coefficients whose functional determinant with respect to the $p_i$'s is not zero (p. 114). [Since $G$ consists of simple points, this is possible, provided one is not required to take the same matrix coefficients at all points.] He then considers the infinitesimal transformations annihilating $f$. They are defined by matrices $C = (c_{\lambda\mu})$ such that

$$(1) \qquad \qquad \sum_{\lambda,\mu} c_{\lambda\mu} x_\mu \frac{\partial f}{\partial x_\nu} = 0.$$

If $C, D$ satisfy that condition, so does their bracket $[C, D] = C.D - D.C$. They form therefore what we call a Lie algebra. I shall say so (and denote it $\mathfrak{g}$), though for Lie and his contemporaries this was an "infinitesimal group" or simply a "group", a basis of which was also called a "complete system of linearly independent differential equations". It is remarked that if $C$ satisfies (1) then the one-parameter subgroup $\{e^{tC}\}$ it generates belongs to $G$, so that the dimension $m'$ of $\mathfrak{g}$ is at least $m$. However, for Maurer it cannot be assumed that around the origin the $p_i$ form a system of parameters for the local Lie group defined by the elements of $G$ close to the origin, and it takes some time before he concludes (in section V) that $m' = m$.

I have used the exponential notation, but Maurer does not. For him, the elements of this group are the solutions $B(t)$ of the equation

$$(2) \qquad \qquad \frac{dB}{dt} = B.C, \quad \text{with initial value } B(0) = \text{Id},$$

and he calls them "elementary transformations".

**4.3.** Section IV is a main point of the paper. Let $C \in \mathfrak{g}$. Let $D$ be a Jordan normal form of $C$. To describe it, I shall follow the notation of Maurer, who refers to [M0]. Let $r_\kappa$ ($\kappa = 1, \ldots, n'$) be the distinct eigenvalues of $C$. Each one gives rise to a certain number of Jordan blocks of the form

$$(1) \qquad \qquad \begin{pmatrix} r_\kappa & & & 0 \\ 1 & \ddots & & \\ & \ddots & \ddots & \\ 0 & & 1 & r_\kappa \end{pmatrix}.$$

We let $\ell_\kappa + 1$ be the number of such blocks, $h$ the index of a block ($0 \leq h \leq \ell_\kappa$) and $e_h^\kappa$ the length of the $h$-th block. Thus the sum of the $e_h^\kappa$ ($0 \leq h \leq \ell_\kappa$) is the multiplicity of $r_\kappa$. There exists an invertible matrix $A$ such that $A.C.A^{-1} = D$. However, Maurer does not write it in that way, but rather

(2) $$A.C = D.A,$$

not in matrix form, though, but coefficient by coefficient. $A$ is itself a juxtaposition of $\sum_1^{n'} (\ell_\kappa + 1)$ matrices, say $A_{\kappa,h}$, indexed by $\kappa$ for the eigenvalue, and by $h$ for one of the blocks with that eigenvalue, where $A_{\kappa,h}$ has $e_h^\kappa$ rows, $n$ columns, and entries

(3) $$[gh\lambda]_\kappa$$

$$(\kappa = 1, \dots, n'; h = 0, \dots, \ell_\kappa; g = 1, \dots, e_h^\kappa, \lambda = 1, \cdots, n)$$

$C$ is said to be of the first kind if the $r_\kappa$ are all zero, i.e. if $C$ is nilpotent, of the second kind if its elementary divisors are of degree 1 (i.e. $D$ is diagonal) *and* the $r_\kappa$ are rational integers, of the third kind otherwise.

If it is of the first kind, then the exponential is a finite sum and the coefficients of $e^{tC}$ are polynomials in the parameter $t$. If it is of the second kind, then the diagonal entries of $e^{tD}$ are integral powers of the parameter $e^t$, and hence the coefficients of $e^{tC}$ are rational functions of a variable running through $\mathbb{C}^*$. Maurer also remarks that the same conclusion holds if the $r_\kappa$ are integral multiples of some number. We shall also call $C$ of the second kind in that case. These are in fact all the algebraic one-dimensional subgroups of $\mathbf{GL}_n(\mathbb{C})$.

**4.4.** Now let $C$ be of the third kind. Then Maurer canonically constructs a subgroup which is a product of one group of the first kind and of some groups of the second kind, contains the one-parameter subgroup $\{e^{tC}\}$ and is minimal for these properties. [In fact, he constructs the algebraic hull of $\{e^{tC}\}$, i.e. the smallest algebraic subgroup of $\mathbf{GL}_n(\mathbb{C})$ containing it.]

Let $D_s$ be the diagonal part and $D^o$ the off-diagonal part of $D$. Thus

$$D = D^o + D_s, \quad [D^o, D_s] = 0, \ D^o \text{ nilpotent}, D_s \text{ semisimple},$$

is the Jordan decomposition of $D$. Let

$$C^o = A^{-1}.D^o.A \quad \text{and} \quad K = A^{-1}.D_s.A,$$

or rather, to follow Maurer more closely, let $(c^o_{\lambda\mu})$ and $(k_{\lambda\mu})$ be the unique solutions of

$$A.C^o = D^o.A, \quad A.K = D_s.A.$$

(As far as I can see, Maurer never writes the inverse of a matrix.) Then $C = C^o + K$ is the Jordan decomposition of $C$, with $C^o$ nilpotent and $K$ semisimple. $K$ is not necessarily of the second kind, and the main point is to represent it in a minimal way as a sum of commuting transformations of the second kind. For this, consider $\mathbb{C}$ as a vector space over $\mathbb{Q}$. The $r_\kappa$ span a finite dimensional $\mathbb{Q}$-subspace. We can choose a basis $\rho_1, \dots, \rho_s$ of that space so that

$$r_\kappa = \sum_{j=1}^{j=s} \rho_j m_{\kappa,j} \quad (m_{\kappa,j} \in \mathbb{Z}, \ i = 1, \dots, n').$$

Then, for $i = 1, \dots, s$, the diagonal matrix $D^i$ with entries $\rho_j.m_{\kappa,j}$, where $\rho_j.m_{\kappa,j}$ is repeated $\ell_\kappa + 1$-times in the $h$-th Jordan block for $r_\kappa$, is of the second kind (in the slightly extended sense mentioned above). Let $(c^i_{\lambda\mu})$ be the unique solution of

(1) $$A.C^i = D^i.A \quad (i = 1, \dots, s).$$

Then $C = C^o + \cdots + C^s$. The matrix $C^o$ is of the first kind, the $C^i$ $(i = 1, \ldots, s)$ are of the second kind, and all the $C^i$'s commute with one another. The groups $B_i = \{e^{tC^i}\}$ are one-dimensional algebraic groups, and $\{e^{tC}\}$ is contained in the product of the $B_i$'s.

The parameters in those one-dimensional groups are $t$ and the exponentials $e^{t \cdot \rho_j}$ $(n \leq j \leq s)$. They are algebraically independent. It follows therefore that if $\{e^{tC}\}$ is in $G$, then so are the $B_i$ $(0 \leq i \leq s)$ (p. 127).

**4.5.** This implies first that if $C \in \mathfrak{g}$, then the semisimple and nilpotent parts of its Jordan decomposition belong to $\mathfrak{g}$. As a further important (and completely new) consequence, the Lie algebra $\mathfrak{g}$ has a basis

$$C^i = (c^i_{\lambda\mu}) \qquad (1 \leq i \leq m; 1 \leq \lambda, \mu \leq n)$$

consisting of transformations of either the first or the second kind. Let $B_i$ be the one-parameter subgroup generated by $C^i$, and $(B_i(p_i))$ its elements. If $C^i$ is of the first kind, they are solutions of

$$(1) \qquad \frac{dB_i}{dp_i} = B_i.C^i \qquad (B_i(0) = \mathrm{Id}),$$

and their coefficients are polynomials in $p_i$. If it is of the second kind, then view it as a solution of

$$(2) \qquad p_i \cdot \frac{dB_i}{dp_i} = B_i.C^i \qquad (B_i(1) = \mathrm{Id}),$$

and then the coefficients are polynomials in $p_i$ and $p_i^{-1}$, hence rational functions of $p_i$, regular for $p_i \neq 0$. Therefore the coefficients of

$$(3) \qquad a(p_1, \ldots, p_m) = b_1(p_1) \cdot \ldots \cdot b_m(p_m)$$

are polynomials (resp. rational functions) in $p_i$ if $C^i$ is of the first (resp. second) kind. Apparently, Maurer views (3) as describing a "general" element of $G$. This is implicit in the last three sections of the paper, and explicit in the summary of it given at the beginning of [M3].

In fact, given two elements

$$a\big((p)\big) \text{ and } a\big((\xi)\big) \qquad ((p) = (p_1, \ldots, p_m), (\xi) = (\xi_1, \ldots, \xi_m)),$$

he claims there are finitely many $(\eta) = (\eta_1, \ldots, \eta_m)$ such that

$$(4) \qquad a\big((\eta)\big) = a\big((\xi)\big).a\big((p)\big).$$

More precisely, let $B = B_1 \times \ldots \times B_m$, and let $\sigma : B \to G$ be the map defined by the product, $A = \sigma(B)$. Then Maurer implies that $\sigma$ is surjective and asserts that the $\eta_i$ are algebraic functions of $(p)$ and $(\xi)$. Now if $G$ is solvable, and the $B_i$ are suitably chosen, this is indeed true (and more precisely $\sigma$ is bijective), but this is not so in general. However the complement of $A$ is contained in a proper algebraic subset (i.e. $A$ contains a non-empty Zariski-open subset). Therefore there exists a Zariski-open subset $U$ of $B$ such that Maurer's assertion is true for $(p)$ and $(\xi)$ in $U$. This is what I gather he proves. Note that the fact that $A$ contains an open Zariski-dense subset already shows that $G$ is a unirational variety.

**4.6.** So far, I have summarized the part of [M1] most relevant in the context of linear algebraic groups. This leaves out the first part, which has been in fact more often quoted and even led to an attribution to him, namely the Maurer equations, later the Maurer-Cartan equations. For the sake of completeness I'll digress to recall them briefly. This now concerns the basic local Lie theory, in particular the second fundamental theorem, and we go back to I, 1.3.

The first fundamental theorem asserts the existence of functions

$$
(1) \qquad \xi_{kj} \qquad (1 \le k \le p, 1 \le i \le n)
$$

in $n$ variables and $\psi_{i,j}$ $(1 \le i, j \le p)$ in $p$ variables such that

$$
(2) \qquad \frac{\partial f_i}{\partial a_j} = \sum_{k=1}^{k=p} \xi_{ki}(f(x,a)) \psi_{kj}(a) \qquad (1 \le i \le n; 1 \le j \le p).
$$

The matrices $(\xi_{ij})$ and $(\psi_{ij})$ have maximal rank. In particular, the latter is invertible. Let $(\eta_{ij})$ be the inverse matrix.

The second fundamental theorem provides first a relation between the $\xi_{ij}$ and their derivatives with respect to the $x_i$, which implies that the $p$ infinitesimal transformations

$$
(3) \qquad X_k = \sum_{i=1}^{n} \xi_{ki} \frac{\partial}{\partial x_i}
$$

form a Lie algebra, namely, they satisfy relations

$$
(4) \qquad [X_i, X_j] = \sum_{k=1}^{p} c_{ij}^k X_k,
$$

where the $c_{ij}^k$ are constants. Maurer's contribution is first to show that the $\psi_{ij}$ satisfy the conditions

$$
(5) \qquad \frac{\partial \psi_{ij}}{\partial a_k} - \frac{\partial \psi_{ik}}{\partial a_j} = \sum_{1 \le \mu, \nu \le p} c_{\mu\nu}^i \psi_{\mu j} \psi_{\nu k}
$$

(equation (J), p. 117).

This implies that the infinitesimal transformations on the space of the parameters

$$
(6) \qquad A_k = \sum_{i=1}^{p} \xi_{ki} \frac{\partial}{\partial a_i}
$$

satisfy the same relations as the $X_i$:

$$
(7) \qquad [A_i, A_j] = - \sum_{k=1}^{p} c_{ij}^k A_k
$$

(equation (J'), p. 118).

They provide an explicit description of the Lie algebra of the parameter group, which Lie and Engel ([LE] p. 805) find of definite interest ("hat nicht geringes Interesse").

**4.7.** The paper [M2] has a substantial overlap with [M1], but does not quote it at all. There, Maurer shows more familiarity with Lie's theory, in particular with the first volume of Lie-Engel. The first part of the paper is again devoted to foundational material in Lie theory. It repeats in part what was discussed in the previous section. He also adds the remark, which I state in the notation of 4.6

(Maurer uses a different one), namely that the $p$ Pfaffian forms on the parameter space

$$(1) \qquad\qquad \sum \eta_{ij}\, da_j \qquad (i = 1, \cdots, p),$$

let us call them $\omega_i$, are right invariant. To my knowledge, this is the first occurrence of the "Maurer-Cartan" forms. Later on, after Frobenius had introduced the notion of "bilinear covariant" of a one-form (the exterior derivative), Cartan wrote the structural equations 4.6(7) in the form

$$(2) \qquad\qquad d\omega_i = \sum c^i_{jk}\omega_j \wedge \omega_k$$

(the "Maurer-Cartan equations"). As pointed out above, Lie and Engel [LE] speak positively of Maurer's contributions to the second fundamental theorem, but are less than enthused by his proof that (locally) a group element is uniquely a product of $p$ elements taken from given one-parameter subgroups whose Lie algebras span the Lie algebra of the group, which, they claim, they handle much more simply.

The main part of the paper is again devoted to algebraic groups, but, at first, not necessarily linear ones. The notion is a direct outgrowth of Lie's definition of finite and continuous group. Such a group $G$ is called *algebraic* if the $f_i$ and $\varphi_j$ are algebraic functions of their arguments. The problem is to find necessary and sufficient conditions on the structure constants for this to be the case. If the group is algebraic in that sense, then so is its adjoint group. The results then are very close to those of [M1] but under different assumptions, and they are proved anew, without reference to [M1].

If $X_1, \ldots, X_n$ is a basis of the Lie algebra $\mathfrak{g}$ then ad $X$ ($X \in \mathfrak{g}$) is given by a matrix whose coefficients are linear combinations of the structural constants. Maurer looks for conditions under which ad $X$ generates a one-parameter group whose coefficients are rational functions of the group parameter. Again, based on the Jordan decomposition (for which he refers to [M0], as in [M1]) he finds that there are two cases: either ad $X$ is nilpotent or it is semisimple with commensurable eigenvalues. Call such transformations regular. For $G$ to be "algebraic" it is necessary for the Lie algebra to have a basis consisting of regular elements. This is also sufficient for the adjoint group. Maurer then claims that this suffices in general, basing himself on the mistaken idea that $G$ is "composed of its center and of its adjoint group" meaning that it is (locally) the direct product of those. This is of course an error, which is pointed out in [LE], pp. 806-7.

**4.8.** We now come to [M3], a direct sequel to [M1] and the most important contribution of Maurer to the theory of linear algebraic groups.

Maurer first summarizes [M1], with some changes in terminology: transformations of the first, second and third kind are now called respectively regular of the first kind, regular of the second kind, and irregular. Moreover, instead of assuming that the defining property of the group is to leave invariant one homogeneous rational function (a "uselessly restrictive assumption") he introduces "regular groups" which are in substance our linear algebraic groups, i.e. a linear group which is the set of all invertible matrices whose coefficients satisfy a number of prescribed algebraic conditions. To reduce this to the previous case, he considers the diagonal map, call it $\mu$, of $G$ into a direct product of sufficiently many copies of $G$, and deduces from a result of Christoffel that this realization of the group is characterized by the rational homogeneous functions it leaves invariant. He points out further that a transformation $C$ is of the first (resp. second, resp. third) kind if and only if $\mu(C)$ is so. It is then clear that the results of the previous paper also hold in this apparently more general situation (section I).

**4.9.** He then proves four lemmas in section II $\bigl($in which $C, C', C'' \in \mathbf{M}_n(\mathbb{C})\bigr)$, preceded by the remark, proved somewhat awkwardly, that the eigenvalues of $\operatorname{ad} C$ are differences of those of $C$.

1) If $C$ is nilpotent, and $[C, C'] = \omega.C', \omega \neq 0$, then $C' = 0$ (since $\operatorname{ad} C$ is also nilpotent).

2) If $C$ is arbitrary and $C'$ is such that $[C, C'] = \omega C'$ with $\omega \neq 0$, then $C'$ is nilpotent (because $[C, C'^m] = m\omega C'^m$ for any natural number $m$, and if $m$ is big enough, $m\omega$ is not an eigenvalue of $\operatorname{ad} C$, hence $C'^m = 0$).

3) If $C$ is diagonalizable and $C', C''$ are such that

$$[C, C'] = \omega C', \qquad [C, C''] = C' + \omega \cdot C'',$$

then $C' = 0$ ($C'$ and $C''$ cannot be linearly independent, since $\operatorname{ad} C$ would not be diagonalisable in their span, and, if $C' \neq 0$, then $\omega = \omega + 1$).

4) A finite set of commuting diagonalizable matrices is diagonalizable.

**4.10.** In present day terms, it can be said that the main goal of the paper is to prove that the variety of a linear algebraic group is rational. This is of course not the way Maurer expresses himself, and I shall also give his formulation.

His method is in essence similar to one used today to prove the theorem over an algebraically closed ground field of arbitrary characteristic, by exhibiting a Zariski open subset which, as a variety, is a product of a solvable subgroup and a unipotent subgroup, for which rationality has already been proved.

By [M1], the Lie algebra $\mathfrak{g}$ of a linear algebraic group $G$ is spanned by regular transformations $C_i$ ($1 \leq i \leq m$). Each one generates a one-parameter subgroup $B_i$. As pointed out above, Maurer appears to view the set $A$ of products $b_1 \cdot \ldots \cdot b_m$ ($b_i \in B_i$) as the group $G$ itself. Since this is not always true, there is some fuzziness and maybe some doubt as to the validity of his main arguments. It turns out that all the proper subgroups he needs to construct are solvable, and he gives enough information about the $C_i$ to make it obvious that we indeed have $A = G$ in those cases. It is only in the last step that $A \neq G$, but this does not matter for the result aimed for, even if he does not formulate it quite correctly.

**4.11.** The Lie algebra $\mathfrak{g}$ of the algebraic group $G$ is said to be of the second class if all its elements are regular of the second kind (i.e. diagonalizable, with integral eigenvalues). This is of course an oversight, since a complex linear combination of such matrices is not in general regular of the second kind. What is meant, and is actually used in the proofs, is that $\mathfrak{g}$ consists of diagonalizable matrices. Then, since it is assumed to be algebraic, it has a basis consisting of regular transformations of the second kind in view of [M1]. His goal in III is to prove that it is commutative, isomorphic to a product of $\mathbb{C}^*$, i.e. is an "algebraic torus" or simply a "torus" in today's terminology.

He first shows that $\mathfrak{g}$ is commutative, by an ingenious use of lemmas 2 and 3. [His argument proves in fact that any linear Lie algebra over an algebraically closed field $K$ of characteristic zero which consists of semisimple elements is commutative.] Being algebraic, it is indeed spanned by regular transformations of the second kind. Since it is commutative, we clearly have $A = G$ in this case. Each $B_i$ is parameterized by a variable $t_i \in \mathbb{C}^*$, so the coefficients of a general transformation $a(t_1, \ldots, t_m)$ of $G$ are rational functions of the $t_i$ and $t_i^{-1}$. The main point of the proof is to show that the $t_i$ are in turn rational functions of the matrix coefficients of $a$, and for this, it suffices to prove that the map $(t, \cdots, t_m) \mapsto b(t_1) \cdot \ldots \cdot b(t_m)$ is a bijection of $\mathbb{C}^{*n}$ onto $G$. The arguments are rather similar to those we would use to show that $G$ can be put in diagonal form with the first $m$ diagonal entries arbitrary in $\mathbb{C}^*$, the remaining ones equal to 1.

**4.12.** Maurer intercalates in IV a general principle, which will repeatedly allow him to prove that certain rational maps (between smooth varieties) are bijective, hence birational.

Let $G, \mathfrak{g}$ and $\{C_i\}$ $(1 \le i \le m)$ be as in 4.8. Assume that the first $q$ (resp. last $m - q$) transformations $C_i$ span the Lie algebra of an algebraic subgroup $A'$ (resp. $A''$). Of course, for him $A' = B_1 \cdot \ldots \cdot B_q$ and $A'' = B_{q+1} \cdot \ldots \cdot B_m$. But in the cases he considers these are actually groups I shall denote $G'$ and $G''$. It is also assumed that the maps

$$(t_1, \ldots, t_q) \mapsto b_1(t_1) \cdot \ldots \cdot b_q(t_q), \qquad (t_{q+1}, \ldots, t_m) \mapsto b_{q+1}(t_{q+1} \ldots t_m)$$

are bijective, which implies that in each case the parameters are rational functions of the matrix coefficients of the image. Finally, assume that if $a \in G'$ is different from the identity, then no power $a^m$ $(m \in \mathbb{Z}, m \neq 0)$ is the identity. Then the map $\mu : G' \times G'' \to G$ defined by the product is bijective, so that the parameters $t_i$ of an element of the product are also rational functions of the coefficients.

Assume $\mu$ is not injective. Then there exist $a'_o \in G'$, $a''_o \in G''$, both $\neq 1$, such that $\mu(a'_o \times a''_o) = 1$. This means that $a'_o = (a''_o)^{-1}$ belongs to $G' \cap G''$. Then so do the $(a'_o)^m = (a''_o)^{-m}$ $(m \in \mathbb{Z})$. Since no power $(a'_o)^m$ $(m \neq 0)$ of $a'_o$ is the identity, $a'_o$ generates an infinite cyclic subgroup of $G' \cap G''$; hence $\mu^{-1}(1)$ is infinite. Given $a' \in G', a'' \in G''$, we have

$$\mu(a' \times a'') = \mu(a'.(a'_o)^m \times (a''_o)^{-m}.a'') \qquad (m \in \mathbb{Z}).$$

Therefore $\mu^{-1}\big(\mu(a' \times a'')\big)$ is infinite, too. But $\mu$ is a rational map between varieties of the same dimension $m$, whose image contains an open set (since the functional determinant at the identity is $\neq 0$). Therefore its general fibers should be finite, a contradiction.

**4.13.** The Lie algebra $\mathfrak{g}$ (or the group $G$) is said to be of the first class if all of its elements are regular of the first kind (i.e. nilpotent). Section V is devoted to those. Maurer deduces from that assumption first that $\mathfrak{g}$ is nilpotent. From a result of Killing, it follows that the $C_h$ can be arranged so that for each $i \le m$ the transformations $C_1, \ldots, C_i$ span a subalgebra stable under all $\operatorname{ad} C_h$ $(1 \le h \le m)$ (i.e., an ideal). From this he deduces by an easy induction, also using 4.12 above, that

$$(t_1, \ldots, t_m) \mapsto b_1(t_1) \ldots b_m(t_m)$$

is bijective, and that $a \neq 1 \Rightarrow a^q \neq 1$ for all non-zero $q \in \mathbb{Z}$. In particular, as a variety, the group is isomorphic to $\mathbb{C}^m$. In other words, he establishes the main properties of algebraic unipotent groups. [In fact, the property $a \neq 1 \Rightarrow a^q \neq 1$ for all $q \in \mathbb{Z} - \{0\}$ is characteristic of groups of the first class, since any other connected linear algebraic group contains a copy of $\mathbb{C}^*$, by [M1].]

**4.14.** The last three sections of the paper are devoted to a group $G$ of the third class (i.e., by definition, one which is neither of the first nor of the second class). By definition, it contains a transformation of the third kind, and hence, by [M1], one of the second kind. Maurer first chooses a maximal set $K_1, \ldots, K_{m_0}$

of commuting transformations of the second kind, spanning a Lie algebra $\mathfrak{a}_o$ and generating a group $A_o$. The transformation ad $K_i$ $(1 \leq j \leq m_o)$ can be simultaneously diagonalized, with integral eigenvalues. Let $K_h$ $(1 \leq h \leq m)$ be a basis of $\mathfrak{g}$ consisting of common eigenvectors of all the $K_i$ $(i \leq m_o)$, the first $m_o$ elements of which are the previously chosen ones. We then have the relations

$$(1) \qquad [K_i, K_h] = \omega_h^{(i)} K_h \qquad (i = 1, \ldots, m_o; \ h = 1, \ldots, m),$$

where the $\omega_h^{(i)}$ are rational integers. By construction $\omega_h^{(i)} = 0$ if $1 \leq i, h \leq m_o$. Then Maurer chooses an integral linear combination $L$ of the $K_i$ $(i \leq m_o)$, whose eigenvalues $\rho_h$ separate the distinct sets $(\omega_h^{(i)}, \ldots, \omega_k^{(m_o)})$. In other words,

$$\rho_h \neq \rho_i \Leftrightarrow (\omega_h^{(1)}, \ldots, \omega_h^{(m_o)}) \neq (\omega_i^{(1)}, \ldots, \omega_i^{(m_o)}).$$

In particular, $\rho_h = 0$ if and only if $\omega_h^{(1)} = \ldots = \omega_h^{(m_o)} = 0$.

The remaining $K_h$ are assumed to be arranged so that

$$(1) \quad \rho_h = 0 \quad \text{for} \ \ h \leq m_o + m_o',$$

$$(2) \quad \rho_h > 0 \quad \text{for} \ \ m_o + m_o' \leq h \leq m_o + m_o' + m_+,$$

$$(3) \quad \begin{cases} \rho_h < 0 \quad \text{for} \ \ h > m_o + m_o' + m_+, \\ \rho_h < 0 \quad \text{for} \ \ h > m_o + m_o' + m_+, \end{cases}$$

and so that the $\rho_h$ are in order of increasing absolute value in the last two cases.

By lemma 2 in 4.9, the $K_h$ for $h > m_o + m_o'$ are regular of the first kind (nilpotent). Maurer deduces from the Jacobi identity that $[K_h, K_i]$ belongs to the eigenvalue $\rho_h + \rho_i$ of $L$, so is in particular zero if $\rho_h + \rho_i$ is not an eigenvalue.

It follows that the $K_h$ for $m_o + m_o' < h \leq m_o + m_o' + m_+$ (resp. for $h > m_o + m_o' + m_+$) span the Lie algebra $\mathfrak{a}_+$ (resp. $\mathfrak{a}_-$) of a group of the first class $A_+$ (resp. $A_-$). Also the $K_h$ $(h \leq m_o + m_o)$ span the Lie algebra of a commutative group $\Gamma$, containing $A_o$, which normalizes $A_+$ and $A_-$.

Section VII is devoted to $\Gamma$. The claim is that it is the direct product of $A_o$ and a group $A_o'$ of the first class. Let us call its Lie algebra $\mathfrak{b}$. By our assumption on $\mathfrak{a}_o$, it cannot contain a regular transformation of the second kind which is linearly independent of the $K_i$ $(i \leq m_o)$. Let $C$ be an irregular transformation in $\mathfrak{b}$. In [M1] it is represented canonically as a sum of regular transformations $C_i$ which belong to $\mathfrak{g}$, since $G$ is algebraic.

Maurer shows that his construction is such that since $C$ commutes with the $K_i$ $(i \leq m_o)$, so do the $C_i$. As a consequence, the basis $K_i$ $(i \leq m_o)$ of $\mathfrak{a}_o$ can be completed to a basis of $\mathfrak{b}$ by adding regular transformations of the first kind. It is then easy to see that these generate a subgroup $A_o'$ of the first class so that $\Gamma$ is the direct product of $A_o$ and $A_o'$.

Maurer has now constructed four groups, namely, $A_o, A_o', A_+, A_-$, the first of which is of the second class, the others of the first class. $A_o.A'$ is the centralizer of $A_o$, and $A_o.A_o'$ normalizes $A_+$ and $A_-$. From 4.12 it follows again that $H = A_o.A_o'.A_+$ is a group, the semidirect product of $A_o.A_o'$. and $A_+$. For both $A_-$ and $H$, the natural parameters are rational coefficients of the matrix coefficients. In particular, $A_-$ and $H$ are rational varieties. The Lie algebra $\mathfrak{g}$ is a direct sum of those of $A_-$ and $H$. A further application of 4.12 then shows that the product map $A \times H \mapsto A.H \subset G$ is bijective and birational; hence $A.H$ (which is a Zariski-open subset) is isomorphic, as a variety, to $A \times H$.

Note that, in today's parlance, $A_o$ is a maximal torus of $G$ and $\Gamma$ the Cartan subgroup of $G$ containing it.

**4.15.** Maurer's next publication [M4] is an unfortunate one, since he sketches what he claims to be a proof of a theorem on the finiteness of invariants for any (connected) linear Lie group, a statement we know to be false. However, it also contains some interesting results, with correct proofs, including one which nowadays is routinely attributed to Weitzenböck (although the latter refers to [M4] for it). In this section and the next, I discuss the contents of [M4] and [M5] postponing to 4.17 some remarks on the history of the invariant problem, which complement those made in Chapter II.

The result announced is then as follows. If $G$ is a Lie subgroup of $\mathbf{GL}_n(\mathbb{C})$, then the polynomials over $\mathbb{C}^n$ invariant under $G$ form a finitely generated algebra. Maurer also claims to prove a more general result, which he needs for induction purposes: the finiteness statement is also true for the regular functions on an affine variety $V$, provided $V$ is defined by equations all invariant under $G$.

The Lie groups are implicitly assumed to be connected. Invariance under $G$ is equivalent to being annihilated by the Lie algebra $\mathfrak{g}$ of $G$, and it is in that form that the problem is treated. Following tradition at that time, I shall also call invariant a polynomial annihilated by $\mathfrak{g}$. Maurer remarks right away that any invariant under $\mathfrak{g}$ is also invariant under the smallest "regular" Lie algebra containing $\mathfrak{g}$, so there is no restriction in assuming that $\mathfrak{g}$ is so, i.e. that it is spanned by regular transformations. The argument is divided into three main steps. The first one deals with one-dimensional groups. Then, Maurer proceeds by induction on $\dim \mathfrak{g}$, assuming the theorem proved for Lie algebras of dimensions $< \dim \mathfrak{g}$. There he distinguishes two cases, depending on whether $\mathfrak{g}$ is simple (3rd step) or not (2nd step).

In the first step, $\mathfrak{g}$ is either nilpotent (regular of the first kind) or spanned by a diagonalizable matrix with integer eigenvalues (regular of the second kind). The latter case is easily taken care of, in view of the fact that all spaces under consideration are direct sums of eigenspaces (which Maurer proves). This works also for the regular functions on an affine variety.

The nilpotent case is more interesting. Let $C$ be a generator of $\mathfrak{g}$. Using the Jordan normal form, Maurer shows that $B$ can be embedded in a "triple" $A, B, C$ of matrices satisfying the commutation relations for $\mathbf{sl}_2$

$$(1) \qquad [A, B] = -2B, \qquad [A, C] = 2C, \qquad [C, B] = A.$$

He then points out that the differential equation $Cf = 0$, with $C$ in normal form, is the same as the one which leads to the determination of covariants for $\mathbf{sl}_2$, and the result in that case follows from a finiteness theorem of P. Gordan (mentioned in II, **6**).

He also deduces a number of relations on the action of $A, B, C$ on eigenvectors of $A$, which are familiar in the discussion of linear representations of $\mathbf{sl}_2$. (For a sample, see II, no 12.) In particular, if $Ax = k.x$ and $q, r$ are the smallest exponents such that $B^q x = C^r x = 0$, then $q = r + k$. [This follows from the fact that the extreme weights of the (irreducible) representation in the space spanned by the $B^i x$ and $C^j x$ are opposite.] This relation also implies that if $x$ is annihilated by any two of $A, B, C$ then it is annihilated by the third one (which is not obvious if one of the two is $A$).

To prepare for the second step, Maurer considers the case of invariants of one linear transformation not just on the polynomials on a vector space, but also on the polynomials on some Lie algebra centralized by the given transformation, and claims to prove the finiteness in that case, too. This is incorrect, since it would for instance immediately imply, by induction on dimension, the theorem for a Lie algebra consisting of nilpotent transformations (and contradict Nagata's example). This of course renders the second step invalid.

If $\mathfrak{g}$ is simple (3rd step), then Maurer leans on the Killing-Cartan structure theory, and specifically uses the $\mathbf{sl}_2$ triples $\mathfrak{s}_\alpha$ associated to the roots, see 5.2 or VI, 2.2. The remark just made on the action of $\mathbf{sl}_2$ implies that the invariants of the maximal solvable subalgebra $\mathfrak{b}$ (see VI, 2.2) are the same as those of $\mathfrak{g}$. So the theorem for $\mathfrak{g}$ would indeed follow by induction if it were true for $\mathfrak{b}$.

The paper [M5] was meant to be the first in a series devoted to the detailed description of the argument outlined in [M4]. The subsequent ones never appeared, to my knowledge. Presumably, Maurer discovered an irreparable flaw in his argument. The paper [M5] is devoted solely to "regular" groups. It reviews what is needed from [M1], [M3] and explains in more detail the new material introduced and used in the above sketch.

**4.16.** As far as I know, this is the last paper written by Maurer on linear algebraic groups. His results resurfaced fifty or sixty years later, with at best perfunctory acknowledgments, quite inadequate, even though the more recent results are usually proved in greater generality. It is true that there are some errors, and that the presentation, in particular the notation, is often awkward, but this should not hide the fact that he had developed a considerable insight into this topic.

**4.17.** As an appendix to this section, it seems appropriate, in view of [M4], to make some historical remarks on the invariant problem, even though they bring us into the 20th century, at variance with the title of this chapter.

As pointed out in Chapter II, an important problem in the 19th century was whether the algebra of polynomials on a vector space which are invariant under a given group of linear transformations is finitely generated, though the problem was formulated differently by the early practitioners of the theory (Aronhold, Clebsch, Gordan, Cayley, Sylvester), at a time when group theory hardly existed (see [P0] for a recent account). As pointed out in II, **6**, remark 1), positive answers had been given for some classical groups, by Gordan, Study, and Hilbert. The latter's paper [H] raised the question as to whether it is true for a general continuous linear group, and was Maurer's starting point. Later, Hilbert incorporated it into a more general one, the 14th problem in his famous list at the ICM Paris 1900, but we limit ourselves here to the version stated above. Hilbert also refers there to a positive answer given by Maurer in [M4]. Apparently, this made little impression and the problem was left in abeyance for a number of years, until Weitzenböck turned to it in [Wz]. The paper [M4] had been completely forgotten: Weitzenböck, who had published a book on invariant theory in 1923, says in [Wz] that he had become aware of it only recently. [Indeed, he makes no mention of Maurer in his announcements in Koninkl. Akad. Wetensch. Amsterdam Proc. Section Sci. **33** (1930), 47-50, 227-231, 232-242).] He views its results as valid. His goal is to give a full proof, rather than just a sketch, as in [M4], even more generally for semi-invariants (polynomials reproduced by group elements up to a constant factor). Although he points out that there is no need to consider "regular" transformations, in Maurer's sense, his argument is similar to Maurer's in several ways. It is also divided into the same three steps, starting with the one-dimensional case. The theorem in that case is nowadays attributed to Weitzenböck, probably beginning with Weyl (see below), but this seems unjustified to me. The proof is quite similar to Maurer's, to which Weitzenböck refers explicitly. In particular, in the most important case of a nilpotent transformation, there is the same reduction to a theorem of P. Gordan. It is true that Maurer limits himself to regular transformations. However, his argument extends trivially to the case where the given Lie algebra is commutative, spanned by one nilpotent transformation and several diagonalizable ones, with integral eigenvalues, and Maurer had proved that the smallest regular algebra containing a given linear transformation is of that form. But, surely, this is not the reason for that misnomer. Simply, [M4] had been overlooked.

From there on, Weitzenböck also proceeds by induction on dim $\mathfrak{g}$. His argument for the third step ($\mathfrak{g}$ simple) is the same as Maurer's. Again, the induction in the second part ($\mathfrak{g}$ not simple) cannot be correct, and the mistake, as in the case of Maurer, lies in an erroneous generalization of the one-dimensional case. In fact, Weyl, in his review of the paper in Zentralblatt der Mathematik **4** (1932), p. 243, after having summarized the main steps, adds that at this time the proof contains a gap ("die Beweisführung enthält vorläufig eine Lücke"), namely, that Weitzenböck uses the theorem for a one-dimensional transformation in the case of "gebundenen Variablen", for which the Gordan finiteness theorem had not been established. [On the other hand, R. Brauer, in his review in Jahresbuch über die Fortschritte der Mathematik, **58**₁ (1932), p. 117, does not question the validity of the proof.] Later on, it became clear, at least to Weyl, that the "vorläufig" had to be erased, and in his "Classical Groups" he writes (p. 275): "we know of no instance where the first main theorem about invariants fails, but a proof holding for all classes is likewise unknown", and in a footnote to that statement ([18], p. 314) he refers to [M4] and [Wz] as "attempts which miscarried". He adds that [Wz] contains a correct proof for one linear transformation, presumably the source of the attribution to Weitzenböck of this result.

In 1958, M. Nagata provided a counterexample (Proc. ICM 1958, 459-462; Amer. J. Math. **81** (1959), 766-772): a 13-dimensional commutative unipotent group acting on a 32-dimensional space. According to his introduction, he was not aware of any earlier work on the 14th problem (in the restricted sense used here) and viewed it as an open problem. Another counterexample, along the same lines, but smaller and simpler, has recently been given by R. Steinberg (*Nagata's example*, in Algebraic Groups and Lie Groups (Papers in Honour of R. W. Richardson), Austral. Math. Soc. Lecture Ser. **9**, Cambridge Univ. Press, 1997, 375-384; Collected Papers, 569-78).

Again, let $\mathfrak{g}$ be one-dimensional and nilpotent. The reduction to a theorem of Gordan alluded to by Maurer and by Weitzenböck may not be as familiar today as it apparently was to them. A proof in the context of algebraic group theory has been given by C.S. Seshadri (J. Math. Kyoto Univ. **1** (1962), 403-9).

As pointed out above, an important point in Maurer's proof for a nilpotent linear transformation $C$ (first step) is the fact that $C$ can be embedded in a subalgebra of $\mathbf{gl}_n(\mathbb{C})$ isomorphic to $\mathbf{sl}_2$. This allows one to also view the theorem for $C$ as a very special case of the following statement, proved by Dž. Hadžiev (*Some questions in the theory of vector invariants*, Math. USSR–Sbornik **1** (1967), 383-396). *Let* $\mathfrak{u}$ *be a maximal nilpotent subalgebra of a complex semisimple Lie algebra* $\mathfrak{g}$ *acting linearly on a finite dimensional complex vector space* $V$. *Then the algebra* *of invariants of* $\mathfrak{u}$ *in* $\mathbb{C}[V]$ *is finitely generated.* This theorem itself has since been generalized in several ways (see F. Grosshans, *The invariants of unipotent radicals of parabolic subgroups*, Invent. Math. **73** (1983), 1-9).

## §5. Élie Cartan

**5.1.** Élie Cartan published one C.R. note on linear algebraic groups [C4]. For him, as for Picard, a linear Lie group is algebraic if the matrix coefficients are algebraic functions of the group parameters, in the sense of Lie's theory. As remarked in 3.3, this is equivalent to Maurer's definition. He points out that, according to Maurer, it may then be arranged that group parameters enter rationally. He is mainly concerned with conditions under which a finite and continuous group of linear transformations is algebraic in the above sense. His three main examples are

a) the derived group of a linear Lie group,

b) the adjoint group of a linear Lie group with a radical of rank 0 (i.e., the Lie algebra of the radical is nilpotent and its image in the adjoint representation consists of nilpotent matrices), and

c) if $G$ is a transitive Lie group of transformations, whose Lie algebra is centerless and has a radical of rank 0, then it can be arranged that the variables and the parameters enter algebraically in the defining equations of the transformation group and that the transformations depend rationally on the parameters.

**5.2.** Cartan does not supply proofs. However, these results follow so directly from the considerations in §§3 and 4 in Chap. VI of his thesis that it may be not too far fetched to believe that the argument outlined below is similar to his.

The "group parameters", say $a_1, \ldots, a_p$, are meant in the sense of Lie. They are coordinates which parametrize the transformations in some open set neighborhood of 1. "To arrange that they enter rationally" no doubt means, as it did for Picard, that the Lie algebra is spanned by elements generating one-parameter linear groups $h_i$ for which the coefficients are rational functions of the parameter, so that the coefficients of the product $h_1(a_1) \ldots h_n(a_n)$ (a "general element of the group") are rational functions of the $a_i$'s. The one-parameter subgroups in question are then the groups $\exp \mathbb{C}X$, where $X$ is either nilpotent or diagonalizable with commensurable eigenvalues (cf. §4.5).

It was well known to Killing and Cartan that the radical $\mathfrak{r}$ of the derived algebra of the given Lie algebra has rank 0. It consists of nilpotent matrices, and the existence of such a basis for the radical is then obvious. If $G$ is semisimple, one is provided by the structure theory developed by Killing and Cartan [C3]: Let $\mathfrak{h}$ be a Cartan subalgebra, and $\Phi$ the set of roots of $\mathfrak{g}$ with respect to $\mathfrak{h}$. Let $X_\alpha$ be root vectors and $H_\alpha = [X_\alpha, X_{-\alpha}]$. Then $H_\alpha, X_\alpha, X_{-\alpha}$ span a copy $\mathfrak{s}_\alpha$ of $\mathfrak{sl}_2$, and the representation theory of $\mathfrak{sl}_2$ shows that $X_{\pm\alpha}$ is nilpotent and $H_\alpha$ diagonalisable with commensurable eigenvalues, so the $X_\alpha$ ($\alpha \in \Phi$) and the $H_\alpha$ ($\alpha$ a simple root) form the desired basis. In his thesis, he had indeed shown, using this basis, that if $\mathfrak{g}$ is semisimple, one can choose group parameters on which the coefficients of the group elements depend rationally ([C3], VIII, §5), and he referred to [M1] for very remarkable related work.

In general, let $\pi : \mathfrak{g} \to \mathfrak{g}/\mathfrak{r}$ be the natural projection. The Lie algebra $\mathfrak{g}/\mathfrak{r}$ is semisimple. Let $\mathfrak{s}_\alpha \subset \mathfrak{g}/\mathfrak{r}$ be as before. By Engel's theorem $\pi^{-1}(\mathfrak{s}_\alpha)$ contains a supplement to $\mathfrak{r}$. As $\alpha$ runs through the roots, this then supplies the desired complement to the basis of $\mathfrak{r}$. Indeed, Cartan developed the representation theory of $\mathbf{sl}_2$, and proved full reducibility, precisely in order to establish Engel's theorem.

The argument is the same for b). In fact, the above proof of a) depends only on the fact that $\mathfrak{r}$ has rank 0, a case which contains b).

The assertion c) is presented as an application. "$G$ transitive" means that (locally) we are dealing with a homogeneous space, acted upon by group translations. It is then clear.

## §6. Karl Carda

I shall conclude this chapter by discussing briefly two papers by K. Carda ([C1], [C2]) which appear to fit into this chapter since they carry "algebraic groups" in their titles, but in fact hardly do so, because Carda's algebraic groups are quite different from the algebraic groups as understood here. (Besides, [C1] only just falls in the 19th century, and [C2] misses by six years.)

**6.1.** This work is a direct outgrowth from Lie's theory. To put it in context, let me come back to some aspects of the notion of (local) transformation groups in Lie's sense (cf I, §1): the transformations depend on parameters $a_1, \ldots, a_p$ varying in some open set of $\mathbb{C}^p$, and to $(a_1, \ldots, a_p)$ there is assigned an analytic isomorphism of a given space $U$, say an open set of $\mathbb{C}^n$, with coordinates $x_i$:

$$x_i \mapsto x_i' = f_i(x_1, \ldots, x_n; a_1, \ldots, a_p),$$

where the $f_i$'s are holomorphic in their arguments. Let $G'$ be a second such group, operating on a space $V$ with coordinates $y_j$. There are two notions of isomorphism attached to that situation. The one we focused on so far (called "Gleichzusammensetzung" by Lie) is an isomorphism of $G$ and $G'$ as local Lie groups, involving only the parameters, not the variables operated upon. But there is also an obvious notion of isomorphism of transformation groups: $G$ and $G'$ are isomorphic as such if there exists a biholomorphic map $\varphi : U \to V$ such that $G = \varphi^{-1} \circ G' \circ \varphi$ (all this of course is meant locally, with suitable restrictions on the domains of the parameters and variables). Clearly, this implies "Gleichzusammensetzung". Two isomorphic groups are viewed as representing the same equivalence class of transformation groups. Lie has determined the equivalence classes of groups operating on the line or the plane, i.e. for $n = 1, 2$, and has given for each class a basis of the Lie algebra of the transformation group $G$, written as vector fields on $U$. For $n = 1$, there are three possibilities, of dimensions 1, 2, 3, all realized within the Lie algebra of $\mathbf{SL}_2$ viewed as the group of fractional transformations of the line ([L], Chap. 12, §4; [LE], p. 6). From $n = 2$, he finds 31 types ([L], Chap. 14, §4; [LE], pp. 71-73). To set up such a classification for $n = 2$ Lie of course needs some features which are preserved under equivalence. He uses the existence of families of curves which are invariant (as a set, not individually) under the group. They are the integral curves of vector fields invariant under the given Lie algebra. If there are none, the group is said to be *primitive* (which implies transitivity), and Lie finds three such classes. If there is at least one, i.e. if the group is *imprimitive*, then he discusses various cases depending on how many such families exist.

**6.2.** I now come to Carda. He says that the above transformation group $G$ is *algebraic* if, for fixed values of the parameters $a_i$, the transformation $g : U \mapsto U$ defined by the $f_i$ is algebraic. No requirement (beyond analyticity) is imposed on the parameters. This notion is quite different from the one we have used. In particular, any linear Lie group would be algebraic in that sense. In the rest of this section, we use Carda's notion. Let $G$ and $G'$ be algebraic and equivalent in Lie's sense. They will be viewed as belonging to the same *algebraic equivalence class* if they are isomorphic by means of a biholomorphism $\varphi : U \to V$ which is algebraic.

Carda's problem is then to determine which of Lie's classes of groups on the line or the plane contain algebraic representatives and, if so, how many algebraic equivalence classes.

In [C1], he starts with one-dimensional groups on the line and finds three classes, represented by the generators of the Lie algebra

$$\partial_x, \quad x.\partial_x, \quad \sqrt{x(1-x)(1-cx)}\,\partial_x \qquad (c \neq 0, 1).$$

They correspond indeed to the three one-dimensional algebraic groups: the additive group of $\mathbb{C}$, the multiplicative group of $\mathbb{C}^*$ and an elliptic curve. He makes essential use of a theorem of H. A. Schwarz (for which he provides a proof first published by E. Picard), asserting that if a one-dimensional group of rational transformations of the plane leaves a curve invariant and is simply transitive on it, then the curve has genus 0 or 1.

The last part of [C1] is devoted to groups acting on the plane. In particular, for the primitive classes (of which there are three) and the imprimitive classes which leave exactly two infinite families of curves invariant (9 classes according to Lie), each contains one algebraic equivalence class of algebraic groups.

**6.3.** The second paper solves this problem for one more class, represented by the Lie algebra with basis

$$\partial_x, \quad 2x\partial_x + y\partial_y, \quad x^2\partial_x + xy\partial_y.$$

It is isomorphic to $\mathbf{sl}_2$. If we pass to homogeneous coordinates $u, v, w$, related to $x$ and $y$ by the relation

$$x = u/w, \qquad y = v/w,$$

then the corresponding group is the subgroup of $\mathbf{SL}_3$ leaving the plane $(u, w)$ stable and fixing the line $u = w = 0$ pointwise. From the point of view of Lie's classification it is transitive (generically of course) and imprimitive, leaving one one-parameter family of curves invariant, namely the curves $x = c\lambda, y = d.\lambda$. It is again shown that any algebraic group in that class is algebraically equivalent to the given one.

# References for Chapter V

[B1]   A. Borel, *Représentations linéaires et espaces homogènes kähleriens des groupes simples compacts*, manuscript, 1954; first published in his Oeuvres/Collected Papers I, Springer, 1993, pp. 392–396.

[B2]   ———, Linear algebraic groups, GTM **126**, Springer, 1991.

[BH]   A. Borel and F. Hirzebruch, *Characteristic classes and homogeneous spaces II*, Amer. J.Math. **81** (1959), 315–382; Borel, Oeuvres/Coll. Papers II, 1–68; Hirzebruch, Ges. Abh./Coll. Papers I, 496–563.

[C1]   K. Carda, *Zur Theorie der algebraischen Gruppen der Geraden und der Ebenen*, Monatshefte Math. Phys. **11** (1900), 31–58.

[C2]   ———, *Ueber eine Schar dreigliedriger algebraischer Gruppen der Ebene*, Monatshefte Math. Phys. **17** (1906), 225–233.

[C3]   É. Cartan, *Sur la structure des groupes de transformations finis et continus*, Thèse, Nony, Paris, 1894; Oeuvres Complètes I₁, 137–287.

[C4]   ———, *Sur certains groupes algébriques*, C.R. Acad. Sci. Paris **120** (1895), 545–548; Oeuvres Complètes, I₁, 289–292.

[G]   P. Gordan, *Ueber Combinanten*, Math. Annalen **5** (1872), 95–122.

[H]   D. Hilbert, *Ueber die vollen Invariantensysteme*, Math. Annalen **42** (1893), 313–373; Ges. Abh. II, 287–344.

[L]   S. Lie, Vorlesungen über continuierliche Gruppen (G. Scheffers ed.), Teubner, Leipzig, 1893.

[LE]   S. Lie and F. Engel, Theorie der Transformationsgruppen III, Teubner, Leipzig, 1893.

[M0]   L. Maurer, *Zur Theorie der linearen Substitutionen*, Thesis, Universität Strassburg, 1887.

[M1]   ———, *Ueber allgemeinere Invarianten-Systeme*, Sitzungsber. Math.-Phys. Kl. Kgl. Bayer. Akad. Wiss. München **18** (1888), 103–150.

[M2]   ———, *Ueber continuierliche Transformationsgruppen*, Math. Annalen **39** (1891), 409–440.

[M3]   ———, *Zur Theorie der continuierlichen, homogenen und linearen Gruppen*, Sitzungsber. Math.-Phys. Kl. Kgl. Bayer. Akad. Wiss. München **24** (1894), 297–341.

[M4]   ———, *Ueber die Endlichkeit der Invariantensysteme*, Sitzungsber. Math.-Phys. Kl. Kgl. Bayer. Akad. Wiss. München **29** (1899), 147–175.

[M5]   ———, *Ueber die Endlichkeit der Invariantensysteme*, Math. Annalen **57** (1903), 265–313.

[P0]   K. Parshall, *Towards a history of nineteenth-century invariant theory*, in The History of Modern Mathematics (D.E. Rowe and J. McCleary, eds), vol. 1, Academic Press, 1989, pp. 157–206.

[P1]   E. Picard, *Sur les groupes de transformation des équations différentielles linéaires*, C.R. Acad. Sci. Paris **96** (1883), 1131–4; Oeuvres Complètes II, 97-100.

[P2]   ———, *Traité d'Analyse, Vol. III*, Gauthier-Villars, Paris, 1896; 2nd ed., 1908.

[S1]   E. Study, Methoden zur Theorie der ternaeren Formen, Teubner, Leipzig, 1889.

[S2]   ———, Letter to S. Lie, Dec. 31, 1890*.

[S3]   ———, Letter to F. Engel, July 31, 1892**.

[W]   E. Weiss, *E. Studys mathematische Werke*, Jahresber. Deutschen Math.-Verein. **43** (1933), 108–124, 211–225.

[Wz]   R. Weitzenböck, *Ueber die Invarianten von linearen Gruppen*, Acta Math. **58** (1932), 231–293.

---

*Brefsamling Nr. 289 of the Department of Manuscripts, National Library of Norway, Oslo division (formerly known as the University Library of Oslo).

**F. Engel Archive, Mathematisches Institut, Justus-Liebig Universität at Giessen, Germany.

# CHAPTER VI

# Linear Algebraic Groups in the 20th Century

The interest in linear algebraic groups was revived in the 1940s by C. Chevalley and E. Kolchin. The most salient features of their contributions are outlined in Chapter VII and VIII. Even though they are put there to suit the broader context, I shall as a rule refer to those chapters, rather than repeat their contents. Some of it will be recalled, however, mainly to round out a narrative which will also take into account, more than there, the work of other authors.

## §1. Linear algebraic groups in characteristic zero. Replicas

**1.1.** As we saw in Chapter V, §4, Ludwig Maurer thoroughly analyzed the properties of the Lie algebra of a complex linear algebraic group. This was Chevalley's starting point. In order to algebraize Maurer's results and extend them to more general ground fields, Chevalley introduced in [C3], 1943, the notion of *replica* of an endomorphism $X$ of a finite dimensional vector space $V$ over a field $k$, which we assume here to be of characteristic zero, since the main use of that notion pertains to that case. In particular, if $X$ is nilpotent, its replicas are only its multiples. If $X$ is semisimple, with eigenvalues $\alpha_1, \ldots, \alpha_n$ (in some extension of $k$), then the replicas of $X$ are the polynomials $p(X)$ in $X$, without constant term, such that any linear relation between the $\alpha_i$ with integral coefficients is also satisfied by the $p(\alpha_i)$. Chevalley calls a linear Lie algebra *algebraic* if it contains the replicas of all its elements. In that terminology, the main result of Maurer's 1888 paper (see V, 4.3-4.5) asserts that the Lie algebra of a complex linear algebraic group is algebraic. [CT1], 1946, also sketches a proof of the converse.

**1.2.** The papers of Chevalley and Chevalley-Tuan [C3], 1943, [C4], 1947, [CT1], [T] quickly gave rise to some work by M. Goto [G1], [G2] and Y. Matsushima [M3], [M4] (1947-8). They provided variants, or alternate proofs, or proofs of results only announced there. As a new point, Goto slightly generalized the notion of algebraic Lie algebra to not necessarily linear ones, by saying that a Lie algebra over the field $k$ is algebraic if the Lie algebra of its inner derivations is so. He shows that it is then isomorphic to a linear algebraic one.

The results of Chevalley and Tuan were first proved in [CT2], 1951, and were later incorporated by Chevalley in a more general theory of linear algebraic groups over a field of characteristic zero [C5], 1951, [C6], 1954 (see VII, **4**).

## §2. Groups over algebraically closed ground fields I

**2.1.** Chevalley's approach to algebraic groups in [C1] and [C2] uses a formal analog of the exponential mapping to set up a correspondence between Lie algebras and Lie groups with the usual properties. It is therefore tied to characteristic zero.

Completely different is the point of view of Kolchin [K1], [K2], 1948, who considered linear algebraic groups over an algebraically closed ground field of arbitrary characteristic and provided proofs insensitive to the characteristic, without any recourse to, nor even any mention of, the Lie algebra (VIII, §4). I cannot help drawing here an analogy with Lie: both were moved by the wish to establish some Galois theory of differential equations. In Lie's case, this hope, labeled Lie's "idée fixe" by Hawkins [H3], led to applications to differential equations which are interesting, but still minor in comparison with the wealth of uses of Lie groups in so many parts of mathematics. Similarly, the case of positive characteristic initiated by Kolchin found hardly any application in his Galois theory of homogeneous linear differential equations, or more generally of differential fields (his motivation), which is confined to characteristic zero; but it underwent a tremendous development in other directions.

We refer to VIII, §4, for some details on Kolchin's contributions. Here, let me only mention: a) Lie's theorem is valid for any linear connected solvable group; b) a connected group consisting of semisimple matrices is commutative and isomorphic to a product of $k^*$ (which generalizes a result of Maurer mentioned in V, §4); c) any group consisting of unipotent matrices can be put in triangular form and is nilpotent (a vast generalization of Engel's theorem on Lie algebras consisting of nilpotent elements); and  d) a commutative group $G$ is the direct product of the subgroups $G_s$ and $G_u$ consisting respectively of the semisimple and unipotent matrices contained in $G$ (a fact which implies the existence of the multiplicative Jordan decomposition in any linear algebraic group).

Kolchin's work on algebraic groups had no echo for several years. Meanwhile, the early fifties saw more and more intrusion of algebraic geometry in the study of complex semisimple groups and certain homogeneous spaces, as well as the beginnings of a systematic theory of algebraic groups (linear or not). Sections 2.2 to 2.5 are devoted to these developments. In 2.2, 2.3, 2.4 the ground field is $\mathbb{C}$.

**2.2.** *Some notation.* To discuss complex semisimple Lie groups, I have to recall some facts, to be found in any textbook on semisimple Lie algebras, and fix notation.

Let $\mathfrak{g}$ be a complex semisimple Lie algebra, $\mathfrak{h}$ a Cartan subalgebra. A root of $\mathfrak{g}$ with respect to $\mathfrak{h}$ is a non-zero linear form $\alpha \in \mathfrak{h}^*$ such that

$$(1) \qquad \mathfrak{g}_\alpha = \{x \in \mathfrak{g} | [h, x] = \alpha(h).x\} \neq 0.$$

The set $\Phi = \Phi(\mathfrak{h}, \mathfrak{g})$ of $\alpha$ for which $\mathfrak{g}_\alpha \neq 0$ is a reduced root system in $\mathfrak{h}^*$, and $\mathfrak{g}$ is a direct sum

$$(2) \qquad \mathfrak{g} = \mathfrak{h} \oplus \bigoplus_{\alpha \in \Phi} \mathfrak{g}_\alpha.$$

Fix an ordering on $\Phi$. Then

$$(3) \qquad \mathfrak{b} = \mathfrak{h} \oplus \bigoplus_{\alpha > 0} \mathfrak{g}_\alpha$$

is the Lie algebra of a maximal solvable subalgebra (later called a Borel subalgebra). All roots are linear combinations with integral coefficients of the same sign of $\ell = \dim \mathfrak{h}$ roots, making up the set $\Delta = \Delta(\mathfrak{h}, \mathfrak{g})$ of simple roots. The subalgebras containing $\mathfrak{b}$ correspond to the subsets $I$ of $\Delta$. The subalgebra $\mathfrak{p}_I$ assigned to $I$ is spanned by $\mathfrak{b}$ and the $\mathfrak{g}_\alpha$, where $\alpha$ is a negative linear combination of the elements in $I$. There was no nice terminology for the subgroups $P_I$ with Lie algebra the $\mathfrak{p}_I$ until R. Godement suggested calling them parabolic subgroups. I shall therefore, anachronistically, call them that.

Fix a scalar product on $\mathfrak{h}$ invariant under the Weyl group $W(\mathfrak{h}, \mathfrak{g})$ of $\Phi$. The fundamental highest weights $\omega_\alpha$ ($\alpha \in \Delta$) are defined by the condition

$$\frac{2(\omega_\alpha, \beta)}{(\beta, \beta)} = \delta_{\alpha\beta} \qquad (\alpha, \beta \in \Delta).$$

The *dominant weights* are the positive integral linear combinations of the $\omega_\alpha$. The isomorphism classes of irreducible representations of $\mathfrak{g}$ correspond bijectively to the dominant weights. We let $P(\Phi)$ be the set of $\lambda \in \mathfrak{h}^*$ such that $2(\lambda, \alpha) \cdot (\alpha, \alpha)^{-1} \in \mathbb{Z}$ for all $\alpha \in \Phi$. It is a lattice, the lattice of *weights* of $\Phi$; it contains the lattice $Q(\Phi)$ spanned by the roots. Once an ordering is chosen, we let $P(\Phi)^+$ be the monoid of dominant weights.

**2.3.** Complex semisimple Lie groups happen to be linear algebraic groups, but this was pretty much ignored in the wake of the global theory of Weyl and Cartan, even though, as we saw in V, §5, Cartan had been well aware of that fact.

Like every homogeneous space, $G/B$ is quasiprojective. On the other hand, it can also be written as $G_u/T$, where $G_u$ is a maximal compact subgroup of $G$ (a compact real form) and $T$ a maximal torus of $G_u$; hence it is compact. Therefore it is a projective variety. This already leads to a very simple derivation of a result of Morozov I had heard about in 1953, with a rather obscure proof of the conjugacy of maximal solvable subalgebras: apply to a projective equivariant embedding of $G/B$ a simple consequence of Lie's theorem, namely, any projective variety invariant under a connected solvable group of projective transformations admits a point fixed under that group.

The projective homogeneous spaces of $G$ became the focus of much attention. If $G/H$ is one, then $H$ contains a conjugate of $B$ by Lie's theorem, hence is conjugate to one of the $P_I$ (see 2.2) and conversely. (It can also be written as a quotient $G_u/H_I$ of $G_u$ by the centralizer $H_I$ of the subalgebra $\mathfrak{h}_I = \bigcap_{\alpha \in I} \ker \alpha$. As such, it is homogeneous kählerian [B3], 1954.)

In [G3], 1954, Goto showed that these spaces are rational varieties, by constructing a Zariski open subset $\Omega$ isomorphic to some affine space. Let $N_I^-$ be the unipotent subgroup whose Lie algebra $\mathfrak{n}_I^-$ is the sum of the $\mathfrak{g}_\alpha$, corresponding to the negative roots, not linear combinations of elements in $I$. Then $\mathfrak{g}$ is the direct sum of $\mathfrak{p}_I$ and $n_I^-$; hence $N_I^- . P_I$ is Zariski open in $G$ and the orbit of the identity under $N_I^-$ is the desired $\Omega$. The construction is similar to—in fact, a special case of—the one used by Maurer (V, §4) to show the rationality of $G$.

For a number of these spaces, such as the complex Grassmannians, C. Ehresmann had shown in his thesis [E], 1934, the existence of a "complex analytic cellular decomposition": a partition into "cells", each isomorphic to some affine space, such that the closure of any cell is the union of cells of smaller dimension. F. Bruhat, in the course of a study of irreducibility of certain infinite dimensional representations, was led to conjecture that this was a general phenomenon—more precisely, that the double cosets $BgB$ would be finite in number and the quotients $BgB/B$ would be cells. He checked it for the classical groups, and it was then proved in general by Harish-Chandra [HC], 1956, and Chevalley [C7], 1955, independently. Harish-Chandra's proof is also valid over the reals, in a form to be discussed later, and Chevalley's argument also holds in positive characteristic (see §3 or VII). More precisely, let $H$ be the Cartan subgroup with Lie algebra $\mathfrak{h}$. The Weyl group $W(\mathfrak{h}, \mathfrak{g})$ can be identified to $\mathcal{N}(H)/H$. Then one has $G = BWB$, the space $G/B$ is the disjoint union of the orbits of the fixed points of $H$ in $G/B$, and each orbit is acted upon simply transitively by a suitable unipotent subgroup. This is the so-called *Bruhat decomposition* of $G$ or $G/B$. It also yields a decomposition of $G/P_I$ into cells indexed by $\mathcal{N}(H)/\mathcal{N}_{P_I}(H)$. The affine subspace exhibited by Goto is then the biggest cell of the Bruhat decomposition.

**2.4.** *Linear representations of complex semisimple groups.* We revert to the notation of 2.2. Let $G$ be the simply connected group with Lie algebra $\mathfrak{g}$, and $B, H, N^+$ the subgroups with Lie algebras $\mathfrak{b}, \mathfrak{h}, \mathfrak{n}^+ = \sum_{\alpha > 0} \mathfrak{g}_\alpha$. The group $B$ is the semidirect product of $H$ and $N^+$. The Lie algebra $\mathfrak{h}$ may be viewed as the universal covering of $H$, and the exponential map as the covering map. Its kernel $\Lambda$ is the lattice in $\mathfrak{h}$ dual to $P(\Phi)$. An element $\lambda \in P(\Phi)$ will be viewed as a character (homomorphism into $\mathbb{C}^*$) of $H$, whose value on $h \in H$, denoted $h^\lambda$, is defined by the rule

$$h^\lambda = e^{2\pi i \lambda(x)}, \quad \text{where } h = \exp x \ (x \in \mathfrak{h}),$$

or as a character of $B$, trivial on $N^+$.

Let $(\sigma, \mathbb{C}^{N+1})$ be an irreducible representation of $\mathfrak{g}$, and $\omega$ its highest weight. We view $\sigma$ as a representation of $G$ in $\mathbf{P}_N(\mathbb{C})$. By Lie's theorem, $B$ has a fixed point, say $z$. It may be fixed under a bigger subgroup $G_z$. It is easily seen that $G_z = P_I$ (cf. 2.2), where

$$I = \{\alpha \in \Delta, (\omega, \alpha) = 0\}.$$

The orbit $G.z$ of $z$ is then isomorphic to $G/P_I$. Since the highest weight has multiplicity one, by Cartan's theory, $z$ is the unique fixed point of $B$, and Lie's theorem implies that $G.z$ is the smallest orbit of $G$, and is contained in the closure of any orbit of $G$. For $G = \mathbf{SL}_3$, we are back to Study's discussion, recalled in Chapter V, §1. This point of view also leads to a geometric construction of $\sigma$. Let $\omega^*$ be the highest weight of the representation contragredient to $\omega$, and

$$(1) \qquad E^\omega = \{f \in \mathbb{C}[G], f(g.b) = f(g) \cdot b^{\omega^*} \ (g \in G, b \in B)\}.$$

Acted upon by left translations, it is a $G$-module, isomorphic to $\sigma$, called the *dual Weyl module*. This is an equivalent formulation of a geometric construction: consider the homogeneous line bundle on $G/B$ defined by $\omega^*$. Then $E^\omega$ may be viewed as the space of regular sections of that bundle. (See [B2], 1954. This theorem is first described in [S3], 1954; a rather similar result was announced in [T1], 1955.)

**2.5.** The notion of linear algebraic group is a special case of that of algebraic group: an algebraic variety $G$, endowed with a group law such that the map $G \times G \to G$ defined by $(x, y) \mapsto x.y^{-1}$ is a morphism of algebraic varieties. There had been nonlinear examples of such algebraic groups for a very long time, well before the birth of Lie theory: first of all of elliptic curves, then the Jacobians of algebraic curves, and after them the quotients $\mathbb{C}^n/\Lambda$, where $\Lambda$ is a lattice of rank $2n$ satisfying Riemann's bilinear relations, i.e. the complex tori admitting a projective embedding, later called abelian varieties. They played a fundamental role in algebraic geometry and number theory all along. This was over $\mathbb{C}$, but the development of algebraic geometry led André Weil to set up the foundations of a theory of abelian varieties over general fields [W1], 1948. By definition, an abelian variety is an irreducible algebraic group admitting a projective embedding. The group law is then necessarily commutative. (The definition assumes, in principle more generally,

André Weil, talking with the author in 1955.

that the group variety is only complete, in Weil's sense, but this anyhow implies
the existence of a projective embedding.) In the early fifties, the notion of algebraic
group appeared to gain in importance so that, there too, some foundations were
in order. They were provided by Weil [W2], [W3], 1955. An important new point
pertains to spaces of cosets. If $H$ is a closed subgroup of an algebraic group, then
the space of left cosets $G/H$ (or of right cosets) admits a unique structure of algebraic variety invariant under the action of $G$, such that the canonical projection
$\pi : G \to G/H$ is separable. It then follows that it has a "universal property": any
morphism of $G$ into an algebraic variety which is constant on the left cosets $xH$
factor through $\pi$. The proof of the existence of the quotient structure was rather
awkward. See VIII, 10.6, for two other more direct constructions.

Since abelian varieties and linear algebraic groups were the only known categories of algebraic groups, the question arose as to whether any algebraic group
was built from those two types. A positive answer was supplied, independently, by
Barsotti and Chevalley (who published his proof only later). Namely: Let $G$ be
a connected algebraic group. Then it contains a unique biggest closed subgroup
$N$, isomorphic to a linear group, which is normal and such that the quotient $G/N$
is an abelian variety. See VII, §10, for references. Another proof was supplied by
M. Rosenlicht [R1], 1956, who also contributed some further foundational material.

**2.6.** In [K1], 1948, one goal of Kolchin was to develop a theory of linear
algebraic groups, suitable for his purposes, independently of the Lie theory, which
he viewed as "far deeper". The same concern for autonomy led him to introduce in
[K3], 1973, his own definition of algebraic group, in the framework of an algebraic
geometry (also of his own making) for subsets of homogeneous spaces. He also
establishes there the Barsotti-Chevalley theorem (see VIII, §§10.2 to 10.6).

This being said, I now revert to linear algebraic groups.

## §3. Groups over an algebraically closed ground field II

*In this section, $k$ is an algebraically closed field.*

**3.1.** Let $f : V \to W$ be a morphism of varieties over $k$, where $V$ is irreducible. A fundamental property of $f$ is the fact that $f(V)$ contains a (non-empty) Zariski-open subset of its Zariski closure $\overline{f(V)}$. In particular, $\overline{f(V)} - f(V)$ is contained in an algebraic set of strictly smaller dimension than $f(V)$. Now let $k = \mathbb{C}$, view $V, W$ as complex analytic spaces. In this case $\overline{f(V)}$ is also the closure of $f(V)$ in the ordinary topology, and so $f(V)$ contains an open dense subset of its closure, in the usual topology. This is a fundamental difference with holomorphic maps: In that case $\overline{f(V)}$ may have strictly bigger dimension than $f(V)$. Clearly, this difference should make the world of complex algebraic groups simpler than that of complex Lie groups, but to use this basic fact, we need some group theoretical consequences. First, there is the fact that if $f : G \to H$ is a morphism of algebraic groups (with $G$ connected), then $f(G)$ is a (closed) algebraic subgroup. The proof is obvious: If $h \in \overline{f(G)}$, then $f(G)$ and $h.f(G)$ contain non-empty Zariski open subsets of $f(G)$, and hence have a non-empty intersection. Therefore there exist $x, y \in G$ such that $f(x) = h.f(y)$, whence $h = f(x.y^{-1}) \in f(G)$.

Again, this is completely false for morphisms of complex Lie groups, as is well known.

A second simple consequence, which generalizes the previous one, concerns transformation groups. Let $G$ be a connected algebraic group and $V$ a variety on which $G$ operates morphically (i.e., there is a morphism $G \times V \to V$ of varieties with the usual properties). *Then $G$ always has closed orbits.* Indeed, let $v \in V$. Its orbit $G.v$ is the image of $G$ under the orbit map: $g \mapsto g.v$. Therefore $G.v$ contains a Zariski-open subset of its Zariski closure $\overline{G.v}$. In fact, by homogeneity, it is open. Clearly, $\overline{G.v} - G.v$ is invariant under $G$. If $w \in \overline{G.v} - Gv$, then $Gw$ has strictly smaller dimension than $G.v$. Therefore, an orbit of minimal dimension in $\overline{G.v}$ is necessarily closed.

**3.2.** This simple remark is the starting point of [B4], 1956, which establishes some basic theorems on linear algebraic groups over $k$. It includes the results of Kolchin, and also extends those proved in [C5] and [C6] for algebraically closed ground fields of characteristic zero.

As in Kolchin, there are no Lie algebras. I was also influenced by some work of Hopf, Samelson and Stiefel on compact connected Lie groups, done in the forties, where they obtained some basic properties of those groups by topological or geometrical methods, without recourse to the infinitesimal theory. The connected diagonalizable subgroups, isomorphic to products of $k^*$, play in [B4] a role similar to that of tori in their theory, and I suggested calling them also tori (although, over $\mathbb{C}$, they are products of $\mathbb{C}^*$, hence not compact). If a distinction needs to be made, one can speak of algebraic tori. For instance, the fact that in a connected group the centralizer of a torus is connected, and its proof, were suggested by a similar statement in [H], 1940-1.

In [C3] Chevalley had introduced a notion of Cartan subgroup of any group (*see* VII, 9). It is shown in [B4] that the Cartan subgroups are algebraic, are the centralizers of maximal tori, and are conjugate (as are the maximal tori). Any semisimple element belongs to some maximal torus.

**3.3.** I gave a copy of that paper to Chevalley in the summer of 1955. Next spring, building on it, he had obtained the structure and classification of semisimple algebraic groups over $k$ ([C8], 1956-8). I refer to Chapter VII, §12, for references and more details.

Here, I shall simply summarize some highlights.

There is first a global version of the structural properties recalled in 2.2, in which the roots are viewed as characters of a maximal torus, rather than as linear forms on a Cartan subalgebra.

Let $T$ be a torus. It is isomorphic to a product of $k^*$, say with coordinates $t_1, \dots, t_s$ ($s = \dim T$). A character of $T$, i.e. a morphism of $T$ into the 1-dimensional torus $\mathbf{GL}_1$, has the form

$$t = (t_1, \dots, t_s) \mapsto t_1^{m_1} \dots t_s^{m_s} \quad (m_i \in \mathbb{Z}, i = 1, \dots, s).$$

The group $X(T)$ of characters of $T$ is thus a free abelian group of rank equal to dim $T$. We let $X(T)_\mathbb{Q} = X(T) \otimes_\mathbb{Z} \mathbb{Q}$. The value of a character $\lambda$ on $t$ will be denoted $t^\lambda$.

Now let $G$ be a simple algebraic group over $k$. Fix a maximal torus $T$ of $G$. It acts on the Lie algebra $\mathfrak{g}$ of $G$ by the adjoint action. This representation is semisimple, and we can again write

$$(3) \qquad \mathfrak{g} = \mathfrak{t} \oplus \bigoplus_\alpha \mathfrak{g}_\alpha, \quad \text{where } \mathfrak{g}_\alpha = \{x \in \mathfrak{g}, \operatorname{Ad} t(x) = t^\alpha.x\},$$

where $\dim \mathfrak{g}_\alpha = 1$ and the $\alpha$ are elements of $X(T)$ which form an irreducible reduced root system $\Phi = \Phi(T, G)$ in $X(T)_\mathbb{Q}$. The group $X(T)$ is a lattice intermediary between the lattice $P(\Phi)$ of weights of $\Phi$ and the lattice $Q(\Phi)$ generated by $\Phi$ (the root lattice). The Weyl group $W(\Phi)$ is induced by the automorphisms of $G$ leaving $T$ invariant, and may be identified to $W(T, G) = \mathcal{N}(T)/T$.

The Lie algebra $\mathfrak{g}_\alpha$ is tangent to a unique one-parameter group $U_\alpha$ invariant under inner automorphisms by $T$, and there exists an isomorphism $\varphi_\alpha : K^* \to U_\alpha$ such that

$$(4) \qquad t.\varphi(k).t^{-1} = \varphi(t^\alpha.k) \quad (k \in K^*).$$

To be more accurate, the roots are introduced via (4) in Exposé 12 of [C8], and (3) is stated later, in Exposé 21; see VII, **12**.

Let us call a root system $\Phi$ in some rational vector space and a lattice between $P(\Phi)$ and $Q(\Phi)$ a *diagram*. Then, if we associate to $G$ the diagram $\big(\Phi(T, G), X(T)\big)$ we get an bijection between isomorphism classes of simple groups over $k$ and isomorphism classes of diagrams $(\Phi, \Gamma)$, where $\Phi$ is irreducible and reduced. This is the main final result of [C8]. It shows that the classification is the same for all $k$. As $\Gamma$ varies between $P(\Phi)$ and $Q(\Phi)$, $G$ describes the group with root system $\Phi$. Let us denote by $G_\Gamma$ the group with diagram $(\Phi, \Gamma)$. If $\Gamma \subset \Gamma'$, there is a natural surjective homomorphism $f_{\Gamma, \Gamma'} : G_\Gamma \to G_{\Gamma'}$ with finite kernel. In characteristic zero the kernel is isomorphic to $\Gamma'/\Gamma$, in characteristic $p$ only to the $p'$-part of $\Gamma'/\Gamma$ (the biggest subgroup of order prime to $p$). In characteristic zero, the differential of $f_{\Gamma, \Gamma'}$ is an isomorphism, but not so in characteristic $p$, if $\Gamma'/\Gamma$ has $p$-torsion, and that $p$-torsion gives information on the kernel of $df_{\Gamma, \Gamma'}$. If $\Gamma = P(\Phi)$, $G$ is the "simply connected" group with root system $\Phi$; if $\Gamma = Q(\Phi)$, it is the adjoint group. The homomorphisms $f_{\Gamma, \Gamma'}$ are central isogenies: isogenies because the kernel is finite, central because global and infinitesimal kernel belong to any maximal torus and its Lie algebra respectively.

[C8] also determines the *special isogenies*, the (rare) cases where there is an isogeny from $G$ to $G'$ with infinitesimal kernel not contained in the Lie algebra of a torus.

Around 1958, Chevalley wrote a paper on the properties of the Bruhat decomposition, not published at the time, but circulated informally, posthumously published in 1990 [C9]. We refer to VII for some details.

**3.4. Remark.** The roots in 2.2, call them infinitesimal roots, are the differentials of the present roots; hence (3) implies 2.2(1) if the $\alpha$ in 2.2(1)(2) are viewed as the differentials of the global roots under consideration here. In characteristic zero, infinitesimal and global roots determine one another, but not always in positive characteristic. It may happen that a global root has zero differential, or that two global roots have the same differential. In that case the spaces $\mathfrak{g}_\alpha$ in 2.2(2) may have dimensions $> 1$.

**3.5.** In [C8], Chevalley also extends the determination of irreducible representations in terms of highest weights. Start from 2.4(1), but now with $\mathbb{C}$ replaced by $k$, again assuming that $G$ is simply connected. As before, $E^\omega$ is a $G$-module. It is not necessarily irreducible, though. It has a unique line $D^\omega$ stable under $B$, on which $B$ acts via $\omega$, which is contained in all $G$-invariant subspaces. Their intersection $L(\omega)$ is then an irreducible $G$-module with highest weight $\omega$. This again establishes a bijective correspondence between dominant weights and equivalence classes of irreducible representations.

In characteristic zero, the character of $E^\omega$ is given by Weyl's formula. This is still true in arbitrary characteristic, but does not describe the character of $L(\omega)$. For this, it is necessary to determine a composition series of $E^\omega$, the subject matter of a number of investigations and conjectures (see, e.g., [J]).

## §4. Rationality properties

*From now on $k$ is a (commutative) field and $K$ an algebraically closed extension.*

**4.1.** The results summarized in §3 brought to some provisory conclusion the theory over an algebraically closed ground field, so attention turned to rationality properties. This topic had a considerable development and is still pursued. Since my goal here is to relate some history rather than give a survey up to the most recent advances, I shall stop at around 1970, or at any rate limit myself to directions already well engaged in by that time.

Recall that an affine or projective variety $V$ over $K$ is defined over $k$ if the ideal $I(V)$ of the variety is generated over $K$ by polynomials (homogeneous in the projective case) with coefficients in $k$. If so, and if $k'$ is an extension of $k$ in $K$, we let $V(k')$ denote the set of points of $V$ with coefficients in $k$, the variety itself being identified with $V(K)$. We shall also say that $V$ is a $k$-variety.

Assume that $V$ is the set of zeroes of a family of polynomials with coefficients in $k$. It is then said to be $k$-closed (in the Zariski topology) in affine or projective space, as the case may be. This does not quite imply that $V$ is defined over $k$, only that $I(V)$ is generated by polynomials with coefficients in a purely inseparable extension $k'$ of $k$, so that $V$ is defined over $k'$. Of course, if $k$ is of characteristic zero, or more generally if $k$ is perfect, there is no distinction between $k$-closed and defined over $k$. Over non-perfect fields, however, this is often a source of considerable difficulties, which sometimes forced one to find new techniques.

**4.2.** If $V = G$ is a linear group, we are interested in properties of $G(k)$, such as knowing whether some type of subgroups defined over $K$ have representatives defined over $k$ and, if so, whether any two which are conjugate under $G(K)$ are already so under $G(k)$.

If $G$ and a closed subgroup $H$ are defined over $k$, then the canonical structure of variety of $G/H$ constructed in [W3] is already defined over $k$. However, $G(k)/H(k)$ may be different from $(G/H)(k)$. The relation between the two is in the domain of Galois cohomology. The latter is also the natural framework for classification problems, such as that of the $k$-forms of a $k$-group $G$, i.e. of the isomorphism classes over $k$ of $k$-groups which are isomorphic over $K$ to the given group.

**4.3.** A first rationality result is due to S. Lang [L], 1956, about groups over finite fields. Let $k = \mathbb{F}_q$ be the field with $q$ elements, where $q = p^s$ is a power of a prime, the characteristic of $k$. Any $k$-variety admits a "Frobenius map" $F_q$, which maps a point $x$ with coordinates $x_i$ (in some chart) to the point $x^{(q)}$ with coordinates $x_i^q$. Its fixed point set is $V(k)$. Its differential is zero.

Let $G$ be a $k$-group. In the language of [L] (which is borrowed from Weil's Foundations), the main point is to show that if $g$ is a generic point over some field of definition, then so is $g^{-1}.g^{(q)}$. In today's language, we have to prove that the morphism of varieties $s : g \mapsto g^{-1}.g^{(q)}$ is dominant and separable; and for this it suffices to show that the differential of $s$ is an isomorphism at all points. This is clear, since the differential of the Frobenius map is zero and that of $g \mapsto g^{-1}$ is an isomorphism. This argument is also valid for the map $s_a : g \mapsto g^{-1}.a.g^{(q)}$ for any $a \in G$, and shows that $G^o \subset s_a(G)$.

Now let $G$ be connected. This implies: a) if $H$ is a closed $k$-subgroup, then the projection $G(k) \rightarrow (G/H)(k)$ is surjective; b) if $G$ acts $k$-morphically on a $k$-variety $V$, then $V(k) \neq \emptyset$; c) if $G'$ is a $k$-group and $f : G \rightarrow G'$ is surjective, with finite kernel, then $G(k)$ and $G'(k)$ have the same cardinality [L]; and finally, as remarked by Serre (see footnote 2, p. 45 of [R2]), d) $G$ has a maximal torus, a Cartan subgroup and a Borel subgroup defined over $k$.

**4.4.** In the structure theory of linear algebraic groups, a basic role is played by maximal tori or, equivalently, by their centralizers, the Cartan subgroups. An important rationality question was therefore the existence of a maximal torus defined over $k$, or, equivalently, as is rather easily shown, the existence of a Cartan subgroup defined over $k$. This being settled for finite fields, the attention turned to infinite fields. This existence was proved first over perfect fields by M. Rosenlicht [R2], 1957, later over general fields, by Rosenlicht for solvable groups [R3], 1963, and by A. Grothendieck in general ([G4], 1970, which was informally available earlier). For solvable groups, it was shown in [R3] that maximal tori defined over $k$ exist and are conjugate under $G(k)$ or, more precisely, under $\mathcal{C}^\infty G(k)$, where $\mathcal{C}^\infty G$ is the last member of the descending central series (or, equivalently, the smallest normal subgroup such that the quotient by it is nilpotent; it is always defined over $k$). $G$ is always the semi-direct product of its unipotent radical $G_u$ by a maximal torus. However, $G_u$, which is clearly $k$-closed, is not always defined over $k$. This may happen even if $G$ is nilpotent, in which case it is the direct product of a unique maximal torus, always defined over $k$, by $G_u$.

The case of an arbitrary field turned out to be more difficult, and in fact appeared at first to require going beyond the usual theory, because Grothendieck's proof was obtained in the context of the theory of group schemes. This was a bit puzzling: the statement of the theorem uses only concepts of the usual theory, so one could not help wondering to what extent is was necessary to appeal to a much broader theory to establish it. In 1965, T.A. Springer and I noticed that very little was needed (see [BS1], and [BS2] for a more detailed account). The general theory of schemes could be bypassed, for the purpose in discussion here, by slightly enlarging the notion of quotient group. So far, we have used only the quotient $G/N$ of a $k$-group by a normal $k$-subgroup. What was needed was the possibility of dividing by an ideal $\mathfrak{m}$ of the Lie algebra $\mathfrak{g}$ of $G$ which is also stable under $G$, acting on $\mathfrak{g}$ by the adjoint representation.[1] Given such an $\mathfrak{m}$, one can define a $k$-group $G/\mathfrak{m}$, endowed with a $k$-morphism $f : G \to G/\mathfrak{m}$ which is bijective on the set of $K$-points, whose differential has $\mathfrak{m}$ as kernel, and such that the comorphism $f^o : K[G/\mathfrak{m}] \to K[G]$ maps $K[G/\mathfrak{m}]$ bijectively onto the subfield of $K[G]$ annihilated by the derivations of $K[G]$ defined by $\mathfrak{m}$. (This quotient and $G/N$ are indeed special cases of the quotient of a group scheme by a normal subgroup subscheme.) The map $G \to G/\mathfrak{m}$ is a so-called purely inseparable isogeny of height one, a notion which had appeared earlier [S4], and had in turn been suggested by the use of purely inseparable extensions of height one of a field, introduced by N. Jacobson in his Galois theory of inseparable extensions. They had also been considered by Chevalley in [C8], as mentioned in VIII, **12**.

It was also necessary to establish that $\mathfrak{g}$ admits an additive Jordan decomposition compatible with morphisms. This was shown first by T.A. Springer, a remark which is in fact at the origin of [BS1], [BS2]. That it had not been pointed out earlier testifies to the very limited use of Lie algebras in the treatment of linear algebraic groups over $K$. However, they play an important role in rationality questions, especially over imperfect fields, even though it is totally different from their use in the usual Lie theory.

Combined with the usual theory (i.e. the theory of "reduced groups", from the point of view of scheme theory), this suffices to show the existence of a maximal torus defined over $k$ ([BS1], 1966, [BS2], 1968, or [B5], §18). Some other facts, also proved by Grothendieck, are established at the same time: the variety of Cartan subgroups is a rational $k$-variety. If $G$ is reductive or if $k$ is perfect, $G$ is unirational over $k$, and in particular $G(k)$ is Zariski dense in $G$. For perfect $k$, the last assertion was also proved in [R2].

**4.5.** *Split solvable groups.* Let $G$ be connected and solvable. Then it has a composition series

$$(1) \qquad\qquad G = N_0 \supset N_1 \supset \cdots \supset N_s = \{1\}$$

such that $N_i$ is closed, connected, normal in $N_{i-1}$ $(1 \le i \le s = \dim G)$ and there exists an isomorphism $f_j$ of $N_j/N_{j+1}$ onto either the additive group $\mathbb{G}_a$ of $K$ or the multiplicative group of $K^*$ $(j = 0, \ldots, s - 1)$.

The group $G$ is said to split over $k$, or be $k$-split, if all the $N_i$ and $f_j$ are defined over $k$. Of course, $G$ always splits over $K$. If $k$ is perfect and $G$ is unipotent, then $G$ splits over $k$. If $G$ is a torus, contained in $\mathbf{GL}_N$, it splits over $k$ if and only if there exists $x \in \mathbf{GL}_N(k)$ such that $x.G.x^{-1}$ is diagonal. If so, we can choose $x$ so that the diagonal entries of $g \in G$ are $(x_1, \ldots, x_s, 1, \ldots, 1)$, where the $x_i \in K^*$ are arbitrary and $s = \dim G$. All the rational characters of $G$ are then defined over $k$. A $k$-torus always splits over a finite separable extension of $k$. The split solvable

---

[1]Moreover, $\mathfrak{m}$ should be a *restricted* Lie subalgebra of $\mathfrak{g}$ (see e.g. [B5], 3.1, for that notion).

groups have some interesting properties, discovered mostly by Rosenlicht (see [R3] or [B5], §15), notably: a) if $V$ is a projective (or complete) $k$-variety on which $G$ operates $k$-morphically and $V(k) \neq \emptyset$, then $V(k)$ contains a point fixed under $G$ (a relative version of the Lie-Kolchin theorem); b) if $V$ is affine and $G$ is transitive, then $V(k) \neq \emptyset$; c) if $H$ is a connected $k$-group and $G$ a connected solvable $k$-split subgroup, then $H(k) \to (H/G)(k)$ is surjective; and d) $G$ is a rational $k$-variety and its unipotent radical is defined over $k$.

A key result needed to prove all that (and more) is the following characterization of $k$-split solvable groups ([R3], Theorem 2):

The connected solvable (resp. unipotent) group $G$ is $k$-split if and only if

$$k[G] \subset k[x_1, \dots, x_n, x_1^{-1}, \dots, x_n^{-1}] \ (\text{resp. } k[x_1, \dots, x_n]),$$

where the $x_i$ are algebraically independent over $k$.

**4.6.** *Semisimple groups.* The heart of the theory of linear algebraic groups is the structure of semisimple groups. A main task was therefore to relativise the theory of Chevalley (3.3 or VII, 12), i.e. to extend it, *mutatis mutandis*, to the rational points. The results are usually stated and proved for the slightly more general class of reductive groups (groups whose radical is a torus), but I shall stick to semisimple groups for simplicity. I shall also assume that $G$ is $k$-simple (i.e., it had no proper normal $k$-subgroup of strictly positive dimension).

This theory was developed by J. Tits and myself, first independently, over perfect fields, and then jointly over arbitrary fields ([BT], 1965). Over perfect fields much of it was also obtained by I. Satake [S1], 1971.

In this theory, the role of maximal tori is played by tori which are $k$-split and maximal for that property. They are shown to be conjugate under $G(k)$, and their common dimension is the $k$-rank $\mathrm{rk}_k(G)$ of $G$. If there are none, i.e. $\mathrm{rk}_k(G) = 0$, the group is said to be *anisotropic over $k$*, and there is no further structure theory. (If $k$ is a locally compact field, this is equivalent to $G(k)$ being compact in the ordinary topology.)

Assume now that $\mathrm{rk}_k(G) > 0$, in which case $G$ is said to be *$k$-isotropic*. Fix a maximal $k$-split torus $S$. Since it is $k$-split, all its characters are defined over $k$, and the image of $S$ in the adjoint representation can be diagonalized over $k$. We can write

$$\mathfrak{g} = \mathcal{Z}(\mathfrak{s}) \oplus \bigoplus_\alpha \mathfrak{g}_\alpha$$

where

$$\mathfrak{g}_\alpha = \{x \in \mathfrak{g} \mid \mathrm{Ad}\, t.x = t^\alpha.x\}$$

and the $\alpha$ for which $\mathfrak{g}_\alpha \neq \{0\}$ form an irreducible root system $_k\Phi(S, G) = {}_k\Phi$ in $\mathfrak{s}^*$. This is very similar to what was recalled in 3.3, but there are important differences: the root system need not be reduced, it may be of type BC in Bourbaki's notation [B6]. The space $\mathfrak{g}_\alpha$ may have dimension $\geq 2$. If $2\alpha$ is not a root, then $\mathfrak{g}_\alpha$ is the Lie algebra of a uniquely determined commutative unipotent $k$-subgroup $U_\alpha$ invariant under $S$. If $2\alpha$ is a root, this is not so, but $\mathfrak{g}_\alpha + \mathfrak{g}_{2\alpha}$ is the Lie algebra of a 2-step unipotent group $U(\alpha)$ invariant under $S$. Of course, $\mathfrak{z}(\mathfrak{s})$ is the Lie algebra of $\mathcal{Z}(S)$, and the zero-eigenspace of $S$. The Weyl group $W(_k\Phi)$ of $_k\Phi$ may be identified with the group of automorphisms of $\mathfrak{s}^*$ defined by $\mathcal{N}(S)/\mathcal{Z}(S)$. Fix an ordering on $_k\Phi$ and let $_k\Delta$ be the set of simple roots. The subsets of $_k\Delta$ parametrize the conjugacy classes of parabolic $k$-subgroups, and any two which are conjugate over $K$ are already so over $k$. Let $\mathfrak{n}^+$ be the sum of the $\mathfrak{g}_\alpha$ for $\alpha > 0$. It is the Lie algebra of a $k$-subgroup $N^+$ normalized by $\mathcal{Z}(S)$ and $_kP_\phi = \mathcal{Z}(S).N^+$ is a minimal parabolic $k$-subgroup. The subgroup $_kP_I$ $(I \subset \Delta)$ is generated by $N^+$ and the centralizer of $S_I$, where $S_I = (\bigcap_{\alpha \in I} \ker \alpha)^o$. Finally, there is a Bruhat decomposition

$$G(k) = \coprod_w P_\phi(k).w.P_\phi(k),$$

whence also a decomposition of $G(k)/_kP_\phi(k)$ by the orbits of $_kP_\phi(k)$, or, equivalently, of $N^+$; every orbit is $k$-isomorphic to an affine space.

**4.7.** For use in §5, I shall add some more information on $_kW$ and its action on $S$ and related spaces. The group $_kW$ acts by inner automorphisms on $S$, and by the adjoint representation on $\mathfrak{s}$ and $\mathfrak{s}^*$. Recall that the group of characters $X(S)$ of $S$ is free abelian of rank $r = \dim S$ and that all its elements are defined over $k$. The dual $\mathbb{Z}$-module $X_*(S)$ in $\mathfrak{s}^*$ may be identified to the group of $k$-morphisms of $\mathbf{GL}_1$ into $S$. If $\lambda$ is such a $k$-morphism and $\mu \in X(S)$, then $\mu \circ \lambda$ is a morphism of $\mathbf{GL}_1$ into itself, hence of the form $t \mapsto t^m$ for some $m \in \mathbb{Z}$. The canonical bilinear form expressing the duality between $X(S)$ and $X_*(S)$ is given by $\langle \mu, \lambda \rangle = m$. The group $_kW$ operates on $X_*(S)$ and $X(S)$, respecting the duality. It also acts on $X_*(S)_K := X_*(S) \otimes_\mathbb{Z} K$, which may be identified to $\mathfrak{s}$, and on the real vector space $X_*(S)_\mathbb{R} = X_*(S) \otimes_\mathbb{Z} \mathbb{R}$. There it is a reflection group, the reflection hyperplanes being the $V_\alpha \otimes_\mathbb{Z} \mathbb{R}$ $(\alpha \in {}_k\Phi)$, where $V_\alpha$ is the submodule of $X_*(S)$ annihilated by the root $\alpha$. The connected components of the complement of the union of the $V_\alpha \otimes_\mathbb{Z} \mathbb{R}$ are the (open) "Weyl chambers". These are permuted in a simply transitive manner by $_kW$. The *positive Weyl chamber* $\mathfrak{a}^+$ (relative to the given ordering) is defined by the inequalities $\alpha > 0$ $(\alpha \in {}_k\Delta)$. Its closure is a fundamental domain for $_kW$.

**4.8.** *k-forms.* A $k$-form of $G$ is a $k$-group $H$ which is isomorphic to $G$ over $K$. Similarly, the $k$-forms of a Lie algebra $\mathfrak{g}$ are the Lie algebras $\mathfrak{h}$ over $k$ such that $\mathfrak{h} \otimes_k K$ is isomorphic to $\mathfrak{g}$. One is interested in classifying the $k$-forms of $G$ (or of $\mathfrak{g}$ in characteristic zero) up to $k$-isomorphisms.

The first result along those lines is the determination by Elie Cartan of the real forms of a complex simple Lie algebra (i.e. $k = \mathbb{R}, K = \mathbb{C}$) in [C0], 1914. Later he gave a global version of his theorem: Let $G$ be a complex simple Lie group, $K$ a maximal compact subgroup. Then the real forms of $G$, up to isomorphism over $\mathbb{R}$, correspond bijectively and naturally to the conjugacy classes of elements of order two in the automorphism group of $G$ (see Chapter IV, 2.5.3).

The classical groups, over $\mathbb{C}$ say, can be viewed as the group of automorphisms of algebras with involution; more precisely, the adjoint simple groups are the identity components of the automorphism groups of algebras with involution. In [W4], 1960, Weil uses this as a basis to classify the $k$-forms of the adjoint groups of classical groups if $k$ has characteristic zero. They appear as the identity component of the automorphism groups of simple algebras with involution defined over $k$.

The most comprehensive treatment so far of classification over general fields is due to Tits [T2], 1966. The main parameters are relative root systems and some properties of $\mathcal{Z}S$, in particular of the derived groups of $\mathcal{Z}S$, which is $k$-anisotropic. Many of these results over a perfect field can also be found in Satake [S1], 1971.

## §5. Algebraic groups and geometry. Tits systems and Tits buildings

**5.1.** In the early 1870's, Klein and Lie arrived at a general principle, codified in Klein's Erlanger program [K0], 1872, giving groups a dominant position in geometry: a geometry on a given space is determined by its group of automorphisms and the subgroups leaving certain configurations stable or pointwise fixed. Two geometries for which the automorphism groups are isomorphic, so that these aggregates of subgroups correspond to one another, should be viewed as equivalent.

The standard example of such a relationship is given by the projective group and projective geometry. Later, similar geometries were developed for symplectic or orthogonal groups, the geometric objects being the totally isotropic subspaces of the given projective space. These are in fact associated to parabolic subgroups, and the development of the theory of semisimple groups allowed Tits to introduce similar geometries for the exceptional groups from 1953 on. This was done first over $\mathbb{C}$ [T1], then gradually over more general fields, and led Tits eventually to what are now called Tits buildings or Tits geometries and Tits systems. I shall start from those, first systematically expounded in [T4], 1974, referring to [T1], 1955, for earlier literature.

This framework has brought to another level the relations between groups and geometry, which have been extensively used to study both. In fact, it puts them on an equal footing. On the one hand, in agreement with the Erlanger program, the essence of a geometry is given by its group of automorphisms and the collection of subgroups leaving certain configurations invariant; but on the other hand the Tits axioms allow one to construct a group of automorphisms starting from the geometry. Both directions have proved to be important. I shall try to give an idea of some of these connections, for semisimple groups over an arbitrary field.

**5.2.** We go back to the situation of 4.6 and denote by $\mathcal{P}_k$ the set of parabolic $k$-subgroups of $G$. Each one is conjugate to a unique standard parabolic $k$-subgroup $_kP_I$ ($I \in {}_k\Delta$). We let $I$ be the type of $P$ and denote by $\mathcal{P}_I$ the set of parabolic $k$-subgroups of type $I$. A parabolic $k$-subgroup is minimal (among parabolic $k$-subgroups) if $I = \varnothing$, and (proper) maximal if $I = {}_k\Delta - (\alpha)$ ($\alpha \in {}_k\Delta$).

The building $\mathcal{T}(G/k)$, or simply $\mathcal{T}$, is a simplicial complex of dimension $r - 1$ ($r = \mathrm{rk}_k G$). Its vertices are the proper maximal parabolic $k$-subgroups. Distinct parabolic $k$-subgroups $P_0, \ldots, P_s$ are the vertices of an $s$-dimensional simplex $\sigma$ if and only if their intersection is a parabolic subgroup.

The type $I$ of their intersection will also be called the type of $\sigma$. Note that the vertices of $\sigma$ are necessarily of distinct types.

Each simplex corresponds in this way to an element of $\mathcal{P}_k$. Clearly, this map is bijective and reverses inclusions. Let $g \in G(k)$. The inner automorphism $i_g$ obviously preserves $\mathcal{P}_k$ and the $\mathcal{P}_{k,I}$. Therefore it induces an automorphism of $\mathcal{T}$ which preserves the types. The stability group $G_\sigma$ of a simplex $\sigma$ is the parabolic subgroup which it represents. Since it preserves the types, it is also the "fixer" of $\sigma$, i.e. the subgroup of $G(k)$ leaving it pointwise fixed.

An outer automorphism of $G$, as an algebraic group over $k$, also preserves $\mathcal{P}_k$ and the inclusion relations between parabolic subgroups, and therefore induces a simplicial automorphism of $\mathcal{T}(G/k)$. However, it may permute the $\mathcal{P}_{k,I}$; hence it does not always preserve the type. The standard conjugacy theorems recalled in 4.6 show that, modulo an inner automorphism, any automorphism may be assumed to preserve a maximal $k$-split torus $S$ and $_k\Delta$. The bijection of $_k\Delta$ so obtained then describes the permutation of the types.

Fix a maximal $k$-split torus $S$, and let $\mathcal{P}_S$ be the set of parabolic $k$-subgroups containing $S$. It is a finite set on which $_kW$ acts naturally by inner automorphisms. The subcomplex $\mathcal{A}$ of $\mathcal{T}$ made up of the simplices attached to the elements of $\mathcal{P}_S$ is called an *apartment*. It may be identified to the simplicial complex cut out on the unit sphere in $X_*(S)_\mathbb{R}$. In particular, it provides a triangulation of that sphere. By construction, the closed Weyl chambers cut out on the sphere simplices of maximal dimension $r-1$, and these are called *chambers* of $\mathcal{T}$. They correspond to the minimal parabolic $k$-subgroups. Since $\mathcal{P}_\varnothing$ is one conjugacy class under $G(k)$, the latter acts transitively on the chambers. More generally, $G(k)$ acts transitively on the set of simplices of a given type $I$, which may be identified to $G(k)/_kP_I(k)$.

**5.3.** In classical cases, this complex can be defined geometrically. We give some examples. If $G = \mathbf{SL}_{n+1}(k)$, it is the *complex of flags* in $k^{n+1}$: a flag in $k^{n+1}$ is a strictly increasing sequence of subspaces

$$(1) \qquad \{0\} \subsetneqq F_1 \subsetneqq \cdots \subsetneqq F_j \neq k^{n+1}.$$

Similarly a flag in $\mathbf{P}_n(k)$ is a strictly increasing sequence of projective subspaces $F_i$ ($\dim F_i \in [0, n-1]$). Obviously, a flag in $k^{n+1}$ determines one in $\mathbf{P}_n(k)$ and conversely. The complex $\mathcal{T}'$ of flags in $k^{n+1}$ or $\mathbf{P}_n(k)$ is a simplicial complex of dimension $n-1$. Its vertices are the proper subspaces, and the vertices $P_0, P_1, \ldots, P_s$ are the vertices of an $s$-simplex if and only if, ordered by increasing dimension, they form a flag. The simplices of $\mathcal{T}'$ therefore correspond bijectively to the flags in $k^{n+1}$ or $\mathbf{P}_n(k)$. The collineations of $\mathbf{P}_n(k)$ are then the automorphisms of $\mathcal{T}'$ which preserve the dimensions, while the correlations are the automorphisms which reverse the dimensions (an $s$-simplex is mapped to an $(n-1-s)$-simplex). So $\mathcal{T}'$ captures the essence of projective geometry.

If $G = \mathbf{Sp}_{2m}(k)$, the complex $\mathcal{T}$ can be similarly identified to the complex of "totally isotropic flags", i.e. flags consisting of subspaces on which the restriction of the standard bilinear form is identically zero (the totally isotropic subspaces).

Now let char $k \neq 2$, and let $f$ be a non-degenerate quadratic form on $k^n$. Its Witt index is the maximal dimension $r$ of totally isotropic subspaces (relative to $f$). We assume it is $\neq 0$, i.e. that $f$ is isotropic. The orthogonal group $O(f)$ is algebraic, defined over $k$, of $k$-rank equal to $r$.

First let $r \neq n/2$. Then $\mathcal{T}$ can again be defined as the complex of totally isotropic flags in $k^n$. If $r = n/2$, the definition has to be modified to take into account special properties of the maximal totally isotropic subspaces: they form two families $V, V'$ stable under $SO(f)(k)$, but permuted by elements of $O(f)(k)$ of determinant $-1$. Two maximal isotropic subspaces $F, F'$ are in one family if their intersection has even codimension in $F$ and $F'$, in different families if it has odd codimension. A totally isotropic subspace of dimension $r-1$ is contained in exactly one maximal subspace of each family. Let us say that two isotropic subspaces are *incident* if either one is strictly contained in the other or if both have dimension $r$ and their intersection has dimension $r-1$. The flags consist of isotropic subspaces of dimensions between 1 and $r$, but different from $r-1$. A chamber corresponds to a maximal set of incident isotropic subspaces. The maximal ones are the maximal sets of distinct incident isotropic subspaces. One such consists of $r$ totally isotropic subspaces

$$F_1 \subset F_2 \subset \cdots \subset F_{r-2} \subset F_r, F_{r'} \in V'$$

where the subscript indicates dimension, $F_r \in V, F'_r \in V'$.

**5.4.** *Spherical buildings.* The complexes $\mathcal{T}(G/k)$ are examples of a geometric notion introduced by Tits under the name of spherical building. It is a special case of a more general notion of building. Another type will be mentioned in Chapter VII. In order to avoid too many preliminaries, I shall limit myself to the spherical ones.

Given a finite reflection group acting on $\mathbb{R}^n$, leaving no line pointwise fixed, we call the simplicial decomposition of the sphere defined by the reflection hyperplanes a *Coxeter complex.*.

A simplicial complex will be called a *chamber complex* if the maximal simplices, to be called chambers, all have the same dimension. A simplicial complex $\mathcal{T}$ is called a *spherical building* if it satisfies the following conditions.

B1.  $\mathcal{T}$ is a chamber complex containing a family $\mathcal{A}$ of subcomplexes called apartments.

B2. The apartments are finite Coxeter complexes.

B3. Any two chambers are contained in an apartment.

B4. Given two apartments $A$ and $A'$ and simplices $\sigma, \sigma' \in A \cap A'$, there exists an isomorphism of $A$ onto $A'$ keeping $\sigma$ and $\sigma'$ pointwise fixed.

$\mathcal{T}$ is *thin* (resp. *thick*) if a simplex of codimension one is contained in exactly two (resp. at least three) chambers.

**5.5.** The complex $\mathcal{T}(G/k)$ constructed in 5.2 is a thick spherical building. This of course reflects some quite special properties of $\mathcal{P}_k$. They, too, were axiomatized by Tits, under the name of $BN$-pairs, but are now usually called Tits systems. At first I limit myself to the Tits systems with finite Weyl groups.

A Tits system in a group $G$ consists of two subgroups $B, N$ with the following properties:

BN 0.  $B$ and $N$ generate $G$.

BN 1.  $H = B \cap N$ is normal in $N$.

BN 2. The group $W = N/H$, called the Weyl group of the Tits system, has a generating set $R$ such that

(BN 2)′. $rBwB \subset BwB \cup BrwB$, and

(BN 2)″. $rBr \neq B$

Expressions such as $BwB$ are a shorthand for $Bw'B$, if $w' \in N$ is a representative of $w$. Since $w'$ is defined modulo $H \subset B$, this is unambiguous. It can be shown that $R$ is completely determined by $B$ and $N$ and consists of elements of order two.

These axioms, and those of the buildings, are amazingly powerful. Here we limit ourselves to the case where $W$ is finite. It is then a finite reflection group.

The example we have considered earlier, as the notation hints, is where $G = G(k)$, $H = S(k)$, $S$ a maximal $k$-split torus, $N = \mathcal{N}(S)(k)$ and $W = {}_kW$ is the relative Weyl group. These axioms are strong enough to prove a Bruhat decomposition and construct the analogue of the parabolic subgroups, namely, a subgroup containing a conjugate of $B$. Given a parabolic subgroup $P$, there exists a unique subset $I \subset R$ such that $P$ is conjugate to $P_I = B.I.B$, and $G$ is the disjoint union of the subsets $BwB$ ($w \in W$). These properties allow one to associate to the Tits system a simplicial complex exactly as in 5.2, which is a thick building in the sense of 5.4.

Conversely, given a building $\mathcal{T}$ of spherical type and rank $\geq 3$, it is possible to define an automorphism group of $\mathcal{T}$ preserving the type, endowed with a canonical

Tits system, in which $B$ is the stability group (or fixer) of a chamber, $N$ the stability group of an apartment and $H$ the fixer of an apartment.

## §6. Abstract automorphisms

**6.0.** The results surveyed here pertain mainly to the following question: to what extent does an automorphism (of abstract group) of the group $G(k)$ of $k$-rational points of a $k$-group $G$ respect the algebraic group structure? The problem does not have an interesting answer unless $G$ is assumed to be semisimple, and we shall limit ourselves to that case. There are three obvious kinds of automorphisms:

a) Restriction to $G(k)$ of an automorphism (of algebraic group) of $G$ defined over $k$.

b) An automorphism induced by an automorphism $\sigma$ of $k$, to be called a field automorphism: if $G \subset \mathbf{GL}_n$, this is the map $g = (g_{ij}) \mapsto g^\sigma = (g_{ij}^\sigma)$.

c) Assume there exists a non-trivial homomorphism $\mu : G(k) \to \mathcal{C}G(k)$. If $\varphi$ is any automorphism of $G(k)$, then the map $g \mapsto \mu(g).\varphi(g)$ is an automorphism of $G(k)$, which is not of type a) or b) even if $\varphi$ is.

Type c) is relatively rare. Obviously, there is no such $\mu \neq 1$ if $G(k)$ is equal to its derived group or if $\mathcal{C}G(k) = \{1\}$.

We shall call *standard* any automorphism of one of these types and, more generally, any composition of such automorphisms. The general results to be surveyed here essentially assert that the automorphisms (of abstract group) of $G(k)$ are all standard. Some extensions to isomorphisms, even homomorphisms, and to certain types of subgroups of $G(k)$ will also be considered.

If $k = \mathbb{R}$ or $\mathbb{Q}_p$, the group $G(k)$ is also a topological group, with the $k$-topology. Moreover, $k$ has no non-trivial automorphism, so there are only automorphisms of type a) or c). A more precise and somewhat more ambitious question is whether in that case the topology of $G(k)$ can be described purely in terms of the abstract group structure.

**6.1.** The first result in that direction, due to O. Schreier and B. L. van der Waerden [SW], 1928, pertains to the group of projective transformations of $n$-dimensional projective space $\mathbf{P}_n(k)$ over $k$, i.e. $G = \mathbf{PGL}_{n+1}$. The restrictions to $G(k)$ of automorphisms of algebraic group of $G$ are of two kinds: the inner automorphisms $X \mapsto A.X.A^{-1}$ and, if $n \geq 2$, the outer automorphisms $X \mapsto A.{}^t X^{-1}.A^{-1} \left( A \in G(k) \right)$. The theorem asserts that all automorphisms of $G(k)$ are generated by those automorphisms and field automorphisms.

View $G$ as the quotient of $\mathbf{GL}_{n+1}$ by its center. Any automorphism of $\mathbf{GL}_{n+1}(k)$ induces one of $G(k)$. We shall discuss mainly the case where $n \geq 2$ and $k$ is infinite. The result is intimately connected with the fundamental theorem of projective geometry (to be referred to as FTPG). Let us give one formulation of it. Recall that a *collineation* of $\mathbf{P}_n(k)$ is a bijection which preserves hyperplanes. This implies that it preserves projective subspaces and their inclusions. Similarly, a correlation is a bijection of $\mathbf{P}_n(k)$ onto the dual projective space $\mathbf{P}_n(k)^*$ of hyperplanes in $\mathbf{P}_n(k)$ which respects inclusion of subspaces. Let $h$ be a collineation. The FTPG asserts that it is induced by a semilinear transformation, i.e., there exist $A \in \mathbf{GL}_{n+1}(k)$ and an automorphism $\sigma$ of $k$ such that

$$(1) \qquad h(x)_i = \sum_{i,j} a_{i,j} x_j^\sigma \qquad (x = x_0, \ldots, x_n).$$

B. L. van der Waerden

Similarly, a correlation is given by a semilinear invertible map of $\mathbb{C}^{n+1}$ onto its dual space. Schreier and van der Waerden first show that the dimension $n$ can be read off the structure of $G(k)$. They then describe a family of unipotent elements (called $t$-elements by them, transvections later and elsewhere (see [D], I, §1)), which fix hyperplanes pointwise. Given an automorphism $\alpha$ of $G(k)$, the idea is then to show that the $t$-elements are permuted by $\alpha$, so that $\alpha$ induces either a collineation or a correlation. The fundamental theorem of projective geometry then yields the result.

Some special considerations I shall not go into are needed to handle the remaining cases ($n = 2$ or $k$ finite). In fact, there is an error in the treatment of $\mathbf{PGL}_2(k)$, corrected by L.K. Hua ([H1], Appendix). For $k$ infinite, the main argument also implies that $\mathbf{PGL}_m(k)$ is isomorphic to $\mathbf{PGL}_n(k')$ if and only if $m = n$ and $k = k'$.

**6.2.** The next contributions concern simple or semisimple real Lie groups. Preliminary to this were some foundational results of von Neumann and Cartan which were recalled in IV, 3.3.

Now let $k = \mathbb{R}$ or $\mathbb{C}$, and let $G$ be a semisimple linear algebraic $k$-group. The group $G(k)$, endowed with the $k$-topology, is a Lie group over $k$. The regular $k$-functions, in the sense of algebraic groups, are exactly the analytic functions whose set of translates (by right or left translations) forms a finite dimensional vector space over $k$. This characterization shows that there is a unique structure of linear algebraic group compatible with the Lie group structure of $G(k)$. As a consequence, any automorphism of $G(k)$ as a Lie group, or any morphism $G(k) \to H(k)$ of Lie groups, where $H$ is another semisimple $k$-group, is automatically rational. In that case, continuity already implies analyticity and rationality.

**6.3.** The next case, chronologically, of the automorphisms problem pertains to compact connected semisimple groups. It was studied, independently and practically simultaneously, by Cartan [C1], 1930, and van der Waerden [W0], 1933. We start with [C1].

The main result of [C1] is the following. Let $G$ be a compact connected semisimple Lie group, $H$ a Lie group and $\varphi : G \to H$ a homomorphism of abstract groups, whose image is bounded. Then $\varphi$ is continuous.

Clearly this implies that any automorphism of $G$ or any bounded linear representation of $G$ is continuous. The most important step is the case where $G = \mathbf{SO}_3$, to which the major part of the paper is devoted. Again, an essential point is a reduction to the FTPG, this time for the real projective plane.

Let $\Gamma$ be the closure of $\varphi(G)$. It is shown to be a compact connected semisimple group with center reduced to $\{1\}$. In $G$, the elements of order 2 form one conjugacy class, say $\mathcal{A}$, the set of $180^o$ rotations around an axis. The proof proceeds with a careful analysis of the conjugacy class $\mathcal{E}$ of elements of order 2 containing $\varphi(\mathcal{A})$. It is a symmetric space for $\Gamma$, and Cartan makes full use of the theory of such spaces, which he had developed in the late twenties (see Chapter IV). Eventually, he shows that $\Gamma$ is a product of copies of $G$, which lets him reduce easily to the case where $\Gamma$ is isomorphic to $G$. Now view $\Gamma$ as the group of proper rotations of a three-dimensional euclidean space $V$. Then the elements of $\mathcal{E}$ are the $180^o$ rotations around some axis going through the origin. Therefore $\varphi$ induces a map from the 1-dimensional subspaces in $\mathbb{R}^3$ to the one-dimensional subspaces in $V$, i.e. from $\mathbf{P}_2(\mathbb{R})$ to the real projective plane $P(V)$ of lines through the origin of $V$. Cartan shows that two distinct (resp. orthogonal) lines go into distinct (resp. orthogonal) lines and that three coplanar lines go into coplanar lines. Hence $\varphi$ yields a *collineation* of $\mathbf{P}_2(\mathbb{R})$ onto $P(V)$. Since $\mathbb{R}$ has no non-trivial automorphism, this collineation is a projective transformation by the FTPG; in particular, it is continuous. This implies the theorem when $G = \mathbf{SO}_3$. Following this, Cartan takes care of the case where $G = \mathbf{SU}_2$, so that the theorem is now proved when $G$ is three-dimensional. In general, the structure theory of compact semisimple groups shows that any such group is generated by three-dimensional simple groups, on which the restriction of $\varphi$ is now known to be continuous, and the theorem follows.

**6.4.** The results of [W0] are similar, slightly sharper, but are obtained from a completely different point of view. Let $G$ be a connected Lie group with discrete center, and $n$ the dimension of $G$. Given $a \in G$, not central, consider the set of products

$$(1) \qquad G(a) = \prod_{j=1}^{n} c_j(b_j.a.b_j^{-1}.a^{-1}).c_j^{-1},$$

where the $b_j$ and $c_j$ run through $G$, or, as the author points out later, only through an arbitrary neighborhood of the identity. Let $\mathcal{M}$ be the set of $G(a)$ ($a \in G$, $a \notin CG$). Van der Waerden proves the following two facts:

a) If $G$ is simple as a real Lie group and $n \geq 2$, then $\mathcal{M}$ contains a neighborhood of the identity.

Hans Freudenthal

b) If $G$ is compact, then, for suitable $a$, $G(a)$ is contained in a prescribed neighborhood of the identity.

From a) he deduces first that if $G$ is simple non-abelian (as a Lie group), the quotient of $G$ by its center is simple as an abstract group. If now $G$ is compact connected semisimple, a) and b) show that $\mathcal{M}$ contains a fundamental system of neighborhoods of the identity. Since $\mathcal{M}$ is obviously invariant under any automorphism $\alpha$ of $G$, this proves that $\alpha$ is continuous.

A further consequence of b) is that if a linear representation of a simple non-abelian compact Lie group $G$ is bounded around the identity, then it is continuous. The author carries out the (easy) extension of this result to any compact connected semisimple group.

**6.5.** In [F], 1941, H. Freudenthal studies isomorphisms of non-compact real semisimple groups. He first refers to an example of a discontinuous automorphism of $\mathbf{SL}_2(\mathbb{C})$ due to von Neumann, pointed out to van der Waerden, mentioned in [W0], obtained by applying a discontinuous automorphism of $\mathbb{C}$ to the coefficients of the elements of the group. (Apparently, this is the first example in the literature of field automorphism 6.0.(b).) However, he points out that $\mathbf{SL}_2(\mathbb{C})$, viewed as a 6-dimensional real Lie group, has a complexification which is not simple (in fact a product of two copies of $\mathbf{SL}_2(\mathbb{C})$). He then considers a simple real Lie group (of dimension $\geq 3$) which is *absolutely simple*, i.e. the complexification of its Lie algebra $\mathfrak{g}$ remains simple as a complex Lie algebra, and he shows that, under that assumption, any automorphism of $G$ is continuous, or rather, slightly more generally, that any abstract isomorphism of $G$ onto a real Lie group is continuous. I shall first assume that $CG = \{1\}$. Freudenthal's goal, following [W0], is to give an abstract group description of a fundamental set of neighborhoods of the identity. In fact, his subsets are those of [W0], but it is much harder to establish that they have that property.

To start with, he shows that if $H$ is a subgroup of a Lie group $L$, each element of which is connected to the identity by an analytic arc in $H$, then $H$ is a Lie subgroup (a result which H. Yamabe was to prove later for any arcwise connected subgroup, Osaka Math. J. **2**, 1950, 13-14). From this he deduces that a Lie group which is simple as a Lie group is simple modulo its center as an abstract group (as was known earlier in [W0], cf. 6.4).

A Cartan subalgebra $\mathfrak{h}$ of $\mathfrak{g}$ is a maximal commutative subalgebra consisting of semisimple elements. Its complexification $\mathfrak{h}_c = \mathfrak{h} \otimes_{\mathbb{R}} \mathbb{C}$ is a Cartan subalgebra of $\mathfrak{g}_c = \mathfrak{g} \otimes_{\mathbb{R}} \mathbb{C}$. The algebra $\mathfrak{g}_c$ is the Lie algebra of the complexification $G_c$ of $G$. We can view $G_c$ as an algebraic group defined over $\mathbb{R}$, and then $G$ is the identity component in the usual topology of $G_c(\mathbb{R})$. However, this is not the point of view of Freudenthal, who views $G$ as a real Lie group, and wants to prove that the abstract group structure determines the Lie group structure; so we shall pursue this discussion in the Lie theoretic framework of [F]. Freudenthal uses the term *"regular group"* for the connected subgroup $H$ of $G$ with Lie algebra a Cartan subalgebra $\mathfrak{h}$. It is a connected commutative Lie group, hence the direct product of a topological torus (product of circle groups) $H_{co}$ by a connected diagonalisable group $H_{sp}$, to be called respectively the compact and the split part of $H$. Their Lie algebras are denoted $\mathfrak{h}_{co}$ and $\mathfrak{h}_{sp}$. The dimension of $\mathfrak{h}$ depends only on $\mathfrak{g}$, but those of $\mathfrak{h}_{co}$ and $\mathfrak{h}_{sp}$ may vary with $\mathfrak{h}$.

The next step (Satz 2) provides a group theoretic description of regular groups, by the following conditions: (a) commutative, infinitely divisible, maximal for these properties; and (b) any subgroup of countable index is of countable index in its normalizer [(b) is a precursor of the condition used by Chevalley in his characterization of Cartan subgroups of linear algebraic groups (see VII, §9)].

The most difficult step is to construct group theoretically a compact neighborhood of the identity. This makes full use of the root structure of $\mathfrak{g}_c$ with respect to $\mathfrak{h}_c$ (see 2.2). As is known, one can choose a basis $x_\alpha$ of $\mathfrak{g}_\alpha$ so that $h_\alpha = [x_\alpha, x_{-\alpha}]$ is characterized by $\beta(h_\alpha) = [h_\alpha, x_\beta] = (\beta, \alpha)x_\beta$. More generally, for $\lambda \in \mathfrak{h}^*$, let $h_\lambda \in \mathfrak{h}$ be defined by the condition $\beta(h_\lambda) = (\beta, \lambda)$ $(\beta \in \mathfrak{h}^*)$.

The set of restrictions of roots to $\mathfrak{h}$ is invariant under complex conjugation. Given a root $\alpha$, Freudenthal considers the intersection, to be denoted here $\mathfrak{h}_{(\alpha)}$, of $\mathfrak{h}$ with the subalgebra of $\mathfrak{g}_c$ generated by $x_\alpha, x_{\bar\alpha}, x_{-\alpha}, x_{-\bar\alpha}$. Let $H_{(\alpha)}$ be the corresponding group. The root $\alpha$ is said to be real if $\alpha = \bar\alpha$, imaginary if $\bar\alpha = -\alpha$, and complex otherwise. The Lie algebra $\mathfrak{h}_{(\alpha)}$ is one dimensional, contained in $\mathfrak{h}_p$ if $\alpha$ is real, in $\mathfrak{h}_{co}$ if $\lambda$ is imaginary, and is two-dimensional if $\alpha$ is complex. Let $H_{(\alpha)}$ be the subgroup of $H$ with Lie algebra $\mathfrak{h}_{(\alpha)}$. Starting from these, Freudenthal constructs a family of subgroups of $H$, containing the $H_{(\alpha)}$, which he shows to be defined in group theoretical terms, and he considers a minimal element $L$. It is one-dimensional, contained either in $H_{co}$ or $H_{sp}$. If $L$ is in $H_{co}$ it is compact. If not, it acts on some group $\exp \mathbb{R}.x_\alpha$ ($\alpha$ real) and it is possible, using this action, to define group theoretically a compact neighborhood of the identity in $L$. Starting with $L$ in the compact case, with a compact neighborhood of the identity in the other case, using conjugates and commutators, he then defines a compact neighborhood $U$ of 1 in abstract group theoretical terms. He then considers the same sets 6.5(1) as van der Waerden, where the $b_j$ and $c_j$ run through $U$. Van der Waerden's argument yields the theorem for $G$ absolutely simple with center reduced to $\{1\}$ if $G = H$. To handle the case where $H$ is isomorphic to $G$ as an abstract group, Freudenthal needs, and establishes, a group theoretical definition of "absolutely simple". This then proves the main theorem when $G$ is absolutely simple with center reduced to 1. The case of a non-trivial center follows easily. Freudenthal also remarks that the case of a semisimple group, all of whose simple factors are absolutely simple, follows as in [W0].

**6.6.** The four papers surveyed above fall into two categories: In one ([SW], and [C1] for $\mathbf{SO}_3$), a reduction to the FTPG is carried out using the standard representation of the group, while in the other ([W0] and [F]), the proof relies on a study of structural properties of the group. Both directions were pursued further. In this and the next subsection, I shall point out some key points of these developments, without aiming at completeness, or trying to cover the latest results.

The first approach was carried out and developed much further for all classical groups, over almost arbitrary fields. A thorough survey of work done up to 1962 is given in [D].

As before, the general approach is to use the standard, defining, representation of the group, and to relate elements of the group, characterized abstractly, to subspaces or geometric configurations attached to the given representation space. I shall not try to duplicate Dieudonné's survey, and shall limit myself to indicating in some examples what kind of elements and geometric objects are involved.

a) $G = \mathbf{GL}_n(k)$. Let $s$ be an involution in $G$. It leaves invariant two subspaces, of complementary dimensions $a, b$ $(a + b = n)$, which span $k^n$, in which case $s$ is said to be of type $(a, b)$. The integer $a$ can be characterized by the fact that $s$ belongs to a maximal commutative family of conjugate involutions of cardinality $\binom{n}{a}$. This in particular singles out the involutions for which $a = 1$, called *extremal involutions*. An automorphism of $G$ then yields a permutation of extremal involutions and eventually a collineation or correlation of $\mathbf{P}_{n-1}(k)$.

b) $G = \mathbf{Sp}_{2m}(k)$ and char $k \neq 2$. The starting points are the involutions of type $(2, 2m - 2)$, called extremal. Two extremal involutions form a *minimal pair* if the intersection of their invariant 2-dimensional subspaces is one-dimensional. It is then shown that every automorphism permutes minimal pairs. Given a one-dimensional subspace $D$ of $k^{2m}$, let $I(D)$ be the set of extremal involutions whose 2-dimensional invariant subspaces contain $D$. It is proved that any automorphism of $G$ permutes the $I(D)$, whence we also get a permutation of the lines in $k^{2m}$.

c) $G = O(f)(k)$ is the orthogonal group on $k^n$ with respect to a non-degenerate quadratic form $f$, and char $k \neq 2$. There is first a characterization of the symmetries with respect to hyperplanes which are not isotropic (i.e., to which the restriction of $f$ is non-degenerate). It is obtained by counting the number of involutions commuting with certain families of involutions. Therefore an automorphism $\alpha$ of $G$ yields a permutation of non-isotropic one-dimensional subspaces. This provides a reduction to the FTPG if $f$ is anisotropic $\big(f(x) \neq 0$ if $x \neq 0\big)$. In the case where $f$ is isotropic $\big(f(x) = 0$ for some $x \neq 0\big)$, the task is to extend that permutation to isotropic lines. Given one, say $D$, let $I(D)$ be the set of non-isotropic planes containing it. One can then show that there exists an isotropic line $D'$ such that

$\alpha$ maps $I(D)$ onto $I(D')$. This extends the correspondence and eventually allows one to use the FTPG also in that case.

Dieudonné's survey also includes unitary groups and special or projective versions of the classical groups. In a number of cases, special considerations are needed in small dimensions, for some small fields and in characteristic two. Classical groups over non-commutative fields are also included.

**6.7.** This direction was pursued by O. T. O'Meara and his school, who considered more generally automorphisms, or isomorphisms, of certain types of subgroups of $G(k)$, the goal being always to show that such automorphisms have the standard form (6.0). One series of works is concerned with the group $G(J)$, where $J$ is an integral domain and $G$ is the general or special linear group, or the symplectic group. Dedekind rings were also considered. We refer to [T3] for a summary and list of references. Another type of subgroups of $\mathbf{GL}_n(k)$ for which a similar conclusion was established is those which are "full of transvections", i.e. contain at least one transvection associated to any given pair consisting of a hyperplane and a line in that plane, both containing the origin. An analogous theorem was also proved in the symplectic case. This was also extended to the case of isomorphisms between subgroups full of transvections of two groups $G(k)$ and $G'(k')$, both either general linear or symplectic [OM 1], [OM 2].

**6.8.** This approach was essentially case by case. In his foreword to [D], Dieudonné referred to the recent work of Chevalley on semisimple algebraic groups, and said he expected that there would eventually be general methods to treat isomorphisms uniformly, including all the cases he had considered.

A step in that direction is made in [BT2], 1973, where such a program is carried out for *isotropic* absolutely simple groups over an arbitrary infinite field, and for a more general problem, namely of homomorphisms $H \to G'(k')$, where $G'$ is semisimple and $H$ is a subgroup of $G(k)$ containing a normal subgroup $G(k)^+$ to be defined below.

To state the result, or at any rate a sample result of [BT2], we need the notion of base change. As before, let $L$ be a $k$-group, $k'$ a field and $\varphi : k \to k'$ a homomorphism. Then one can define a $k'$-group $^\varphi L$ and a canonical homomorphism $\varphi^c : L(k) \to {}^\varphi L(k')$. [View $L$ as a matrix group. Then $^\varphi L$ is the $k'$-group whose equations are obtained by applying $\varphi$ to those defining $L$. The elements of $^\varphi L(k')$ are the matrices $(\varphi(g_{ij}))$, where $g = (g_{ij})$ runs through $L(k)$ and $\varphi^o$ is the homomorphism

$$g = (g_{ij}) \mapsto (\varphi(g_{ij})).]$$

Assume now that $L$ is simple over $k$, and isotropic. Then it has proper parabolic $k$-subgroups. The group $L(k)^+$ is, by definition, generated by the subgroup $U(k)$, where $U$ runs through the unipotent radicals of the parabolic $k$-subgroups of $L$. It is perfect, and normal in $L(k)$. If $L$ is simply connected and split over $k$, then the subgroup $L(k)^+$ is equal to $L(k)$, but in general it is a proper subgroup.

A special case of the main theorem of [BT2] is then the following: Let $G$ be absolutely simple over $k$, let $G'$ be a simple $k'$-group, $H$ a subgroup of $G(k)$ containing $G(k)^+$ and $\alpha : H \to G'(k')$ a homomorphism whose image is Zariski dense. Assume either $G$ simply connected or $G'$ of adjoint type. Then there exist a homomorphism $\varphi : k \to k'$, a $k$-isogeny $\beta : {}^\varphi G \to G'$ with non-zero differential, and a homomorphism $\gamma : H \to \mathcal{C}G'(k')$ such that $x(h) = \gamma(h)\, \beta(\varphi^0(h))\ (h \in H)$ (see [BT2] §8 for more general results).

This type of homomorphism is the natural extension to homomorphisms of the standard isomorphisms of 6.0.

The result extends, *mutatis mutandis*, to semisimple $G'$. In order to formulate it, we have to use one further notion, that of *restriction of scalars*. Let $L$ be a $k$-group. Assume it is simple over $k$, but not absolutely simple. Then there exist a finite separable extension $k'$ of $k$ and a $k'$-group $L'$ such that $L = \mathcal{R}_{k'/k}L'$ is obtained from $L'$ by restriction of scalars. This notion, introduced by A. Weil, is an extension of the idea of viewing an $n$-dimensional vector space $V$ over $k'$ as a $d.n$-dimensional vector space over $k$, where $d$ is the degree of $k'$ over $k$. In particular, there is a natural isomorphism $L(k) = L'(k')$.

Now assume $G'$ to be $k$-simple, but not absolutely $k'$-simple. We can write $G' = \mathcal{R}_{\ell.k'}G''$, where $\ell$ is a finite separable extension of $k'$ and $G''$ an absolutely simple $\ell$-group. In particular, $G'(k') = G''(\ell)$. The result is then very similar, $\beta(\varphi(k))$ being replaced by $\beta(\mathcal{R}_{\ell/k'}(\varphi^0(k)))$. If $G'$ is semisimple, then the isogeny factor is replaced by a product of factors corresponding to the $k'$-simple factors of $G'$ ([BT2], 8.11).

An incentive to consider homomorphisms was to prove a conjecture of R. Steinberg's on an "abstract" projective representation of $H$ whose restriction to $G(k)^+$ is absolutely irreducible, and where $k'$ is algebraically closed. It is obtained by applying the previous result to the case where $G'$ is a projective group.

If $k$ and $k'$ are locally compact non-discrete fields, then $G(k)$ and $G'(k')$, endowed respectively with the $k$- and $k'$-topology, are topological groups, and the problem is to know whether an abstract homomorphism is continuous. This is handled in §9 of [BT2]. In particular, it includes the theorem of Freudenthal (see 6.6) and extends it to other fields.

The general idea in [BT2] is to start from opposite parabolic $k$-subgroups in $G$ and try to show that the images of their rational points have similar properties in $G'(k')$. The proof is long and technical, in particular when $k$ and $k'$ are non-perfect fields of characteristic two, the treatment of which substantially increases the length of the paper. We shall not attempt to give more details.

**6.9.** *Automorphisms of buildings.* As we saw, the proofs of the first results on automorphisms ([SW] and [C1]) depend very much on the FTPG. Recall that a collineation of $\mathbf{P}_n(k)$ may be viewed as an automorphism of the Tits building of $\mathbf{SL}_{n+1}(k)$ which preserves the types, and a correlation as one which reverses the types. The FTPG then says that any automorphism of the building is induced by a semi-linear isomorphism of $k^{n+1}$ onto itself or onto its dual.

The results of [BT2] imply a converse ([T4], Theorem 5.8), also valid for isomorphisms. For $i = 1, 2$, let $G_i$ be an absolutely simple group over a field $k_i$ of adjoint type and of $k_i$-rank $\geq 2$. Let $\mathcal{T}_i$ be the Tits building of parabolic $k$-subgroups of $G_i$. Let $\pi$ be an order preserving bijection of $\mathcal{T}_1$ onto $\mathcal{T}_2$. Then there exist an isomorphism $\sigma$ of $k$ onto $k'$ and a special isogeny of $G_1$ onto $G_2$ inducing $\pi$.

This also generalizes various analogues of the FTPG for many classical groups due to W.L. Chow (see [D], §§2, 3, 4).

**6.10.** As has been pointed out, the big difference between [BT2] and the results surveyed in [D] is that the proofs are case by case in [D], but uniform in [BT2]. However, there is a drawback in [BT2]: It is restricted to isotropic groups, while [D] handles both isotropic and anisotropic ones. This of course makes it likely that the main results of [BT2] have a counterpart for anisotropic groups. Whether this is really the case is still an open question, as far as I know. The most substantial result in that direction has been obtained by B. Weisfeiler in [W5], 1981, the last of several papers by him on the subject. The main result of [BT2] is extended to the case where $G$ and $G'$ are anisotropic, but become split groups over a quadratic extension of the ground field. See [HJW] for an extensive survey of this topic.

## §7. Merger

The transcendental and algebraico-geometric approaches to Lie theory proceeded rather independently for a long time, but interesting relations between the two emerged, grew, and led to a considerable increase in the scope of the theory. Except for some brief indications and references below, I shall not attempt a survey and shall limit myself to a first example relating the Cartan root system and Weyl group of a negatively curved symmetric space (IV, 2.4.3, 2.4.4) to the relative root system and Weyl group in the sense of VI, 4.6.

**7.1.** Let $G$ be a connected absolutely simple non-compact Lie group in adjoint form, $K$ a maximal compact subgroup, and $\mathfrak{g} = \mathfrak{k} \oplus \mathfrak{p}$ the corresponding Cartan decomposition. The space $\mathfrak{p}$ may be identified to the tangent space $T(G/K)_o$ at the origin of the symmetric space $G/K$, of which a model is the "space $\mathcal{E}$", $P = \exp \mathfrak{p}$.

Let $\mathfrak{a}$ be a maximal commutative subalgebra of $\mathfrak{p}$. Then $A = \exp \mathfrak{a}$ is a *maximal flat*, i.e. a totally geodesic submanifold of $P$, isometric to euclidean space and maximal for that property. $G$ operates transitively on the set of maximal flats, and $K$ is transitive on those which contain the origin and whose tangent space at the origin belongs to $\mathfrak{p}$. The centralizer $\mathfrak{z}(\mathfrak{a})$ of $\mathfrak{a}$ in $G$ is the direct product of $A$ and its intersection $M$ with $K$. The roots of $\mathfrak{g}$ with respect to $\mathfrak{a}$ are the non-zero weights of the adjoint representation of $\mathfrak{a}$ in $\mathfrak{g}$. They form an irreducible root system $\Phi(\mathfrak{a}, \mathfrak{g})$, and its Weyl group $W(\mathfrak{a}, \mathfrak{g})$ may be identified with $\mathcal{N}_K(\mathfrak{a})/\mathcal{Z}_K(\mathfrak{a})$. But, in fact, $\mathcal{N}_K(a)$ is also the full normalizer of $\mathfrak{a}$ in $G$.

**7.2.** We claim that $\Phi(\mathfrak{a}, \mathfrak{g})$ can also be viewed as a relative root system $_k\Phi$, in the sense of VI, 4.6, where $k = \mathbb{R}$.

To see this, first note that $G = \mathcal{G}(\mathbb{R})^o$ is the identity component, in ordinary topology, of the group of real points of an algebraic group $\mathcal{G}$ defined over $\mathbb{R}$, namely the group Aut $\mathfrak{g}_c$ of automorphisms of the complexification of $\mathfrak{g}$. Coordinates on $\mathfrak{g}$ were chosen so that $\mathfrak{k}$ (resp. $\mathfrak{p}$) was represented in the adjoint representation by skew-symmetric (resp. symmetric) matrices (IV, 2.4.4). Therefore $A$ is a commutative group of real symmetric matrices, hence diagonalizable over $\mathbb{R}$. It is not difficult to see that $A = S(\mathbb{R})^o$ is the identity component, in ordinary topology, of the group of real points of a maximal $\mathbb{R}$-split torus $S$ of $\mathcal{G}$. Then the conjugacy of maximal flats containing the origin is related to the conjugacy under $G$ of the maximal $\mathbb{R}$-split tori. $\Phi(\mathfrak{a}, \mathfrak{g})$ may be identified to the relative root system $_\mathbb{R}\Phi(S, \mathcal{G})$. Also, the normalizers (resp. centralizers) in $G$ of $\mathfrak{a}$ and $S$ are the same; hence so are the Weyl groups in both theories.

In short, the group $G$ is on one hand a real Lie group, in the topology inherited from that of the real numbers, and on the other a real algebraic group (or of finite index in one), endowed with the Zariski topology. The definition of $_\mathbb{R}\Phi(S, \mathcal{G})$ uses the latter, while the symmetric space interpretation is based on the former, as is the study of maximal compact subgroups.

**7.3.** This comparison is certainly suggestive, but does not bring really new results. However, it did for groups over other local fields, in particular $p$-adic fields. Let $k$ be a $p$-adic field. If $\mathcal{G}$ is an algebraic group defined over $k$, then $\mathcal{G}(k)$

is of course endowed with the relative Zariski topology, but also with the ordinary topology defined by that of $k$, and as such it is a Lie group over $k$. For semisimple (or reductive) $\mathcal{G}$, Bruhat and Tits, generalizing earlier work of Iwahori and Matsumoto, developed a striking theory of maximal compact subgroups of $\mathcal{G}(k)$, based on the use of a simplicial complex, a building in the sense of §5, but where the apartments are euclidean spaces and the Weyl groups are infinite reflection groups, which plays a role remarkably similar to that of the symmetric spaces of maximal compact subgroups in the real case. Besides the Bruhat-Tits papers, Inst. Hautes Études Sci. Publ. Math. **41** (1972), 5-252; **60** (1984), 5-184; J. Fac. Sci. Univ. Tokyo **34** (1987), 671-688, see the books by K. Brown, Buildings, Springer, 1989, P. Garrett, Buildings and classical groups, Chapman and Hall, 1997, or M. Ronan, Lectures on buildings, Academic Press, 1989.

# References for Chapter VI

[B1]    I. Barsotti, *Un teorema di struttura per le varietà gruppali*, Atti Accad. Naz. Lincei Rend Cl. Sci. Fis. Mat. Nat. (8) **18** (1955), 43–50.

[B2]    A. Borel, *Représentations linéaires et espaces homogènes kähleriens des groupes de Lie compacts*, manuscript, 1954; first published in his Oeuvres/Collected Papers I, 392–396.

[B3]    _____, *Kählerian coset spaces of semi-simple Lie groups*, Proc. Nat. Acad. Sci. USA **40** (1954), 1147–1151; Oeuvres I, 397–401.

[B4]    _____, *Groupes linéaires algébriques*, Annals of Math. (2) **64** (1956), 20–82, Oeuvres I, 490–552.

[B5]    _____, Linear algebraic groups, 2nd ed., GTM **126**, Springer, 1991.

[BS]    A. Borel et J-P. Serre, *Théorèmes de finitude en chomologie galoisienne*, Comment. Math. Helv. **39** (1964), 111–164; Borel, Oeuvres II, 362–415.

[BS1]    A. Borel and T.A. Springer, *Rationality properties of linear algebraic groups*, Proc. Sympos. Pure Math. **9**, Amer. Math. Soc., 1966, 26–32; Borel, Oeuvres II, 697–703.

[BS2]    _____, *Rationality properties of linear algebraic groups II*, Tôhoku Math. J. (2) **20** (1968), 443–497; Borel, Oeuvres II, 720–774.

[BT]    A. Borel et J. Tits, *Groupes réductifs*, Inst. Hautes Études Sci. Publ. Math. **27** (1965), 55–150; *Compléments*, ibid. **41** (1972), 253–276; Borel, Oeuvres II, 424–520; III 129–152.

[BT2]    _____, *Homomorphismes "abstraits" de groupes algébriques simples*, Annals of Math. (2) **97** (1973), 499–571; Borel, Oeuvres III, 171–243.

[B6]    N. Bourbaki, Groupes et algèbres de Lie, Chaps. 4, 5, 6, Masson, Paris, 1981.

[C0]    É. Cartan, *Les groupes réels simples, finis et continus*, Ann. Sci. École Norm. Sup. (3) **31** (1914), 263–355; Oeuvres Complètes I$_1$, 399–491.

[C1]    _____, *Sur les représentations linéaires des groupes clos*, Comment. Math. Helv. **2** (1930), 269–283; Oeuvres Complètes I$_2$, 1149–1163.

[C2]    _____, *La théorie des groupes finis et continus et l'analysis situs*, Mém. Sci. Math. XLII, Gauthier-Villars, Paris, 1930; Oeuvres Complètes I$_2$, 1165–1224.

[C3]    C. Chevalley, *A new kind of relationship between matrices*, Amer. J. Math. **65** (1943), 521–531.

[C4]    _____, *Algebraic Lie algebras*, Ann. of Math. (2) **48** (1947), 91–100.

[C5]    _____, Théorie des groupes de Lie II. Groupes algébriques, Hermann, Paris, 1951.

[C6]    _____, Théorie des groupes de Lie III, Groupes algébriques, Hermann, Paris, 1954.

[C7]    _____, *Sur certains groupes simples*, Tôhoku Math. J. (2) **7** (1955), 14–66.

[C8]    _____, Classification des groupes algébriques, 2 vols., Mimeographed notes, Inst. H. Poincaré, 1956-58.

[C9]    _____, *Sur les décompositions cellulaires des espaces G/B*, in Algebraic Groups and Their Generalizations, Proc. Sympos. Pure Math. **56**, Part 1, Amer. Math. Soc., 1994, 1–23.

[CT1]    C. Chevalley and H.F. Tuan, *On algebraic Lie algebras*, Proc. Nat. Acad. Sci. USA **51** (1946), 195–196.

[CT2]    _____, *Algebraic Lie algebras and their invariants*, Acta Math. Sinica* **1** (1951), 215–242.

[D]    J. Dieudonné, La géométrie des groupes classiques, Springer, 1963.

[E]    C. Ehresmann, *Sur la topologie de certains espaces homogènes*, Ann. of Math. (2) **35** (1934), 396–443; Oeuvres Complètes et Commentées, 3–53.

[F]    H. Freudenthal, *Die Topologie der Lieschen Gruppen als algebraisches Phänomenon I*, Annals of Math. (2) **42** (1941), 1051–1074; erratum, ibid. **47** (1946), 829–830.

[G1]    M. Goto, *On replica of nilpotent matrices*, Proc. Japan Acad. **23** (1947), 39–41.

[G2]    _____, *On algebraic Lie algebras*, J. Math. Soc. Japan **1** (1948), 29–45.

[G3]    _____, *On algebraic homogeneous spaces*, Amer. J. Math. **76** (1954), 811–818.

[G4]    A. Grothendieck, Exposés XII, XIII, XIV in Séminaire de Géométrie Algébrique du Bois Marie 1962/63 (SGA3), Lecture Notes in Math. **152**, Springer, 1970.

[HJW]    A.J. Hahn, D.G. James and B. Weisfeiler, *Homomorphisms of algebraic and classical groups: a survey*, Canadian Math. Soc. Conf. Proc. **4** (1984), 249–296.

[HC]    Harish-Chandra, *On a lemma of F. Bruhat*, J. Math. Pures Appl. (9) **35** (1956), 203–210; Coll. Papers II, 223–230.

[H]    H. Hopf, *Ueber den Rang geschlossener Liescher Gruppen*, Comment. Math. Helv. **13** (1940-1), 119–143; Selecta, 152–174.

---

*Reviewed by Math. Reviews as "J. Chinese Math. Soc. (N.S.)".

[H1]    L.K. Hua, *On the automorphisms of the symplectic group over any field*, Ann. of Math. (2) **49** (1948), 739–759; Sel. Papers, 433–453.

[J]     J.-C. Jantzen, Representations of algebraic groups, Pure Appl. Math. **131**, Academic Press, 1987.

[K0]    F. Klein, *Vergleichende Betrachtungen über neuere geometrische Forschungen* (Erlangen, 1872), Math. Annalen **43** (1893), 63–100; Ges. Abh. I, 460–497.

[K1]    E. Kolchin, *Algebraic matric groupes and the Picard-Vessiot theory of homogeneous differential equations*, Ann. of Math. (2) **49** (1948), 1–42; Sel. Wks., 87–128.

[K2]    ———, *On certain concepts in the theory of algebraic matric groups*, Ann. of Math. (2) **49** (1948), 774–789; Sel. Wks., 129–144.

[K3]    ———, Differential algebra and algebraic groups, Academic Press, 1973.

[L]     S. Lang, *Algebraic groups over finite fields*, Amer. J. Math. **78** (1956), 555–563.

[M1]    Y. Matsushima, *Note on the replicas of matrices*, Proc. Japan Acad. **23** (1947), 42–49; Coll. Papers, 28–35.

[M2]    ———, *On algebraic Lie groups and algebras*, J. Math. Soc. Japan **1** (1948), 46–57; Coll. Papers, 39–50.

[OM1]   O.T. O'Meara, Lectures on linear groups, Conf. Board Math. Sci. Regional Conf. Ser. Math. **22**, Amer. Math. Soc., 1974.

[OM2]   ———, Symplectic groups, Math. Surveys **16**, Amer. Math. Soc., 1978.

[OM3]   ———, *General isomorphism theorem for linear groups*, J. Algebra **44** (1977), 93–142.

[R1]    M. Rosenlicht, *Some basic theorems on algebraic groups*, Amer. J. Math. **78** (1956), 401–443.

[R2]    ———, *Some rationality questions on algebraic groups*, Annali di Mat. Pura Appl. (4) **43** (1957), 27–50.

[R3]    ———, *Questions of rationality for solvable algebraic groups over nonperfect fields*, Annali di Mat. Pura Appl. (4) **61** (1963), 97–120.

[S1]    I. Satake, Classification theory of semi-simple algebraic groups, Lecture Notes in Pure Appl. Math. **3**, Marcel Dekker, 1971.

[SW]    O. Schreier und B.L. van der Waerden, *Die Automorphismen der projektiven Gruppen*, Abh. Math. Sem. Hamburg **6** (1928), 303–322.

[S2]    F. Schur, *Ueber den analytischen Charakter der eine endliche continuirliche Transformationsgruppe darstellenden Funktionen*, Math. Annalen **41** (1893), 503–538.

[S3]    J-P. Serre, *Représentations linéaires et espaces homogènes kählériens des groupes de Lie compacts*, Exp. **100**, Sém. Bourbaki 1953-54.

[S4]    ———, *Quelques properiétés des variétés abéliennes en charactéristique p*, Amer. J. Math. **80** (1958), 715–739; Oeuvres I, 544–568.

[S5]    ———, Cohomologie galoisienne, Lecture Notes in Math. **5**, 5ème éd., Springer, 1994.

[T1]    J. Tits, *Sur certaines classes d'espaces homogènes de groupes de Lie*, Mém. Acad. Roy. Belg. **29** (1955).

[T2]    ———, *Classification of algebraic semisimple groups*, Algebraic Groups and Discontinuous Subgroups, Proc. Sympos. Pure Math. **9**, Amer. Math. Soc., 1966, 33–62.

[T3]    ———, *Homomorphismes et automorphismes "abstraits" de groupes algébriques et arithmétiques*, Actes Internat. Congr. Math. (Nice, 1970), Vol. 2, Gauthier-Villars, Paris, 1971, 349–355.

[T4]    ———, Buildings of spherical type and finite BN-paires, Lecture Notes in Math. **386**, Springer, 1974.

[W0]    B.L. van der Waerden, *Stetigkeitssätze für halbeinfache Liesche Gruppen*, Math. Zeitschr. **36** (1933), 780–786.

[W1]    A. Weil, Variétés abéliennes et courbes algébriques, Actualités Sci. Indust. **1064**, Hermann, 1948.

[W2]    ———, *On algebraic groups of transformations*, Amer. J. Math. **77** (1955), 355–391; Oeuvres Sci. II, 197–233.

[W3]    ———, *On algebraic groups and homogeneous spaces*, Amer. J. Math **77** (1955), 493–512; Oeuvres Sci. II, 235–254.

[W4]    ———, *Algebras with involution and the classical groups*, J. Indian Math. Soc. **24** (1960), 589–623; Oeuvres Sci. II, 413–447.

[W5]    B. Weisfeiler, *Abstract isomorphisms of simple algebraic groups split by a quadratic extension*, J. Algebra **68** (1981), 335–368.

CHAPTER VII

# The Work of Chevalley
# in Lie Groups and Algebraic Groups

The first major research interests of Chevalley pertained to algebraic number theory, in particular class field theory. He had set for himself two main goals, to develop first a local class field theory, independently from the global theory, and second an algebraic treatment of the latter, free from analysis. His thesis [C1] realized the first one and made a first step towards the second one, which was achieved later in [C2] [C3]. It is on this occasion that he introduced the concept of idele, which soon became fundamental in algebraic number theory.

## §1. Lie groups, 1941–1946

**1.** In the thirties, the Julia Seminar in Paris, started in 1934 as a successor to the Hadamard Seminar, was a meeting ground for a number of younger mathematicians, including some of those who had founded Bourbaki in 1934. At the suggestion of A. Weil, it was agreed to devote it in 1936-37 to the work of Élie Cartan on Lie groups and Lie algebras. Chevalley participated, giving one lecture on the representation theory of Lie algebras. This seminar was the starting point of his interest in Lie groups, which soon grew to a major one. Already in 1938, some Bourbaki projects included a report by Chevalley on Lie groups, which was to follow one by C. Ehresmann on manifolds, Chevalley promising to "algebraize it to death". Soon the war came, Bourbaki dispersed, Chevalley went to Princeton, and these plans were left in abeyance as far as Bourbaki was concerned, but Chevalley apparently decided to pursue them in part on his own. In 1941 he published his first two papers on Lie groups. In one, [C4], he determined the topological structure of a connected solvable group, showing, in analogy with a well-known result of Cartan on semisimple Lie groups, that a connected solvable group is topologically (or even differentiably) isomorphic to the product of a compact subgroup (a torus) by a euclidean space. Later A. Malcev and K. Iwasawa proved independently a similar result (and the conjugacy of maximal compact subgroups) for a general connected Lie group. (The paper of Iwasawa [I] also introduces what is now known as the *Iwasawa decomposition*, but Iwasawa himself points out that his original proof was valid only for complex semisimple groups and that the one he gives is due to Chevalley.)

**2.** The paper [C5] is much more a harbinger of the future. Let $\mathfrak{g}$ be a complex Lie algebra and $X$ a regular element (i.e. the multiplicity of the eigenvalue 0 of $\operatorname{ad} X$ is the smallest possible for $X \in \mathfrak{g}, X \neq 0$), and let $\mathfrak{n}_0$ be the nilspace of $\operatorname{ad} X$ (the space of elements of $\mathfrak{g}$ annihilated by some power of $\operatorname{ad} X$). It is a nilpotent

---

Modified and updated version of a paper published under the same title in the Proceedings of the Hyderabad Conference on Algebraic Groups, Manoj Prakashan, Madras 1991, 1-22.

Claude Chevalley

subalgebra which plays a basic role in Cartan's study of $\mathfrak{g}$. Chevalley proposed calling it a *Cartan subalgebra* of $\mathfrak{g}$, and his main result in that paper is the conjugacy of Cartan subalgebras. It had already been proved for $\mathfrak{g}$ semisimple by Weyl in his Math. Zeitschrift papers [Wy1]. There the conjugacy was under the adjoint group $\mathrm{Ad}\,\mathfrak{g}$ of inner automorphisms of $\mathfrak{g}$. However, Chevalley assumes more generally that the ground field is algebraically closed, of characteristic zero, so that he does not have a ready-made analogue of the adjoint group to perform the conjugacy. Let $n$ and $\ell$ be the dimension of $\mathfrak{g}$ and $\mathfrak{n}_0$ respectively. In the Grassmannian of $\ell$-planes in $\mathfrak{g}$, Chevalley constructs an irreducible variety $\mathcal{N}$, stable under any automorphism of $\mathfrak{g}$, containing a Zariski-open subset whose points represent the Cartan subalgebras of $\mathfrak{g}$. The generalized eigenspaces of $\mathrm{ad}\,X$ corresponding to the non-zero eigenvalues consist of nilpotent elements. If $Y$ belongs to one of them (or is any nilpotent element of $\mathfrak{g}$, for that matter), then the exponential series

$$(1) \qquad\qquad \exp \mathrm{ad}\,Y = \sum_{k \geq 0} (k!)^{-1}.(\mathrm{ad}\,Y)^k$$

is a finite sum and represents an automorphism of $\mathfrak{g}$. Using a functional determinant argument, Chevalley shows that $\mathcal{N}$ has dimension $n-\ell$ and that the conjugates of $\mathfrak{n}_0$ in $\mathcal{N}$ under products of such elements fill a Zariski-open subset. The conjugacy then follows from the fact that any two non-empty Zariski open subsets of $\mathcal{N}$ have a non-empty intersection.[1] Therefore conjugacy is achieved under the group generated by the elements (1) for $Y$ nilpotent. Chevalley himself saw in this argument the beginning of his work in algebraic geometry. It steers away from the traditional analytic approach to Lie groups, towards an algebraico-geometric setting, so it is hardly surprising that he soon turned his attention to linear algebraic groups.

Chevalley's published work on Lie groups and algebraic groups extends from 1941 to 1961. It divides rather naturally into two parts, with 1954 as a cutoff date. The second period is devoted entirely to algebraic groups, and consists of a few landmark papers. The first period is more tentative. In it, Lie groups and algebraic groups are closely intertwined, the former playing a dominant role at first, but gradually receding into the background, in favor of algebraic groups, which became more and more the primary objects. In this account, I shall try to follow the chronological order, but since Chevalley often pursued various goals concurrently, in different directions, I will at times have to deviate from it and to backtrack.

**3.** Chevalley's first major achievement in the direction of Lie groups is his Princeton book [C10], to be referred to as Lie I, which became quickly and for many years the basic reference, not only for Lie groups, but also for some foundational material on real analytic manifolds. The latter included in particular a new presentation of tangent spaces, exterior differential forms, Cartan's exterior differential calculus, and a global discussion of integral manifolds of a completely integrable system of Pfaffian equations. Although there are hardly any forerunners in the literature, it should not be surmised that all of it was due to him. It is rather likely that some of this material had emerged in discussions and writings within Bourbaki. But the strong algebraic flavor bears his imprint; in particular, the definition of a tangent vector at a point as a derivation of the local ring at that point is most probably his. The exposition of Lie groups is systematically global. The theory of Lie groups had taken a global turn from Weyl's papers [Wy1] on, and there had been some expositions of various global aspects of the theory, in particular in [Ca3], [Ca4], and the last chapters of Pontryagin's book [P], but this was the first textbook exposition. The last chapter contains a proof of the Peter-Weyl theorem and also a version of the Tannaka duality which offers a link between Lie groups and algebraic groups: Given a compact Lie group $G$, Chevalley introduces the "representative ring" $\mathcal{R}(G)$ of $G$, i.e. the ring of complex valued functions on $G$ generated by the coefficients of the finite dimensional representations of $G$. He then shows that the set of homomorphisms of $\mathcal{R}(G)$ into $\mathbb{C}$ (the maximal spectrum of $\mathcal{R}(G)$) admits a natural structure of complex Lie group $G_c$ containing $G$ as a maximal compact subgroup. $G$ appears as the fixed point set of an involution of $G_c$ (for the underlying real group structure) defined by complex conjugation. He also recovers the decomposition of $G_c$ as the product of $G$ by a euclidean space.

---

[1] For brevity, I use the language of the Zariski topology, though it was not available at the time.

## §2. Linear algebraic groups, 1943–1951

**4.** During these years, Chevalley had also been involved with linear algebraic groups. To put his work into context, let me first say a few words about earlier contributions to this topic, all dating from the nineteenth century and due to L. Maurer and E. Picard (except for two papers, Cartan [Ca1] and G. Fano [F], which were not known to Chevalley). (See Chapter V.) Picard had chiefly in mind a Galois theory of linear differential equations, in which the Galois group would be a linear algebraic group, whence his interest in those, to which he devoted some short papers and a chapter in Vol. 3 of his Traité d'Analyse. This led to the Picard-Vessiot theory and later to the work of Ritt and Kolchin (see Chapter VIII). Of more direct interest for us is the work of Maurer (essentially four papers published between 1888 and 1894, the last one [M] being the main one). Initially, Maurer had considered a subgroup $G$ of $\mathbf{GL}_n(\mathbb{C})$ defined by one invariant, i.e. consisting of all the invertible linear transformations of $\mathbb{C}^n$ leaving invariant a given rational homogeneous function on $\mathbb{C}^n$. This is a special case of a linear algebraic group, and Maurer soon realized he might as well study any such group, which he called a "regular group", whether it was characterized by one invariant or not. Given a linear algebraic group $G$, Maurer first investigated the Lie algebra $\mathfrak{g}$ of $G$. If $X \in \mathfrak{g}$, and

$$(2) \qquad X = X_s + X_n \quad (X_s \text{ semisimple }, X_n \text{ nilpotent }, X_s.X_n = X_n.X_s)$$

is its Jordan decomposition, he noted first that $X_s, X_n$ also belong to $\mathfrak{g}$. Now let $X$ be semisimple, put in diagonal form, and let $\lambda_i$ $(1 \leq i \leq n)$ be its eigenvalues. Set

$$C(X) = \left\{ z = (z_i) \in \mathbb{Z}^n \,\middle|\, \sum_i z_i.\lambda_i = 0 \right\}$$

and let $\mathfrak{h}_X$ be the space of diagonal matrices $\mathrm{diag}(\mu_i)$, where $(\mu_i)$ runs through the $n$-tuples of complex numbers satisfying the relations

$$(3) \qquad\qquad \sum_i z_i.\mu_i = 0 \qquad \big( z = (z_i) \in C(X) \big).$$

This is a diagonal subalgebra having a basis consisting of matrices with integral eigenvalues. Maurer shows that $\mathfrak{h}_X$ also belongs to $\mathfrak{g}$. (In present day terminology, he had characterized the Lie algebra of the smallest (algebraic) torus in $\mathbf{GL}_n(\mathbb{C})$ whose Lie algebra contains $X$.) In a later part of the paper, he shows that $G$ is a rational variety (see V, §4), a point I shall come back to later.

However, Maurer's work was almost fifty years old, some proofs were not complete, and some natural questions had not been addressed, so another presentation was in order. Chevalley, first by himself, then in collaboration with H.-F. Tuan, proceeded to provide one and to generalize the theory to an arbitrary ground field of characteristic zero. To that effect he introduced in [C6] a new algebraic notion, that of a *replica* of a linear transformation $X$ of a finite dimensional vector space $V$ over a field $K$. Let $V_{p,q}$ be the tensor product of $p$ copies of $V$ and $q$ copies of its dual $V'$ $(p, q \in \mathbb{N})$. Extended as a derivation of the tensor algebra of $V$, the transformation $X$ defines an endomorphism $X_{p,q}$ of $V_{p,q}$. Then $Y \in \mathrm{End}\, V$ is a replica of $X$ if for any $p, q \in \mathbb{N}$, any element of $V_{p,q}$ annihilated by $X_{p,q}$ is also annihilated by $Y_{p,q}$. Now let $K$ be perfect. Then the Jordan decomposition (2) holds over $K$. It is proved that $X_s, X_n$ belong to the space $\mathrm{Rep}\, X$ of replicas of $X$

and that $\operatorname{Rep} X$ is the sum of $\operatorname{Rep} X_s$ and $\operatorname{Rep} X_n$. If $X$ is semisimple and diagonalized over some extension of $K$, Chevalley shows that $\operatorname{Rep} X$ is the space denoted $\mathfrak{h}_X$ above. If $K$ is of characteristic zero and $X$ is nilpotent, then $\operatorname{Rep} X = K.X$. Thus Maurer's result is equivalent to the statement that if $\mathfrak{g}$ is the Lie algebra of a complex linear algebraic group, then it contains the replicas of all of its elements. In [C9] Chevalley and Tuan announced a converse, the proof of which is given in [C16]. The converse had been formulated over $\mathbb{C}$ only because there was at that time no theory of algebraic groups over other fields, but Chevalley soon began to develop one, to which he devoted [C19]. The title of [C19] certainly conveys the idea that it is meant as a sequel to Lie I, but, in fact, it is hardly so, as can be gathered from the beginning of the introduction:

> The present work constitutes to some extent a sequel to my work 'Theory of Lie groups I', published by Princeton University Press. However, the topics treated here are very different from those considered in 'Theory of Lie Groups', and the proofs of the main theorems contained in this volume do not depend on the general theory of Lie groups.

The field $K$ is always assumed to be infinite. Chevalley defines the notion of algebraic subgroups of $GL(V)$, without any reference to an algebraically closed extension of $K$. As a result, this notion is not quite the same as that of the group of $K$-rational points of an algebraic group defined over $K$, in current terminology, and this has been a source of some confusion. To be more precise, in [C19] a subgroup $G$ of $GL(V)$ is *algebraic* if there exists a set $\mathcal{P}$ of polynomials on $\operatorname{End} V$ such that $G$ is the set of all elements of $GL(V)$ whose coefficients annihilate the elements of $\mathcal{P}$. The ideal $I(G)$ of $G$ is the set of all polynomials on $\operatorname{End} V$ which are zero on $G$. The group $G$ is irreducible as a variety if $I(G)$ is a prime ideal, in which case the quotient field of $K[V]/I(G)$ is by definition the field $K(G)$ of rational functions on $G$. Now let $L$ be an algebraically closed extension of $K$. Let $V^L = V \otimes_K L$ and $G^L$ the smallest algebraic subgroup of $GL(V^L)$ containing $G$. Its ideal is $I(G) \otimes_K L$, so $G^L$ is in the usual terminology an algebraic group defined over $K$; its group of rational $K$-points is indeed $G$, and, by definition, $G$ is Zariski dense in $G^L$. It also follows that $L(G^L)$ is a regular extension of $K$. Conversely, start from an algebraic group $H \subset GL(V^L)$ which is "defined over $K$" (i.e., $I(H)$ is generated by elements with coefficients in $K$); then $H(K) = H \cap GL(V)$ is not necessarily Zariski-dense in $H$. It can be viewed as an algebraic group in its own right, but its ideal $I(H(K))$ may be strictly bigger than $I(H) \cap K[V]$; hence $H(K)^L$ may be of strictly smaller dimension than $H$. While reading Lie II, or any paper written in the framework of that book, one has to remember that $G$ is by definition Zariski-dense in $G^L$, which runs against the usual conventions in algebraic geometry. One drawback of this point of view is that it does not allow for a convenient definition of quotients: If $G$ is as before and $N$ is an algebraic subgroup of $GL(V)$ which is normal in $G$, then the quotient group $G/N$ cannot be identified in general with an algebraic subgroup of $GL(W)$, where $W$ is some vector space over $K$, in Chevalley's sense. [If $H \subset GL(V^L)$ is as above, defined over $K$, and $Q$ is normal in $H$, also defined over $K$, then $H/Q$ is in a natural way an algebraic group defined over $K$, but in general, $(H/Q)(K)$ is not equal to $H(K)/Q(K)$.] The difficulty would be even worse if one tried to define in his context an algebraic structure on a coset space, and no attempt is made in Lie II to do so, even in the case of a quotient group.[2]

A number of generalities are developed over any (infinite) $K$. They include the characterization of an algebraic group by a finite set of semi-invariant polynomials with the same weight, or of rational invariants (proved first in [C18]), and the definition of the Lie algebra $L(G)$ of $G$ as the space of $X \in \mathfrak{gl}(V)$ for which the derivation of the field of rational functions on End $V$ associated to the right translation by $X$ leaves the ideal $I(G)$ of $G$ stable. This is so if and only if for any extension $L$ of $K$, any $s \in G^L$ and any $P \in I(G)$, the differential of $P$ at $s$, in the direction $s.X$, is zero. The notion of a rational representation $\rho : G \to GL(U)$, where $U$ is a finite dimensional vector space over $K$, is defined, as well as the differential $d\rho : L(G) \to L(H)$ for any algebraic subgroup $H$ of $GL(U)$ containing $\rho(G)$.

The main results, however, are established for $K$ of characteristic zero. This restriction is forced upon Chevalley because his main tool is a formal exponential, defined only in characteristic zero, which allows him to go from $L(G)$ to $G$ in more or less the familiar way. The smallest algebraic subgroup of $GL(V)$ whose Lie algebra contains a given diagonal matrix $X \in \mathfrak{gl}(V)$ is described. Its Lie algebra is the space of replicas of $X$. By definition a Lie algebra is *algebraic* if it is the Lie algebra of an algebraic group. The Chevalley-Tuan characterization in terms of replicas is proved in this more general context. It is also shown that the derived algebra of any Lie subalgebra of $\mathfrak{gl}(n, K)$ is algebraic (a result which, unknown to Chevalley, had already been announced by Cartan over $\mathbb{C}$ in [Ca1]; see V, §5). The very last theorem of Lie II is the existence of a multiplicative Jordan decomposition in $G$: If $g \in G$ and

$$(4) \qquad g = g_s \cdot g_u \qquad (g_s \text{ semisimple, } g_u \text{ unipotent, } g_s \cdot g_u = g_u \cdot g_s)$$

is its multiplicative Jordan decomposition, then $g_s, g_u \in G$.

## §3. Lie groups, 1948–1955

**5.** Now I have to backtrack again to cover the contributions of Chevalley to Lie groups in the late forties. The joint paper with S. Eilenberg [C12] introduces the notion of cohomology of Lie algebras and discusses some applications and open problems. Its origin is a paper by Cartan [Ca2] that shows how to reduce the determination of the Betti numbers of the coset space $G/H$ of a compact connected Lie group $G$ modulo a closed subgroup $H$ to an algebraic problem, assuming some theorems he conjectured on that occasion and which were soon proved by G. de Rham. Granted those, Cartan had in substance proved (without the terminology) that the cohomology ring $H^{\cdot}(G/H; \mathbb{R})$ is isomorphic to the cohomology algebra of the algebra of $G$-invariant differential forms on $G/H$, with respect to exterior differentiation. Since the latter algebra may be identified to the algebra $\left(\Lambda^{\cdot} T^*(G/H)_0\right)^H$ of elements in the exterior algebra of the cotangent space $T^*(G/H)_0$ at the origin of $G/H$ which are invariant under $H$, acting via the isotropy representation, this was indeed a reduction to an algebraic problem. It was further simplified for

---

[2]In §7 of [B2], I recover most of the results of Lie II in a much shorter way, the main reason being, it seems to me, that I can avail myself of the existence of a canonical structure of algebraic variety on a coset space $G/H$. To see concretely how this works, compare the proofs of the fact that if $M$ and $N$ are algebraic subgroups of $G$, then the Lie algebra of $M \cap N$ is the intersection of those of $M$ and $N$.

symmetric spaces, the main case of interest to Cartan, because then the exterior differential is identically zero on $G$-invariant forms, so $H^{\cdot}(G/H; \mathbb{R})$ may be identified to $\left(\Lambda^{\cdot}T^*(G/H)_0\right)^H$ itself. Therefore Cartan had had no need to describe the differential in the general case. This is carried out in [C12], and leads to the definition of the cohomology space $H^{\cdot}(\mathfrak{g}; V)$ of a Lie algebra $\mathfrak{g}$ over a field $K$ of characteristic zero, with coefficients in a $\mathfrak{g}$-module $V$, where $V$ is a vector space over $K$, and more generally of the relative cohomology space $H^{\cdot}(\mathfrak{g}, \mathfrak{h}; V)$ of $\mathfrak{g}$ modulo a subalgebra $\mathfrak{h}$, with coefficients in $V$. Both spaces are algebras if $V = K$ is the trivial one-dimensional $\mathfrak{g}$-module. Another incentive to develop this theory was the realization that if $\mathfrak{g}$ is semisimple, then the vanishing of $H^1(\mathfrak{g}; V)$ implies the complete reducibility of the finite dimensional representations of $\mathfrak{g}$, and the vanishing of $H^2(\mathfrak{g}; V)$ is equivalent to a lemma proved by J.H.C. Whitehead to give a new proof of the Levi theorem. The theorems of H. Hopf and H. Samelson on the cohomology of a compact connected Lie group $G$ could be translated into properties of $H^{\cdot}(\mathfrak{g}; \mathbb{R})$, where $\mathfrak{g}$ is the Lie algebra of $G$, but there remained the problem of finding direct algebraic proofs. This program, as well a more systematic approach to the cohomology of $G/H$ via cohomology of Lie algebras, was soon carried out by J-L. Koszul [Kz], and then in joint work of H. Cartan, Chevalley, Koszul and Weil. I shall not pursue this,[3] except to discuss one problem, namely the determination of the Betti numbers of a compact connected Lie group $G$. In that case, Cartan's method shows that $H^{\cdot}(G; \mathbb{R})$ may be identified to the invariants of $G$, acting by the adjoint representation on $\Lambda^{\cdot}\mathfrak{g}$. The determination of those had been carried out for the classical groups by R. Brauer. On the other hand, É. Cartan [Ca2] had already found the Poincaré polynomial[4] of the exceptional group $\mathbf{G}_2$. There still remained the four exceptional groups $\mathbf{E}_6$, $\mathbf{E}_7$, $\mathbf{E}_8$ and $\mathbf{F}_4$. Meanwhile Hopf had shown that $H^{\cdot}(G; \mathbb{R})$ is an exterior algebra on $\ell = \mathrm{rank}\, G$ generators, say of degrees $2m_i - 1$ $(i = 1, \ldots, \ell)$, so that the problem was reduced to finding the $m_i$'s, often called the *exponents* of $G$. This was first done by Yen Chih Ta [Y], in a rather roundabout way, by computing the cohomology of certain symmetric spaces of $G$ and then getting back to $G$ by means of a formula of Hirsch (still somewhat conjectural at that time) expressing the Poincaré polynomial of $G/H$ in terms of those of $G$ and $H$ when $G$ and $H$ have the same rank. In [C14], Chevalley proposed another method: Let $S^*(\mathfrak{g})$ be the algebra of polynomials over $\mathfrak{g}$. It is operated upon by $G$ via the adjoint representation. The joint work alluded to above implies that the algebra $I_G$ of invariants of $G$ in $S^*(\mathfrak{g})$ has $\ell$ algebraically independent homogeneous generators, of degrees $2m_i$ $(i = 1, \ldots, \ell)$, with $m_i$ as above, so that the Poincaré polynomial of $G$ is in principle determined by the structure of $I_G$. Now let $T$ be a maximal torus of $G$, $\mathfrak{t}$ its Lie algebra, $N$ the normalizer of $T$ in $G$ and $W = N/T$ the Weyl group of $G$. (By the way, Chevalley asserts there that the terminology "Weyl group" had been proposed by him. I presume this was orally,

---

[3]A detailed exposition may be found in [GHV]. To the references to earlier publications given there, one should now add the letters of A. Weil, written in 1949 and published for the first time in his Collected Papers [Wi].

[4]Recall that the Poincaré polynomial of a space $X$ with finite dimensional real homology is the polynomial

$$P(X, t) = \sum_{i \geq 0} b_i(X).t^i,$$

where $b_i(X)$ is the $i$-th Betti number of $X$.

because, as far as I know, this is indeed its first occurrence in print.) Then Chevalley states that the restriction to t induces an isomorphism of $I_G$ onto the algebra $I_W$ of invariants of $W$ acting on $S^*(\mathfrak{t})$. In this particular case, this implies that $I_W$ has $\ell$ algebraically independent generators, but Chevalley also claims this can be established more generally for any finite Euclidean reflection group. The proof was published only later [C27].[5] This then reduces the determination of the exponents to the determination of the invariants of a finite group. In that paper, Chevalley announced that it can be carried out and confirmed the results of Yen Chih Ta. Neither [Y] nor [C14] gives all the details. In order to fill this gap, Chevalley and I wrote a paper [C26] in which the exponents are determined with a minimum of computations, using the Hirsch formula and also, for $\mathbf{E}_8$, invariants of the Weyl group of low degree.

**6.** In a completely different direction, [C13] gives algebraic *a priori* proofs of two known theorems pertaining to a complex simple Lie algebra $\mathfrak{g}$. First the existence of a simple Lie algebra corresponding to a given Cartan matrix (the existence part of the classification, which had been until then carried out case by case), and second the existence of an irreducible representation with a given highest weight. (Weyl had already given a general proof [Wy1], [Wy2], but it used transcendental methods. Earlier, Cartan had relied on an explicit construction of $\ell = \mathrm{rank}\,\mathfrak{g}$ fundamental representations, which had to be done case by case.) [C13] gives only a sketch of the proofs, but enough indications to make one realize that Chevalley's main constructions, or related ones, occur in many later treatments. (For instance, the infinite dimensional Lie algebra $\tilde{\mathcal{L}}$ and the $e$-extreme $\tilde{\mathcal{L}}$-modules in Jacobson's book on Lie algebras are already there.) It also brings somehow to mind infinite dimensional representations and even Kac-Moody algebras. It is therefore rather intriguing that, probably around that time, Chevalley had asserted that the exceptional Lie algebras also belonged to infinite series of simple Lie algebras, like the classical ones, except that they were the only finite dimensional members of those series. I heard this from H. Cartan, around 1950 I think, and later from A. Weil. In retrospect, it is a bit puzzling that nobody took him up on that and asked for more details. But apparently nobody did, and I do not know what he had in mind.

**7.** At about that time, more precisely during the academic year 1947-48, Chevalley saw in the audience of his course on Lie groups and Lie algebras a new

---

[5]Without Chevalley's knowledge, under rather unusual, if not unique, circumstances. Chevalley had written down his proof in detail, but his manuscript did not have all the trimmings of a full-fledged paper and could not be published *ne varietur*. The theorem is proved in characteristic zero, but I needed a complement over a perfect field in some work on the cohomology mod $p$ of compact Lie groups, eventually published in [B0]. In order to prove it, I had to quote not only the statement, but also the proof of his theorem and so needed a publication to refer to. When apprised of this, Chevalley still did not agree to publish a paper and told me I could do whatever I wanted with his manuscript ("Fais-en ce que tu veux"). A. Weil was at the time one of the managing editors of the *American Journal of Mathematics*. We agreed to interpret Chevalley's authorization broadly, as allowing me to complete his manuscript under his name and submit it for publication in the *American Journal of Mathematics*. There was a theoretical possibility that Chevalley might become aware of it since, as an associate editor, he regularly received reports on the activities of the Journal, but we were rather confident that he hardly ever looked at them. Indeed, he learned about the paper only when he saw it in the issue of the Journal containing it. Baffled at first, he quickly realized what had happened, and did not mind. I had included the refinement I needed directly in the paper, as a lemma to be used by A. Borel in a forthcoming paper, and I warmly thanked Chevalley in [B0] for having shown his manuscript to me.

student coming from England, who did not seem to know much at first but who caught his undivided attention after some months, when he produced a new proof of Ado's theorem. This was his first contact with Harish-Chandra, who had come to Princeton to study Lie groups under him. In 1948, Harish-Chandra had also found an algebraic proof of the second theorem of [C13] stated above. Under the influence of [C13], he modified his argument to incorporate a proof of the first one as well. It was published in [HC].

## §4. Linear algebraic groups, 1954

**8.** At the end of the introduction to Lie II, Chevalley had announced that Lie I and II were to be part of an exposition of Lie groups and algebraic groups, which would include four more books, but he published only one more, [C24]. Besides the general theory and the classification of semisimple Lie algebras, this treatise was to include the cohomology of Lie algebras and the topology of Lie groups. As far as I know, those chapters were never written, but it may be as a preparation that he had made two further contributions of an expository nature to the global theory of Lie groups. One is a proof of the surjectivity of the exponential map in a compact connected Lie group, which is self-contained and simpler than any other proof I know. The other is a proof of the conjugacy of maximal compact subgroups in a semisimple Lie group not using any Riemannian geometry (in contrast with Cartan's famous argument). Both were included in various internal manuscripts of Bourbaki. The former was eventually published in [Bk], and the latter in ([B3], VII, 3.4).

**9.** Lie III [C24], notwithstanding its title "Algebraic Groups", is a mixture of algebraic groups (in characteristic zero, essentially a standing assumption), Lie groups and Lie algebras. Lie groups and algebraic groups are studied concurrently, the former being however often subsumed to the latter, for instance by reducing questions on Lie algebras or linear Lie groups to the algebraic case by going over to the algebraic hulls or Zariski closures. Apart from some generalities on linear representations, there are two main parts. The first (Chapters III, IV, V) is concerned with general properties of Lie algebras. It presupposes Lie I and II. Furthermore, Chevalley was at the time of the opinion that the theory of semisimple algebras should precede that of solvable ones. Strict adherence to such principles is not always conducive to an economical presentation. For instance, his proof of Lie's theorem on solvable Lie algebras presupposes the theory of reductive Lie algebras and Cartan's characterization of semisimple algebras. Nevertheless, the main structure theorems are covered, including Ado, Levi-Malcev, a discussion of the universal enveloping algebra, and invariant differential operators.

The second main part (Chapter VI) is devoted to Cartan subalgebras and Cartan subgroups, and begins with the introduction of the Zariski topology, which is used systematically. A *Cartan subalgebra* of the Lie algebra $\mathfrak{g}$ is, by definition here, a nilpotent subalgebra which is equal to its normalizer. The equivalence with the original definition [C5] is proved; this implies in particular that Cartan subalgebras so defined do exist. As in [C5], a main goal is the conjugacy theorem over an algebraically closed field. The basic idea is the same as there, but Chevalley takes advantage of the fact that he now has a natural framework to express it. Instead of a group generated by the exponentials of nilpotent inner derivations, he uses the irreducible algebraic subgroup of Aut $\mathfrak{g}$ whose Lie algebra is the derived algebra of ad $\mathfrak{g}$ (which he knows to be algebraic by Lie II). He also shows that the conjugacy holds for $\mathfrak{g}$ solvable, even if the ground field is not algebraically closed. A completely novel element is a definition of Cartan subgroups which is meaningful in any group $G^{.}$. A subgroup $C$ of $G$ is a *Cartan subgroup* if it is maximal nilpotent and every subgroup of finite index of $C$ is of finite index in its normalizer in $G$. If $G$ is algebraic, these groups are shown to exist and to be the irreducible algebraic subgroups whose Lie algebras are the Cartan subalgebras. If $G$ is a connected Lie group, these are closed subgroups, but they are not necessarily connected. In [C22] Chevalley says he expects that such a characterization of Cartan subgroups will be useful to study the properties of an algebraic group as an abstract group, an expectation fully confirmed by subsequent developments. The case of compact connected Lie groups, where the Cartan subgroups are the maximal tori, is also considered, and the conjugacy of Cartan subgroups and the surjectivity of the exponential mapping are proved. It is also shown that a compact Lie group is real algebraic, a fact which had been proved in substance in Lie I, in the discussion of Tannaka duality, but had not been stated explicitly there.

**10.** From now on we shall be concerned only with algebraic groups. The paper [C23] investigates the structure of the field of rational functions $K(G)$ of an irreducible linear algebraic group $G$ over $K$, where $K$ is of characteristic zero. It is shown that $K(G)$ is contained in a purely transcendental extension of $K$ (i.e., $G$ is unirational) and is itself purely transcendental (i.e., $G$ is a rational variety) if $K$ is algebraically closed. In his Traité d'Analyse, Picard had claimed the last result (over $\mathbb{C}$, of course), but Chevalley points out that Picard's argument only yields the unirationality (as we saw in Chapter V, §3). In this paper, Chevalley in fact proves more: He constructs the variety $M$ of Cartan subgroups, shows that it is rational, that $G$ is birationally isomorphic to the product of $M$ by a Cartan subgroup $C$, and that $C$ is unirational. Moreover, $C$ is rational if all the elements ad $c$ $(c \in C)$ have their eigenvalues in $K$.[6]

## §5. Algebraic groups, 1955–1961

From the point of view of algebraic geometry, linear algebraic groups over an algebraically closed ground field are special case of "algebraic group varieties", i.e., algebraic varieties endowed with a product structure such that product and inverse are morphisms of algebraic varieties. One class, besides the linear groups, was well-known, namely the abelian varieties, which can be characterized as the irreducible algebraic groups whose underlying variety is complete. In 1953, as pointed out in [R], Chevalley showed that any irreducible algebraic group $G$ contains a biggest normal linear algebraic subgroup, which is the kernel of a morphism of $G$ onto an abelian variety. Thus abelian varieties and linear algebraic groups are the two building blocks of general algebraic groups. He did not publish it at that time, only later [C33]. The argument is in principle the one alluded to in [R]. It uses a theory of the Albanese variety and of linear systems of divisors in a non-projective situation. It may be that Chevalley waited until more foundational material was available.

---

[6]As described in III, §4, Maurer essentially proved in [M] that a complex irreducible linear algebraic group is a rational variety. Since that paper is about the only reference to earlier work given in Lie II, I once asked Chevalley why he did not quote it in [C23]. His answer was that since there were mistakes in the first part of the paper he had never bothered to look at the rest. In fact, Maurer's approach is in a way more direct than Chevalley's, in the sense that Maurer does not use Levi's theorem (which was anyhow not available at the time).

He does indeed refer to later papers for it. Independently, I. Barsotti [Ba2] and
M. Rosenlicht [R] published proofs different from Chevalley's. [R] also contains
a number of important results on algebraic groups. [C29] is also concerned with
general group varieties. It gives a new proof of a theorem of Barsotti's stating that
a group variety is quasi-projective [Ba1]. Shortly afterwards, W.L. Chow proved it
more generally for homogeneous varieties [Ch].

**11.** In the notice on his own work [C22], written in Japan probably during
the winter 1953-54, Chevalley gives the following motivation for studying algebraic
groups over fields other than the complex numbers:

> The principal interest of the algebraic groups seems to me to be
> that they establish a synthesis, at least partial, between the two
> main parts of group theory, namely the theory of Lie groups and
> the theory of finite groups.

Whether this view was already his at the beginning I do not know, but it became
foremost in his mind in the late forties. The model here was L. Dickson, who, taking
advantage of the fact that the classical groups have an algebraic definition valid over
general fields, had constructed new series of finite simple groups over finite fields.
He had also done that for the exceptional group $\mathbf{G}_2$ (as well as for $\mathbf{E}_6$, but this was
pretty much forgotten at the time and unknown to Chevalley). The task Chevalley
set for himself was then to find models of the four other exceptional groups which
would make sense over arbitrary fields and again lead to new simple groups. His
joint paper with R.D. Schafer [C15] on $\mathbf{F}_4$ and $\mathbf{E}_6$ and his Comptes Rendus notes
on $\mathbf{E}_6$ [C20, 21] are first steps in that direction. In [C22] Chevalley asserts that in
the summer of 1953 he had found new algebraic definitions of $\mathbf{F}_4$, $\mathbf{E}_6$ and $\mathbf{E}_7$, by
making use of the triality principle, and that these groups generate infinite series of
simple groups, the first new ones in fifty years. No doubt he intended at that time
to publish the proofs. In fact, he states in [C22] that in his small book on spinors
[C25] he carries out a synthesis of the methods developed by Weyl and Cartan, and
generalizes their results over arbitrary ground fields, a "generalization which was
necessary in view of the study of the new finite groups I have discovered". But he
never did. There was apparently a breakthrough shortly afterwards, and Chevalley
saw how to carry this out in a uniform, classification-free manner. This leads us to
the first major achievement in the second part of Chevalley's work (in the division
proposed above), the very influential and justly famous "Tôhoku" paper [C28].

Let $\mathfrak{g}$ be a complex semisimple Lie algebra, $\mathfrak{h}$ a Cartan subalgebra of $\mathfrak{g}$, $R$ the
set of roots of $\mathfrak{g}$ with respect to $\mathfrak{h}$, and $S$ the set of simple roots with respect to some
ordering of $R$. Let $P \subset \mathfrak{h}'$ be the lattice of weights, $Q_r$ the sublattice generated
by the roots, $P^\vee$ and $Q_r^\vee$ their duals in $\mathfrak{h}$ (the lattices of coweights and coroots
respectively). First, by a careful, largely new, analysis of the constants of structure
of $\mathfrak{g}$, Chevalley shows that one can choose roots vectors $X_r$ $(r \in R)$ and a basis
$H_s$ $(s \in S)$ of $\mathfrak{g}$ which span a $\mathbb{Z}$-form $\mathfrak{g}_\mathbb{Z}$ of $\mathfrak{g}$ so that, for any field $K$, the elements
$\exp \operatorname{ad} t X_r$ $(t \in K)$ form a unipotent one-parameter group $x_r(K)$ of automorphisms
of $\mathfrak{g}_K = \mathfrak{g}_\mathbb{Z} \otimes_\mathbb{Z} K$. Chevalley also defines a group of automorphisms $H_K$ (resp. $H'_K$)
of $\mathfrak{g}_K$ associated to $\mathfrak{h}$ and to the homomorphisms of $Q$ (resp. $P$) into $K^*$. Let $G_K$
(resp. $G'_K$) be the subgroup of $\operatorname{Aut} \mathfrak{g}_K$ generated by the $x_r(K)$ and $H_K$ (resp. the
$x_r(K)$). In $G_K$, the quotient by $H_K$ of the normalizer $\mathcal{N}(H_K)$ of $H_K$ is the Weyl
group $W$ of $\mathfrak{g}$ with respect to $\mathfrak{h}$. Similarly, $G'_K$ contains $H'_K$ and $\mathcal{N}(H'_K)/H'_K = W$.
The group $G'_K$ is normal in $G_K$, and $G_K/G'_K = H_K/H'_K$. Both $G_K$ and $G'_K$ are

shown to admit Bruhat decompositions. Now let $\mathfrak{g}$ be simple. The main result of the paper is that, except in a few cases where $K$ has two or three elements, $G'_K$ is the derived group of $G_K$ and is a simple group.

Formally, there are no algebraic groups in that paper, though, at the end, Chevalley conjectures that, for $\mathfrak{g}$ classical, $G_K$ is isomorphic to the corresponding (split) classical group modulo its center, and asks whether $G_K$ is algebraic in general. A first answer was supplied by T. Ono [O]. It became clear later, in the light of [C33], that [C28] gives a construction of the scheme over $\mathbb{Z}$ for the adjoint group of type $\mathfrak{g}$ and of the image in the adjoint group of the group of rational points of the simply connected group of the given split type. Altogether, this paper provides a striking illustration for the programmatic statement of the notice [C22] quoted above.

**12.** The next publication of Chevalley is the no less famous Paris Seminar [C30]. There, the framework and point of view are completely different from those of Lie II, III. The Lie algebra appears only briefly, in the last two lectures, and there is no exponential mapping. They are replaced by global arguments in algebraic geometry valid over an algebraically closed ground field of arbitrary characteristic. Since I am responsible for this change of scenery, I'll digress a little and discuss briefly my own work at that time and how it relates to Chevalley's.

I had been aware of [C9] and of Lie II early on, but from a distance. I got closer to algebraic groups during the first AMS Summer Institute in 1953, devoted to Lie groups and Lie algebras, through my joint work with G.D. Mostow [BM] and a series of lectures by Chevalley on Cartan subalgebras and Cartan subgroups of algebraic groups (the future Chapter VI of Lie III). He was not pleased with it, though, and toward the end said he felt it was too complicated and there should be a more natural approach valid in any characteristic. Another topic which came up in discussions was a claim by V.V. Morozov, to the effect that maximal solvable subalgebras of a complex semisimple Lie algebra are conjugate. Nobody understood his argument, but I found a simple global proof, using Lie's theorem and the flag variety. In 1954-55, in Chicago, I made a deliberate effort to get away from characteristic zero. Two papers by Kolchin [K1, 2], the first ones to prove substantial results on linear algebraic groups by methods insensitive to the characteristic of the ground field, led quickly to a structure theory of connected solvable groups (see VIII). Then I saw how to extend my conjugacy proof of maximal connected solvable subgroups to arbitrary characteristics. That was the decisive step. From then on, it was comparatively smooth sailing, and the other results of [B1] followed rather quickly. I lectured on this work and talked about it with Chevalley shortly afterwards, in February 1955 I think, at a conference in Urbana, Illinois. In summer 1955, before leaving the States, I gave him a copy of the manuscript of my forthcoming Annals paper. We did not discuss the subject until the Summer 1956 Bourbaki Congress, where I was told (not by him) that he had classified the simple algebraic groups over any algebraically closed ground field. When asked, he confirmed it and agreed to give us an informal lecture about this work, at which time he announced he would propose the name "Borel subgroup" for a maximal closed connected solvable subgroup of a linear algebraic group. He also told me that, after having read my paper, his first goal had been to prove the normalizer theorem: "A Borel subgroup of a connected linear algebraic group is its own normalizer", after which, "the rest followed by analytic continuation".

The first part of [C30] covers some foundational material and [B1]. The next goal is the normalizer theorem. It is followed by a rather long discussion of various types of subtori, in particular subtori of codimension one of a given maximal torus $T$ which are singular, i.e. whose centralizers have dimension strictly greater than that of the centralizer $T$ (a Cartan subgroup of $G$ by [B1]), which allows Chevalley to prove that the unipotent radical of the identity component of the intersection of the Borel subgroups containing $T$ is the unipotent radical of $G$.[7] Now assume $G$ to be semisimple, i.e. $\mathcal{R}G = \{1\}$. The next goal is to associate to $G$ and $T$ a root system in $X(T)_{\mathbb{Q}}$, where $X(T)$ is the group of rational characters of $T$ and $X(T)_{\mathbb{Q}} = X(T) \otimes_{\mathbb{Z}} \mathbb{Q}$, whose Weyl group may be identified to $\mathcal{N}(T)/T$ acting on $X(T)_{\mathbb{Q}}$ via inner automorphisms. The previous results already imply that the centralizer $\mathcal{Z}(S)$ of a singular torus $S$ of codimension one, modulo its radical, is simple of dimension three. The existence of a reflection amounts to showing that $\mathcal{N}(T) \neq T$ in $\mathcal{Z}(S)$, and follows from a fixed point theorem: A torus acting on a projective variety of dimension $\geq 1$ has at least two fixed points. On the other hand, a one-dimensional connected unipotent group $U$ may be identified with the additive group of the given (universal) algebraically closed ground field $K$, and then any automorphism of $U$ is multiplication by an element of $K^*$. In particular, if such a subgroup of $G$ is invariant under $T$, then the action of $T$ on $U$ by inner automorphisms is described by a rational character. The roots are introduced in this way, and the identity components of their kernels are the maximal singular subtori contained in $T$. Chevalley proves that the roots so defined form a reduced root system in $X(T)_{\mathbb{Q}}$. He associates to $(G, T)$ the pair $(\Phi, \Gamma)$, which I shall call a root *diagram*, where $\Phi \subset X(T)_{\mathbb{Q}}$ is a root system and $\Gamma$ is a lattice in $X(T)_{\mathbb{Q}}$ intermediary between the lattice $P(\Phi)$ of weights of $\Phi$ and the lattice $Q(\Phi)$ generated by the roots, namely $X(T)$ itself. Moreover, $\Phi$ is shown to be irreducible if and only if $G$ is simple as an algebraic group. Next the irreducible rational representations of $G$ are constructed. Let $B$ be a Borel subgroup of $G$ containing $T$. It corresponds to the negative roots for a suitable ordering on $\Phi$. Let $\lambda$ be a dominant weight contained in $X(T)$. In characteristic zero, the irreducible representation of $G$ with highest weight $\lambda$ is afforded by the space of regular sections of the line bundle of $G/B$ defined by $\lambda$ (viewed as a character of $B$). This space can also be defined in arbitrary characteristic and is again a finite dimensional vector space on which $G$ acts canonically. It is shown to be non-zero (because $\lambda$ is dominant), but it is not necessarily irreducible as a $G$-module. However, Chevalley shows that it contains a unique invariant irreducible subspace, and establishes in this way again a bijection between dominant weights and irreducible representations. The last seven chapters of the Seminar are chiefly devoted to a uniqueness theorem: An isomorphism of the diagrams $(\Phi', \Gamma')$ and $(\Phi, \Gamma)$ associated to two groups $G'$, $G$ and maximal tori $T'$, $T$ is induced by an isomorphism of $G$ onto $G'$. The extremely difficult proof consists in first checking this for $\Phi$ of type $\mathbf{A}_n$ or of rank two. The proof in general is then reduced to the rank one and two cases by consideration of the centralizers of singular tori of codimension one and two. In fact, Chevalley proves a more functorial statement. Assume that $\mu$ is an isomorphism of $X(T')_{\mathbb{Q}}$ onto $X(T)_{\mathbb{Q}}$ which maps $\Gamma'$ into $\Gamma$ and induces a bijection $\mu_0$ of $\Phi'$ onto $\Phi$. It is said to be *special* if moreover there exists a bijection $\nu : \Phi \to \Phi'$ such that

---

[7]A much simpler argument has been given recently by D. Luna, Retour sur un théorème de Chevalley, Enseignement Mathématique, **45**, 1999, 317-9.

$\mu_0(\nu(\alpha)) = q(\alpha)\alpha$ ($\alpha \in \Phi$), where $q(\alpha)$ is a power of the characteristic exponent of $K$. This condition is easily seen necessary for $\mu$ to be associated to an isogeny, and Chevalley proves that, conversely, any special $\mu$ is associated to an isogeny $\iota(\mu) : G \to G'$. If the $q(\alpha)$ are equal to one (that is, if $\mu$ is an isomorphism of root systems), then $\iota(\mu)$ is a central isogeny. Using his results on representations, Chevalley shows that if $(\Phi, \Gamma)$ is the root-diagram associated to $(G, T)$, then there is always $\tilde{G}$ associated to $(\Phi, P(\Phi))$, which admits therefore a central isogeny onto $G$. (On the other hand, the adjoint group of $G$ corresponds to $\Gamma = Q(\Phi)$.) There are however a few cases, in characteristic two or three, in which $G$ and $G'$ are simple and the function $q$ takes two distinct values. The existence theorem then produces "exceptional isogenies" which, except in the case of $\mathbf{B}_n$ and $\mathbf{C}_n$ in characteristic two, were unknown before.[8] The Lie algebras of algebraic groups play no role, are not even mentioned, in [K1], [K2], [B1]. Chevalley had defined them in [C19], see **4**, and they do appear in [C30], but in a very minor way, in two places: in Exp. 16, p. 2, to define purely inseparable isogenies of height one (see VI, 4.1), and in Exp. 21, §3, to identify the roots of a semisimple group with the non-zero weights of the adjoint representation, thus arriving at the definition VI, 13 (3) of the roots, the more natural one nowadays.

**13.** The undated paper [C31] is no doubt contemporary with the seminar. Although it was unpublished until 1994 and copies were hard to come by, its subject matter and main results were rather well-known early on, and became part of the literature after M. Demazure had published complete proofs [D]. It is concerned with the closures of the orbits of $B$ in $G/B$, the "Schubert cells" of $G/B$. The orbits themselves are affine subspaces $C(w)$ ($w \in W$), the "Bruhat cells", parametrized by the Weyl group $W$ of $G$. Chevalley first gives a combinatorial description of the cells contained in the closure $\overline{C(w)}$ of $C(w)$ in terms of a reduced decomposition of $w$. He also shows that $\overline{C(w)}$ is non-singular in codimension one, and then conjectures that it is always non-singular, a somewhat surprising conjecture in view of the fact that examples of singular classical Schubert cells in Grassmannians were known; apparently Chevalley was not aware of it. At any rate, these singularities have since been studied, and remarkable connections with representation theory, via the Kazhdan-Lusztig polynomials, have been uncovered. Next Chevalley makes an important step towards the determination of the Chow ring $A(G/B)$ of $G/B$ by computing the intersection of $\overline{C(w)}$ with any Schubert cell of codimension one. The formulas are given in terms of roots and weights, and depend in fact only on the root datum $(\Phi, \Gamma)$ associated to $G$ in [C30]. This suffices to show that $A(G/B)$ is independent of the characteristic of the ground field, so that it may be identified with the similar object attached to the complex analogue of $G$.

Over $\mathbb{C}$, it is well-known that the map $\mu : (G, T) \mapsto (\Phi(G, T), X(T))$ yields a bijection between isomorphism classes of complex semisimple groups and isomorphism classes of root diagrams. The root system characterizes the Lie algebra of $G$, and $X(T)$ characterizes the various locally isomorphic groups with the given Lie algebra. The group is simply connected if $X(T) = P(\Phi)$, and of adjoint type if $X(T) = Q(\Phi)$. In general, $X(T)/Q(\Phi)$ is isomorphic to the center of $G$ and $P(\Phi)/X(T)$ to its fundamental group. As pointed out earlier, Chevalley had set up

---

[8]Recently, R. Steinberg found a surprisingly simple proof of the isogeny theorem (J, Algebra **216** (1999), 366–383). Earlier, M. Takeuchi had given the first classification free proof, also valid for exceptional isogenies. For central isogenies, it is presented in [J].

in [C30] a similar mapping in any characteristic and shown that it is also injective. It remained to prove the surjectivity, in order to complete the classification. In the last lecture of [C30], Chevalley remarks that [C28] gives the existence of $G$ of adjoint type, i.e. when $\Gamma = Q(\Phi)$, and adds that with the methods developed to prove the isogeny theorem, it would be easy to associate to any $\Gamma$ a covering group of the adjoint group, thus establishing the surjectivity of $\mu$. But in [C34] he comes back to this question and handles it in a deeper manner, which has opened new vistas, by a direct generalization of the construction of [C28], now expressed in the language of schemes. This procedure also has a natural built-in explanation of why the classification is independent of the characteristic of the ground field.

**14.** Given $(\Phi, \Gamma)$, let $G$ be the associated complex group, $\mathfrak{g}$ its Lie algebra and $T$ a maximal torus of $G$. The construction of $\mathfrak{g}_{\mathbb{Z}}$ in [C28] allows one to introduce a $\mathbb{Q}$-structure of split group on $G$. Fix a faithful linear representation $(\sigma, E)$ of $G$. It may also be defined over $\mathbb{Q}$. Chevalley asserts the existence of an "admissible" lattice in $E(\mathbb{Q})$, i.e. a lattice spanned by eigenvectors of $T$ and stable under the automorphisms $\exp t\sigma(X_r)\,(t \in \mathbb{Z})$, where the $X_r$ are part of the basis of $\mathfrak{g}$ constructed in [C28]. The subring of the coordinate ring $\mathbb{C}[G]$ generated by the coefficients of $\sigma$ with respect to a basis of an admissible lattice is shown to generate a $\mathbb{Z}$-form $\mathbb{Z}[G]$ of $\mathbb{C}[G]$. It represents a scheme over $\mathbb{Z}$ associated to $(\Phi, \Gamma)$. It is then claimed that $\mathbb{Z}[G] \otimes_{\mathbb{Z}} K$ is the coordinate ring of an irreducible $K$-group $G_{(K)}$ defined and split over the prime field of $K$, with diagram $(\Phi, \Gamma)$, whence the existence.[9]

This fundamental paper, [C28], turned out to be the last (published) research paper of Chevalley's. In the following years, Chevalley devoted his seminar to finite groups and developed an active and successful school in that area, but did not publish anything himself.

---

[9]It was pointed out to me several years ago by J. Tits and D. Verma, independently, that this last claim, which amounts to saying that the scheme over $\mathbb{Z}$ defined by $\mathbb{Z}[G]$ has good reduction everywhere, is not quite true. But it becomes so if $\sigma$ is replaced by its direct sum with the adjoint representation. See also 3.9.3 in F. Bruhat and J. Tits, Inst. Hautes Études Sci. Publ. Math. **60** (1983), 197–376. For a systematic exposition of split groups with (resp. without) schemes, see [DG] (resp. [S]).

# References for Chapter VII

[Ba1]  I. Barsotti, *A note on abelian varieties*, Rend. Circ. Mat. Palermo (2) **2** (1953), 236–257.

[Ba2]  _____, *Un teorema di struttura per les varietà gruppali*, Atti Accad. Naz. Lincei Rend Cl. Sci. Fis. Mat. Nat. (8) **18** (1955), 43–50.

[B0]   A. Borel, *Sur la torsion des groupes de Lie*, J. Math. Pures Appl. (9) **35** (1955), 127–139; Oeuvres I, 477–489.

[B1]   _____, *Groupes linéaires algébriques*, Ann. of Math. (2) **64** (1956), 20–82; Oeuvres I, 490–552.

[B2]   _____, Linear algebraic groups (Notes by H. Bass), Benjamin, 1969; 2nd aug. ed.,, GTM **126**, Springer, 1991.

[B3]   _____, Semisimple groups and Riemannian symmetric spaces, Texts and Readings in Math. **16**, Hindustan Book Agency, New Delhi, 1998.

[BM]   A. Borel and G. D. Mostow, *On semi-simple automorphisms of Lie algebras*, Ann. of Math. (2) **61** (1955), 389 - 405; Oeuvres I, 460–476.

[Bk]   N. Bourbaki, Groupes et algèbres de Lie, Chap. 9: Groupes de Lie réels compacts, Masson, Paris, 1982.

[Ca1]  E. Cartan, *Sur certains groupes algébriques*, C.R. Acad. Sci. Paris **120** (1895), 544–548; Oeuvres Complètes I$_1$, 289–292.

[Ca2]  _____, *Sur les invariants intégraux de certains espaces homogènes clos et les propriétés topologiques de ces espaces*, Ann. Soc. Polonaise Math. **8** (1929), 181–225; Oeuvres Complètes I$_2$, 1081–1125.

[Ca3]  _____, La théorie des groupes finis et continus et l'Analysis Situs, Mém. Sc. Math. XLII, Gauthier-Villars, Paris, 1930; Oeuvres Complètes I$_2$, 1165–1224.

[Ca4]  _____, *La topologie des espaces représentatifs des groupes de Lie*, Enseignement Math. **35** (1936), 177–200; Exposés de Géométrie VIII¡ Hermann, Paris, 1936; Oeuvres Complètes I$_2$, 1307–1330.

—  —  —  —  —

## Books and Papers by C. Chevalley*

[C1]   C. Chevalley, *Sur la théorie du corps de classes dans les corps finis et les corps locaux*, J. Fac. Sci. Tokyo Univ. **2** (1933), 364–476.

[C2]   _____, *Sur les démonstrations arithmétiques dans la théorie du corps de classes* (with H. Nehrkorn), Math. Annalen **111** (1935), 364–371.

[C3]   _____, *La théorie du corps de classes*, Ann. of Math. (2) **41** (1940), 394–417.

[C4]   _____, *On the topological structure of solvable groups*, Ann. of Math (2) **42** (1941), 668–675.

[C5]   _____, *An algebraic proof of a property of Lie groups*, Amer. J. Math **63** (1941), 785–793.

[C6]   *A new kind of relationship between matrices*, Amer. J. Math **65** (1943), 521–531.

[C7]   _____, *On groups of automorphisms of Lie groups*, Proc. Nat. Acad. Sci USA **30** (1944), 274–275.

[C8]   _____, *La théorie des groupes de Lie*, Proc. First Canadian Math. Congr. 1945, Univ. of Toronto Press (1946), 338–354.

[C9]   _____, *On algebraic Lie algebras* (with H.F. Tuan), Proc. Nat. Acad. Sci. USA **51** (1946), 195–196.

[C10]  _____, Theory of Lie groups I, Princeton Univ. Press (1946).

[C11]  _____, *Algebraic Lie algebras*, Ann. of Math. (2) **48** (1947), 91–100.

[C12]  _____, *Cohomology theory of Lie groups and Lie algebras* (With S. Eilenberg), Trans. Amer. Math. Soc. **63** (1948), 85–124.

[C13]  _____, *Sur la classification des algèbres de Lie simples et de leurs représentations*, C.R. Acad. Sci. Paris **227** (1948), 1136–1138.

[C14]  _____, *The Betti numbers of the exceptional simple Lie groups*, Proc. Internat. Congr. Math. (Cambridge, Mass., 1950) **2**, Amer. Math. Soc., 1952,21–24.

[C15]  _____, *The exceptional simple Lie algebras $F_4$ and $E_6$* (with R.D. Schafer), Proc. Nat. Acad. Sci. USA **36** (1950), 137–141.

_____

*Chevalley's Collected Works are still being prepared, except for Volume II (Springer, 1997).

[C16]     _____ , *Algebraic Lie algebras and their invariants* (with H.F. Tuan), Acta Math. Sinica** **1** (1951), 215–242.

[C17]     _____ , *On a theorem of Gleason*, Proc. Amer. Math. Soc **2** (1951), 122–125.

[C18]     _____ , *Two proofs of a theorem on algebraic groups* (with E.R. Kolchin), Proc. Amer. Math. Soc. **2** (1951), 126–137; Kolchin, Sel. Wks., 159–167.

[C19]     _____ , Théorie des groupes de Lie II. Groupes algébriques, Hermann, Paris, 1951.

[C20]     _____ , *Sur le groupe exceptionnel* ($E_6$), C.R. Acad. Sci. Paris **232** (1951), 1991–1993.

[C21]     _____ , *Sur une variété algébrique liée à l'étude du groupe* ($E_6$), C.R. Acad. Sci. Paris **232** (1951), 2168 - 2170.

[C22]     _____ , *Notice sur les travaux scientifiques de Claude Chevalley*, (unpublished).

[C23]     _____ , *On algebraic group varieties*, J. Math. Soc. Japan **6** (1954), 303–324.

[C24]     _____ , Théorie des groupes de Lie III. Groupes algébriques, Hermann, Paris, 1954.

[C25]     _____ , The algebraic theory of spinors, Columbia University Press, New York, 1954; Collected Works II, 65–192.

[C26]     _____ , *The Betti numbers of the exceptional Lie groups* (with A. Borel), Memoirs Amer. Math. Soc **14** (1955), 1–9; Borel, Oeuvres I, 451–459.

[C27]     _____ , *Invariants of finite groups generated by reflections*, Amer. J. Math **77** (1955), 778–782.

[C28]     _____ , *Sur certains groupes simples*, Tôhoku Math. J. (2) **7** (1955), 14–66.

[C29]     _____ , *Plongement projectif d'une variété de groupe*, Proc. Internat. Sympos. Algebraic Number Theory (Tokyo and Nikko, 1955), Sci. Council of Japan, Tokyo, 1956, 131–138.

[C30]     _____ , Classification des groupes de Lie algébriques, 2 vols., Notes polycopiées, Inst. H. Poincaré, 1956-58.

[C31]     _____ , *Sur les décompositions cellulaires des espaces G/B*, in Algebraic Groups and Their Generalizations: Classical Methods, Proc. Sympos. Pure Math. **56**, Part 1, Amer. Math. Soc., 1994, 1–23.

[C32]     _____ , *La théorie des groupes algébriques*, Proc. Internat. Congr. Math. (Edinburgh, 1958), Cambridge Univ. Press, 1960, 53–68.

[C33]     _____ , *Une démonstration d'un théorème sur les groupes algébriques*, J. Math. Pures Appl. (9) **39** (1960), 307–317.

[C34]     _____ , *Certains schémas de groupes semi-simples*, Sém. Bourbaki 1960/61, Exp. **219**, May 1961.

—  —  —  —  —

[Ch]     W. L. Chow, *On the projective embeddings of homogeneous varieties*, in Algebraic Geometry and Topology, A Symposium in Honor of S. Lefschetz, Princeton University Press, 1957, 122–128.

[D]      M. Demazure, *Désingularisation des variétés de Schubert généralisées*, Ann.Sci. École Norm. Sup. (4) **1** (1974), 53–88.

[DG]     M. Demazure and A. Grothendieck (editors), Schémas en groupes I, II, III (SGA3), Lecture Notes in Math. **151–153**, Springer, 1970.

[GHV]    W. H. Greub, S. Halperin and J.R. Vanstone, Connections, curvature and cohomology, Vol 3, Academic Press, 1976.

[HC]     Harish-Chandra, *On some applications of the universal algebra of a semi-simple Lie algebra*, Trans. Amer. Math. Soc. **70** (1951), 28–96; Coll. Papers I, 292–360.

[I]      K. Iwasawa, *On some types of topological groups*, Ann. of Math. (2) **50** (1949), 507–558.

[J]      J.C. Jantzen, Representations of algebraic groups, Academic Press, 1987.

[K1]     E.R. Kolchin, *Algebraic matric groups and the Picard-Vessiot theory of homogeneous linear differential equations*, Ann. of Math. (2) **49** (1948), 1–42; Sel. Wks., 87–128.

[K2]     _____ , *On certain concepts in the theory of algebraic matric groups*, Ann. of Math. (2) **49** (1948, 774–789; Sel. Wks., 129–144).

[Kz]     J.-L. Koszul, *Homologie et cohomologie des algèbres de Lie*, Bull. Soc. Math. France **78** (1950), 65–127; Sel. Papers, 7–69.

[M]      L. Maurer, *Zur Theorie der continuirlichen, homogenen und linearen Gruppen*, Sitzungsber. Math.-Phys. Kl. Kgl. Bayer. Akad. Wiss. München (8) **24** (1894), 297–341.

[O]      T. Ono, *Sur les groupes de Chevalley*, J. Math. Soc. Japan **10** (1958), 307–313.

**Reviewed in Math. Reviews as "J. Chinese Math. Soc. (N.S.)."

[P]     L. Pontrjagin, Topological groups (translated by E. Lehmer), Princeton University Press, 1939.

[R]     M. Rosenlicht, *Some basic theorems on algebraic groups*, Amer. J. Math **78** (1956), 401–443.

[S]     R. Steinberg, Lectures on Chevalley groups (Notes by J. Faulkner and R. Wilson), Yale University, 1967; new version, to appear.

[T]     M. Takeuchi, *A hyperalgebraic proof of the isomorphism and isogeny theorems for reductive groups*, J. Algebra **85** (1983), 179–196.

[Wi]    A. Weil, *Géométrie différentielle des espaces fibrés*, manuscript, 1949; first published in his Oeuvres Scientifiques/Collected Papers, Vol I, Springer, 1979, 422 - 436.

[Wy1]   H. Weyl, *Theorie des Darstellung kontinuerlicher halbeinfacher Gruppen durch lineare Transformationen*, I. II. III and Nachtrag, Math. Zeitschr. **23** (1925), 271–309; **24** (1926), 377–395, 377–395, 789–791; Ges. Abh. II, 543–647; Selecta, 262–366.

[Wy2]   ———, *Die Vollständigkeit der primitiven Darstellungen einer geschlossenen kontinuerlichen Gruppe* (mit F. Peter), Math. Annalen **97** (1927), 737–755; Ges. Abh. III, 58–75; Selecta, 387–404.

[Y]     Yen Chih Ta, *Sur les polynomes de Poincaré des groupes simples exceptionnels*, C.R. Acad. Sci. Paris **228** (1949), 628–630.

# CHAPTER VIII

# Algebraic Groups and Galois Theory in the Work of Ellis R. Kolchin

The Galois theory of differential fields, a generalization of the Picard-Vessiot theory, was a major concern of E. Kolchin's during the first thirty years or so of his scientific life. From the beginning, it appeared that these Galois groups would be algebraic groups, or, rather, naturally isomorphic to such. However, the theory of algebraic groups was not suitably developed in Kolchin's view for his purpose when he first needed it, so that a minor, but persistent and essential, theme in his work is the theory of algebraic groups.

As implied by the title, this lecture will emphasize both themes, but I still do not claim to provide an even-handed discussion of all of Kolchin's work in this area. More specifically, I shall take for granted whatever is needed from the theory of differential fields, or more generally from differential algebra, to which Kolchin had to add in no small measure, in order to proceed as directly as possible to the two main topics and their relationships.

My starting point in describing Kolchin's work will be his Annals paper on algebraic matric groups and Picard-Vessiot theory [K10]. It already contains important results, but also a program which was to occupy him for about 25 years. In order to put this paper in context, I would like first to make some historical remarks, to indicate how the theory presented itself to him then.

## §1. The Picard-Vessiot theory

**1.** The idea of using Lie groups to develop some sort of Galois theory for differential equations is as old as Lie theory itself. It was in the mind of S. Lie when he first developed it, as can already be seen from an 1874 letter to A. Mayer.[1] He had noticed that several classical methods for finding the solutions of certain differential equations could be given a unified treatment by using a one-parameter subgroup leaving the equation invariant. This gave him the idea that if a Lie group leaves invariant a given system of ordinary differential equations this should help one to reduce the search for solutions to the study of a system of lower degree. The existence of such a group would imply certain properties of the solutions, reminding one of the role of the Galois group of an algebraic equation. But the analogy is rather weak: every algebraic equation has a Galois group, the structure of which gives essential insight into the properties of the roots, whereas most differential equations do not admit a non-trivial Lie group of transformations.

---

First Kolchin Memorial Lecture, given at Columbia University on April 16, 1993, published in [K], 505-525, and reproduced here with minor modifications.

[1]Cf. the letter of February 3, 1874 in Vol. V of Lie's Collected Papers, p. 586.

Ellis R. Kolchin

**2.** A point of view closer to the model of Galois theory was proposed by Emile Picard for homogeneous linear differential equations, in a series of C.R. notes and papers between 1883 and 1898 (see [P2]), and in the third volume of his "Traité d'Analyse" [P1]. Consider the ordinary homogeneous differential equation

$$(\mathcal{E}) \qquad \frac{d^n y}{dx^n} + p_1(x)\frac{d^{n-1}y}{dx^{n-1}} + \cdots + p_n(x)y = 0,$$

where the $p_i$'s belong to some field $F$ of functions meromorphic in a given region of the plane, stable under the taking of derivatives and containing $\mathbb{C}$. It has a "fundamental system of solutions" $y_1, \ldots, y_n$, i.e. a set of solutions linearly independent over $\mathbb{C}$. Then every other solution is a linear combination of the $y_i$'s with constant coefficients. As was well-known, the condition for linear independence is that the

Wronskian

$$W(y_1, \ldots, y_n) = \begin{vmatrix} y_1 & \cdots & y_n \\ y_1^{(1)} & \cdots & y_n^{(1)} \\ & \vdots & \\ y_1^{(n-1)} & \cdots & y_n^{(n-1)} \end{vmatrix}$$

be $\neq 0$. Here and below, $y_i^{(j)}$ stands for $\frac{d^j y_i}{dx^j}$. Now let

$$A = (a_i^j) \in \mathbf{GL}_n(\mathbb{C}).$$

Then

$$\overline{y}_j = \sum_i a_j^i y_i$$

is another fundamental set of solutions. We want to impose on $A$ certain conditions analogous to those defining the Galois group of an algebraic equation. In that case it can be required that $A$ is an automorphism of the splitting field of the equation or, equivalently, without using that notion, that it respects all algebraic relations between the roots. Here we use an analogue of that second formulation. Let

$$Y_i^{(j)} \qquad (i = 1, \ldots, n; j = 0, 1, 2, \ldots)$$

be indeterminates. Let $\mathcal{P}(\mathcal{E})$ be the set of polynomials $P(Y_i^{(j)})$ in the $Y_i^{(j)}$, with complex coefficients, such that $P(y_i^{(j)})$, viewed as a function of $x$, belongs to $F$. The condition imposed on $A$ is then that $P(\overline{y}_j^{(j)})$ is the same function of $x$ as $P(y_i^{(j)})$ for all $P \in \mathcal{P}(\mathcal{E})$. These $A$'s form what Picard called the "transformation group of $(\mathcal{E})$", to be denoted $G(\mathcal{E})$; F. Klein called it the "rationality group of $(\mathcal{E})$". Klein also pointed out that if $n = 2$ and the equation has regular singular points, then $G(\mathcal{E})$ is the smallest algebraic group containing the monodromy group of the equation.[2]

To justify the use of "group", Picard pointed out only that $G(\mathcal{E})$ contains the product of any two of its elements. That $A \in G(\mathcal{E})$ implies $A^{-1} \in G(\mathcal{E})$ was shown later by A. Loewy [L].[3]

The group $G(\mathcal{E})$ is first of all a Lie group, but Picard pointed out that it is moreover "algebraic" in the sense that the matrix coefficients are algebraic functions of suitably chosen parameters (rather that merely analytic), which led him to study some properties of such groups. In particular, he determined the structure of those of dimensions 1 or 2 (cf. [96] in [P2]) and stated that the previous parameters can be chosen so that the matrix coefficients are rational functions of them ([P1], XVII, n° 12) (i.e., that the group variety is unirational, in present-day language) (see Chapter V, §3).

The theory was further developed by E. Vessiot over the years 1892–1904; see his account in [V], where earlier references are also given. His main result says, in substance, that $G(\mathcal{E})$ is connected solvable ("integrable" in Lie's sense) if and only if $(\mathcal{E})$ is solvable by "quadratures". That notion is not precisely or consistently defined, as pointed out by Kolchin in [K10], and we shall come back to it later (see

---

[2]This is also true of a linear homogeneous equation with regular singular points of any degree; cf. [Bk], Theorem 2.5.1. Here it is understood that $F = \mathbb{C}(z)$.

[3]There were precedents in Lie theory where this condition had been forgotten. However, Picard includes it when it comes to defining finite groups of substitutions (cf. [P1], p. 452).

J. F. Ritt

§6). Of course, it means adjoining a primitive; but, apparently, the exponential of a primitive was also allowed.

**3.** The whole theory was somewhat obscure, and the necessity of various clarifications or reformulations was felt already at the time. In fact, in giving the definition of $G(\mathcal{E})$ above, I have followed [L] and [V] rather than Picard himself. There was little further development until [K10]. There, after summarizing the history of the subject and pointing out various ways in which it was lacking in precision or rigor, Kolchin states:

> the purposes of the present paper are, first to develop a set of theorems on algebraic matric groups at least adequate to meet the demands of the Picard-Vessiot theory and second, to algebraize, rigorize, round out, and augment that theory.

The framework for the algebraization needed to discuss differential equations was already available, namely the differential algebra, developed by his teacher J. F. Ritt. On the group side, Kolchin felt the need to build a theory of algebraic matric groups, rather than rely on the theory of Lie groups, which he viewed as far deeper than the one to be developed. His paper then has three main parts.

(A) Algebraic matric groups (Chapter I).

(B) Galois theory of differential fields (Chapter III).

(C) The Picard-Vessiot theory (Chapters IV, V).

(Chapter II was devoted to preliminaries on differential algebra.)

## §2. Linear algebraic groups

Before coming to (A), let me mention that there was more at the time to the theory of algebraic matrix groups than he implied. In particular, L. Maurer had published several interesting papers (1890-94) in which he notably characterized the Lie algebras of algebraic groups and essentially proved that a group variety is rational (rather than unirational); see V, §4. His theory had been revived and generalized by Chevalley and Tuan, and then by Chevalley (see VI, 1.1; VII, 4).

Since (B) and (C) are in characteristic zero, this might have filled Kolchin's needs. Whether he was aware of it or not I do not know, but the fact that it was not taken into account was ultimately a "good thing" because Kolchin decided that:

> To emphasize the purely algebraic nature of the subject matter the
> proofs are carried out in manner valid for fields of non-zero as well
> as zero characteristic.

so that part (A) and the companion paper [K11] constitute in fact the birth certificate of the theory of linear algebraic groups over algebraically closed fields of *arbitrary* characteristic. In the following summary, I shall discuss both together.

**4.** Let $C$ be an algebraically closed commutative field of characteristic $p$. The notion of algebraic matrix group over $C$ introduced by Kolchin is now the standard one: a subgroup $G$ of $\mathbf{GL}_n(C)$ is algebraic if it is the set of all invertible matrices whose coefficients annihilate a given set of polynomials in $n^2$ indeterminates with coefficients in $C$, i.e. $G$ is the intersection of $\mathbf{GL}_n(C)$ with a closed subvariety of $\mathbf{M}_n(C)$, called by Kolchin the underlying manifold of $G$.

**4.1.** A first main result is that if $G$ is connected and solvable, it can be put in triangular form. Since then this result has been known as the Lie-Kolchin theorem. Kolchin also shows that $G$ is solvable, as an abstract group, if and only if it is solvable as an algebraic group: the series of the smallest algebraic groups containing the successive derived groups stops at $\{1\}$. (In fact, it was proved later that the abstract derived groups are automatically algebraic subgroups, cf. e.g. [B].)

**4.2.** If $G$ is commutative, then it is the direct product of two subgroups $G_s$ and $G_u$ consisting respectively of the semisimple and unipotent elements in $G$. [An immediate consequence, not drawn by Kolchin, is that any linear algebraic group is stable under the (multiplicative) Jordan decomposition.]

**4.3.** Kolchin also introduces the notions of "anticompact" and "quasicompact" groups of matrices, i.e. the groups consisting of unipotent and semisimple matrices respectively. He shows that any anticompact group can be put in triangular form. If $G$ is quasicompact, algebraic, and connected, then it can be put in diagonal form (i.e., it is a torus in the usual terminology). Also he shows that if $G$ is connected, the $k$-th power map $x \mapsto x^k$ is a dominant morphism, provided that $(k, p) = 1$ in case $p \neq 0$.

**4.4.** These two papers, [K10] and [K11], also contain some general comments about connected components, Jordan-Hölder series, and algebraic subgroups. Altogether, they present the first body of theorems on linear algebraic groups with proofs insensitive to the characteristic of the ground field, carried out in the framework of algebraic geometry, without any recourse to Lie algebras and the usual mechanism of Lie theory. They were influential on my own work, as can be seen from my first paper on this topic (Annals of Math. **64** (1956), 20–82), in which these results are all proved again. They are also incorporated in [B] and belong to any systematic exposition of the theory.

## §3. Generalization of the Picard-Vessiot theory

**5.1.** For (B) and (C) the basic notion is that of a *differential field*. A (commutative) field $F$ is differential if it is endowed with a derivation,[4] i.e. a self-map $\delta$ satisfying

$$\delta(a+b) = \delta(a) + \delta(b), \quad \delta(a.b) = \delta(a).b + a.\delta(b) \qquad (a, b \in F).$$

Then

$$C_F = \{a \in F, \delta(a) = 0\}$$

is a subfield, the *field of constants* of $F$. The $n$-th derivative of $a$ is $\delta^n.a$ and is written $a^{(n)}$. An *automorphism* of $F$ is an automorphism of the underlying field commuting with the derivation.

In fact, this is an *ordinary* differential field. Kolchin develops the theory more generally for *partial* differential fields, i.e. fields endowed with a finite set of commuting derivations.

*In the sequel, unless otherwise stated, $F$ is a differential field of characteristic zero with algebraically closed field of constants $C_F$ or $C$. The field $F$ is partial from §7 on, ordinary before, unless otherwise stated.*

**5.2.** Let $E \supset F$ be a field containing $F$ as a subfield. It is a *differential extension* of $F$ if it is a differential field with derivation leaving $F$ stable and inducing on $F$ its given derivation. It is a *finitely generated differential extension* if there exist moreover finitely many elements $\eta_1, \ldots, \eta_1$ in $E$ such that $E$ is generated over $F$ by the $\eta_i$'s and their derivatives of all orders, in which case we write

$$E = F\langle \eta_1, \ldots, \eta_n \rangle.$$

It is not necessarily a finitely generated field extension of $F$, i.e. its transcendence degree $\partial^0(E/F)$ over $F$ may be infinite.

In the present context, the equation $(\mathcal{E})$ takes the form

$$(\mathcal{E}) \qquad L(y) = y^{(n)} + p_1.y^{(n-1)} + \cdots + p_n.y = 0 \quad (p_i \in F, i = 1, \ldots, n)$$

and we are looking for solutions $y$ in some differential extension of $F$. The standard facts about the solutions of $(\mathcal{E})$ in the classical case recalled earlier generalize to the following theorem (see [K4] and [K10], §§14, 15).

THEOREM. *There exists a finitely generated differential extension $E$ of $F$ containing $n$ solutions $\eta_1, \ldots, \eta_n$ of $(\mathcal{E})$ such that*

$$W(\eta_1, \ldots, \eta_n) = \begin{vmatrix} \eta_1 & \cdots & \eta_n \\ \eta_1^{(1)} & \cdots & \eta_n^{(1)} \\ & \vdots & \\ \eta_1^{(n-1)} & \cdots & \eta_n^{(n-1)} \end{vmatrix} \neq 0.$$

---

[4]Called in [K10] a derivative. From [K17] on, Kolchin switched to *derivation operator* and let *derivative operators* be the operators on $F$ defined by the powers of $\delta$ (similarly for partial differential fields). I shall use derivation and derivatives.

*In any differential extension $E'$ of $E$, the $\eta_i$ are linearly independent over $C_{E'}$, and span over $C_{E'}$ the space of solutions of $(\mathcal{E})$ in $E'$. No differential extension $F'$ of $F$ contains more than $n$ solutions of $(\mathcal{E})$ linearly independent over the constants.*

A set of $n$ linearly independent solutions of $(\mathcal{E})$ is called a *fundamental set of solutions* of $(\mathcal{E})$.

**5.3.** A differential extension $E$ of $F$ is a *Picard-Vessiot extension* (a P.V. extension for short) of $F$ if it is of the form

$$E = F\langle \eta_1, \ldots, \eta_n \rangle,$$

where the $\eta_i$ form a fundamental set of solutions of an equation $(\mathcal{E})$, *and if* $C_E = C_F$.

It is clear from $(\mathcal{E})$ that the derivatives of all orders of the $\eta_i$'s are linear combinations with coefficients in $F$ of the first $(n-1)$ derivatives. Therefore $\partial^0(E/F)$ is finite.

Kolchin proved later that given $(\mathcal{E})$, one can find a fundamental set of solutions $\eta_1, \ldots, \eta_n$ in some extension such that $F\langle \eta_1, \ldots, \eta_n \rangle$ has the same field of constants as $F$ (cf. [K12] or [K29], Prop. 13, p. 412). This is under our standing assumption that $C_F$ is algebraically closed (of characteristic 0). If not, then Seidenberg has produced an equation $(\mathcal{E})$ such that $C_E \neq C_F$ for all differential field extensions $E$ generated over $F$ by a fundamental set of solutions of that equation (cf. [S] or [K29], Exercise 1, p. 413).

**5.4.** Let $E$ be a P.V. extension of $F$. Then, by definition, its *Galois group over $F$* is the group $G = G(E/F) = \mathrm{Aut}(E/F)$ of automorphisms of $E$ (as a differential field) leaving fixed each element of $F$. Let $g \in G$. Then

$$g.\eta_i = \sum_j c_i^j(g).\eta_j,$$

where the $c_i^j(g)$ belong to $C_E$ at first, hence to $C_F$ since both are assumed to be equal. The map

$$g \mapsto c(g) = (c_i^j(g))$$

is a faithful linear representation of $G$ into $GL_n(C_F)$.

Before stating the main theorem concerning $G(E/F)$, we note that if $E_1$ is a differential field intermediate between $E$ and $F$, then its field of constants is equal to $C_F$, since $C_E = C_F$. Therefore, it is clear from the definition that $E$ is a P.V. extension of $E_1$ and $G(E/E_1)$ is defined as before.

**5.5.** THEOREM. *Let $E$ be a P.V. extension of $F$. We keep the previous notation.*

(i)  *The group $c(G)$ is the group of rational points over $C_F$ of an algebraic subgroup of $\mathbf{GL}_n(C_F)$ defined over $C_F$, of dimension equal to $\partial^0(E/F)$.*

(ii)  *The map which assigns to an algebraic subgroup $H$ of $c(G)$ the subfield $E^H$ of elements fixed under $H$ is a bijection between algebraic subgroups and intermediate differential subfields. The inverse bijection assigns to a differential subfield $E_1$ the group $G(E/E_1)$ of automorphisms of $E$ over $E_1$.*

(iii)  *$H$ is a normal subgroup if and only if $\sigma(E^H) = E^H$ for all $\sigma \in G(E/F)$. In that case, the restriction of an automorphism to $E^H$ defines an isomorphism of $G/H$ onto $\mathrm{Aut}(E^H/F)$.*

See §§17–20 in [K10]. There $F$ is ordinary. The generalization to partial differential fields is carried out in [K15]. We indicate briefly how Kolchin defined P.V.

extensions in the general case. To this effect he first remarks that $p_i = \pm W_i/W_0$, where

$$W_i = \det(\eta_k^{(j)}) \qquad (1 \le i, k \le n; 0 \le j \le n, j \ne n - i)$$

by Cramer's rule. Given now $n$ elements $y_1, \ldots, y_n$ in a partial differential field $E$ and partial derivatives $\theta_1, \ldots, \theta_n$, let

$$W(\theta_1, \ldots, \theta_n, y_1, \ldots, y_n) = \det(\theta_j y_i).$$

If the $y_i$'s are linearly independent over constants, there is always a choice of partial derivatives of orders $< n$, say $\sigma_1, \ldots, \sigma_n$, such that

$$(1) \qquad W(\sigma_1, \ldots, \sigma_n; y_1, \ldots, y_n) \ne 0$$

([K15], Lemma 1). Assume now that $F$ is a partial differential field. A differentiable extension of $E$ of $F$ is a P.V. extension if

$$C_E = C_F, \qquad E = F\langle \eta_1, \ldots, \eta_n \rangle,$$

where the $\eta_i$ are linearly independent over constants and

$$W(\theta_1, \ldots, \theta_n; \eta_1, \ldots, \eta_n)/W(\sigma_1, \ldots, \sigma_n; \eta_1, \ldots, \eta_n) \in F$$

for all choices $\theta_1, \ldots, \theta_n$ of partial derivatives of orders $\le n$, the $\sigma_i$'s being a set of partial derivatives of orders $< n$ such that (1) is satisfied by the $\eta_i$'s.

The initial remark above shows readily that this definition is equivalent to the original one in the ordinary case. For other characterizations and further discussion of P.V. extensions, see [K29], VI, n° 6, and the exercises following it, pp. 409–418.

In [K10], Kolchin does not raise the question of whether $E^H$ in (iii) is a P.V. extension of $F$. Later on, he provided a positive answer in [K18], III, n° 3, Corollary to Theorem 2 (see also 9.2, below).

**6.** The last sections of the paper (§§23–27) are devoted to a precise version and generalization of Vessiot's theorem about equations solvable by "quadratures". Consider first two special cases:

**6.1.** The primitives of $a \in F$, i.e. the solutions of $y' = a$, differ by a constant. The equation $y' = a$ is not homogeneous, but any solution $\eta_2$ and the constant $\eta_1 = 1$ span the space of solutions of

$$(1) \qquad y'' - (a'/a).y' = 0.$$

The extension $E = F\langle \eta_2 \rangle$ is P.V. If $g \in G(E/F)$, then

$$g.\eta_1 = \eta_1, \qquad g.\eta_2 = \eta_2 + c = \eta_2 + c.\eta_1,$$

so that $G$ can be identified to the group of upper triangular unipotent matrices in $\mathbf{GL}_2(C_F)$, i.e. to the additive group $\mathbf{G}_a$ of $C_F$.

**6.2.** Now let $E = F\langle \eta \rangle$, where $\eta$ is a solution of

$$y' - a.y = 0.$$

Then $\eta = \exp b$, where $b$ is a primitive of $a$. Any other fundamental solution is a non-zero multiple of $\eta$; hence $G(E/F)$ is isomorphic to a subgroup of $\mathbf{GL}_1$. There are then two cases:

    a) $\eta$ is transcendental, which is equivalent to $G(E/F) = \mathbf{GL}_1$,

    b) $\eta$ is algebraic. Then $G(E/F)$ is finite cyclic, of some order $h$, and $\eta$ is an $h$-th root of an element of $F$.

The results of §27 imply conversely that if $E$ is a P.V. extension with a Galois group of one of the three previous types, then we are in the corresponding case.

**6.3.** From this it follows first that a P.V. extension has a solvable Galois group if and only if it can be obtained from $F$ by a succession of steps 6.1, 6.2. In fact, Kolchin carries out the discussion a bit more generally so as to allow for more general algebraic extensions.

An extension $E = F\langle \alpha_1, \ldots, \alpha_q \rangle$ of $F$ is said to be *Liouvillian* if $C_E = C_F$ and $\alpha_{i+1}$ is either algebraic over $F\langle \alpha_1, \ldots, \alpha_i \rangle$ or a primitive or an exponential of a primitive of an element in that field $(i = 0, \ldots, q - 1)$. Now let $E$ be a P.V. extension of $F$. Then (§25, Theorem), if $E$ is contained in a Liouvillian extension, then $G(E/F)$ has a solvable identity component. Conversely, the latter property implies that $E$ is Liouvillian. More precisely, Kolchin distinguishes ten types of Liouvillian extensions and ten types of algebraic groups, and shows that they match in case of P.V. extensions contained in a Liouvillian extension.

**7.** The above sections 5,6 summarize part (C) in my initial division of [K10]. Part (B) is Kolchin's first attempt at a Galois theory of differential fields, not necessarily associated to differential equations, and is much more tentative. To this effect, he has to introduce a notion of normal extension of $F$, but can prove only part of the desired Galois correspondence. See the introduction of [K16] for a discussion of the "blemishes" of that theory. There he defines the concept of "strongly normal extension", which was to be the basis of all his further work on Galois theory, so I shall go over to it directly. The main results are proved first in [K16], [K18] and in the paper [K19], written jointly with Serge Lang. The whole theory was given a comprehensive exposition in [K29].

## §4. Galois theory of strongly normal extensions

**8.1.** Kolchin first introduces a *universal field* in analogy with the universal field in Weil's Foundations of Algebraic Geometry [W].

A differential extension $U$ of $F$ is a *universal extension* if, given a finitely generated differential extension $F_1$ of $F$ in $U$ and a finitely generated differential extension $E$ of $F_1$, there exists an isomorphism of $E$ into $U$ over $F_1$. It is algebraically closed, and its field of constants $C_U$ is a universal field for $C_F$ in Weil's sense (i.e., algebraically closed, of infinite transcendence degree over $C_F$). Then nothing is lost by restricting oneself to finitely generated extensions of $F$ contained in $U$, and we shall do so. (Cf. [K16], I, n° 5, and [K29], III, n° 7, for an extension to arbitrary characteristics.)

**8.2.** Now let $E$ be a finitely generated differential extension of $F$. What we want is a condition of normality. The most obvious one is that $E$ be stable under any automorphism of $U$ over $F$. However, Kolchin noticed that this forces $E$ to be algebraic over $F$, and so it is too strong, so that isomorphisms of $E$ into $U$ over $F$ have to be allowed. In order to avoid endless repetitions of "into $U$", I shall follow Kolchin's convention and call this simply an isomorphism of $E$ over $F$ (whereas an "automorphism" of $E$ over $F$ is of course an isomorphism of $E$ onto itself fixing $F$ pointwise). However, they must be restricted in some fashion. On the other hand, whatever the definition of normal extension is, a P.V. extension has to be one. Assume then that $E = F\langle \eta_1, \ldots, \eta_n \rangle$ is a P.V. extension, where the $\eta_i$ are a fundamental set of solutions of the underlying equation $L(y) = 0$. Let $\sigma$ be an isomorphism of $E$ over $F$. Then the $\sigma.\eta_i$ form a fundamental set of solutions in

$\sigma(E)$, and therefore

$$\sigma.\eta_i = \sum c_i^j \eta_j \qquad (i = 1, \ldots, n),$$

where the $c_i^j$ belong to the constant field of $\sigma(E)$, hence also to $C_U$. Therefore

$$\sigma(E) \subset E.C_U.$$

Kolchin saw that this was the decisive restriction, and he introduced the following definitions:

DEFINITIONS. (i) An isomorphism $\sigma$ of $E$ over $F$ is *strong* it satisfies the following two conditions:
    (1) $\sigma \mid C_E = \mathrm{Id}$.
    (2) $\sigma(E) \subset E.C_U$ and $E \subset \sigma(E).C_U$, i.e., $E.C_U = \sigma(E).C_U$.
    (ii) $E$ is a *strongly normal extension* of $F$ if it is finitely generated over $F$ (as a differentiable field) and every isomorphism of $E$ over $F$ is strong.

See [K29], VI, n$^{os}$ 2,3. These definitions appear first in [K16], except that (1) is not needed since $C_E = C_F$ by a standing assumption there (on p. 772).
    If $E$ is strongly normal over $F$, then it is clearly so over every differential subfield containing $F$.
    In the sequel, $C(\sigma)$ denotes the field of constants of the field $E.\sigma(E)$. The condition (2) may also be written

(3)                    $$E.C(\sigma) = E.\sigma(E) = \sigma(E).C(\sigma).$$

If $E$ is strongly normal, then $C_E = C_F$ ([K29], V, Prop. 9, p. 393) and $E$ has finite transcendence degree over $F$ ([K29], VI, Prop. 11, p. 394; the latter statement is also contained in [K16], III, n$^\circ$ 2, Theorem 2).
    **8.3.** *The Galois group.* If $E$ is any finitely generated differential extension of $F$, then it is linearly disjoint from $C_U$ over its own field of constants $C_E$ of $E$ (see [K16], I, n$^\circ$ 4, Prop. 3 and the comment following it, or [K29], II, n$^\circ$ 1, Cor. 1, p. 87). As a consequence, a strong isomorphism of $E$ over $F$ extends uniquely to an automorphism of $E.C_U$ fixing $C_U$ pointwise, i.e. to an automorphism of $E.C_U$ over $F.C_U$. Conversely, if $\tau$ is an automorphism of $E.C_U$ over $F.C_U$, then its restriction to $E$ is a strong isomorphism of $E$ over $F$ which determines $\tau$ completely. Now assume $E$ to be strongly normal. Then every isomorphism of $E$ over $F$ is strong, by definition, and so we get a canonical bijection

$$\{\mathrm{Isom}\, E/F\} = \mathrm{Aut}(E.C_U/F.C_U),$$

which allows one to view the left-hand side as a *group*. By definition, this is the Galois group $G(E/F)$ of $E$ over $F$.
    **8.4.** In the sequel we write $C$ for $C_F$, and $C_U$ is viewed as the universal field for $C$, i.e. algebraic varieties defined over $C$ are assumed to have their points in $C_U$. If $M$ is an irreducible one, then the residue field $C(x)$ of $x \in M$ at $x$ is the subfield of $C_U$ generated over $C$ by the values at $x$ of the rational functions on $M$ which are defined at $x$.
    We now give a first formulation, provisory as far as (i) is concerned, of the fundamental theorem of Kolchin's Galois theory of strongly normal extensions:

THEOREM. *Let $E$ be a strongly normal differential extension of $F$.*

(i) *There exists an algebraic group* **G** *defined over $C$ such that $G(E/F)$ may be identified with* $\mathbf{G}(C_U)$ *so that $C(\sigma)$ is the residue field of $\sigma \in G(E/F)$. For a finitely generated extension $K$ of $C$ in $C_U$, this isomorphism identifies $\mathbf{G}(K)$ with* $\{\sigma \in G(E/F) \mid \sigma(E) \subset E.K\}$. *In particular,* $\mathbf{G}(C_F) = \mathrm{Aut}(E/F)$. *The dimension of* **G** *is equal to $\partial^0(E/F)$.*

(ii) *The map which assigns to an algebraic subgroup $H$ of $G(E/F)$ defined over $C$ the field $E^H$ of invariants of $H$ establishes a bijection between algebraic $C$-subgroups of $G(E/F)$ and intermediate differential fields between $F$ and $E$. The inverse bijection assigns to such a field $F_1$ the subgroup of $G(E/F)$ of elements of $G(E/F)$ fixing $F_1$ pointwise.*

(iii) *The $C$-subgroup $H$ is normal if and only if $E^H$ is strongly normal over $F$. In that case, the restriction of $\sigma \in G(E/F)$ to $E^H$ defines an isomorphism of algebraic groups of $G(E/F)/H$ onto $G(E^H/F)$.*

For (ii) and (iii), see [K29], VI, n° 4, Theorems 3 and 4, respectively. It is also shown in the latter that the conditions in (iii) are equivalent to either of:

(a) For each element $\alpha$ of $E^H$ not contained in $F$, there exists a strong isomorphism of $E^H$ over $F$ not leaving $\alpha$ fixed.

(b) $\sigma(E^H) \subset E^H C_U$ for every $\sigma \in G(E/F)$.

We shall soon come to (i) and to what is already proved in [K16] and [K18]. Before doing so, we list some complements to the fundamental theorem. We write $G$ for $G(E/F)$ and view it as an algebraic group over $C$.

**9.1.** The group $G(E/F)$ is not necessarily connected. In fact the fixed point set of the identity component $G^0$ of $G$ is the algebraic closure $F^0$ of $F$ in $E$, and we have

$$[F^0 : F] = [G : G^0].$$

**9.2.** In this theorem, the algebraic groups are not necessarily linear. In fact, Kolchin proves that $E$ is P.V. if and only if $G(E/F)$ is linear. The new implication is of course that $G(E/F)$ linear implies $E$ to be P.V. (see [K18], III, n° 3, Theorem 2, p. 891, or [K29], Cor. 2 to Theorem 8, p. 427). Now, given a linear algebraic group $G$ and a closed normal subgroup $N$, there always exists a rational representation of $G$ with kernel $N$ (a by now standard fact, proved first in [K14]); it follows from this characterization that if $E$ is P.V., any strongly normal extension of $F$ in $E$ is also P.V. ([K18], Cor. to Theorem 2, p. 891). A slightly more direct way to derive the corollary from the theorem is also hinted at in exercise 2 on p. 427 of [K29].

**9.3.** Assume now that $G$ is connected of dimension one. The two cases where it is linear were discussed in 6.1 and 6.2. There remains the one where $G$ is an elliptic curve. This corresponds to the case where $E$ is *Weierstrassian* over $F$, first discussed in [K16], III, n°6, then also in [K18], III, n° 3, and in [K29], VI, n° 5. Briefly, $E = F\langle\alpha\rangle$, with $\alpha$ transcendental, and there exist $g_2, g_3 \in C_F$ and $a \in F$ such that

$$g_2^3 - 27g_3^2 \neq 0, \qquad (\alpha')^2 = a^2(4\alpha^3 - g_2\alpha - g_3).$$

**9.4.** By the theorem, every Galois group is a group variety over $C$. As a step towards a converse, Kolchin shows that given a connected group variety over $C$, there exists a differential field with field of constants $C$ and a strongly normal extension of that field with Galois group $G$. This is still far from a solution to the "inverse problem" of Galois theory, namely, given $F$, to determine the possible Galois groups of strongly normal extensions of $F$.[5]

**9.5.** Let $G$ be connected. Then by the Chevalley-Barsotti structure theorem $G$ has a biggest normal linear subgroup $N$ such that $G/N$ is an abelian variety (see e.g. [C], and VII, 10; a proof is also given in [K29], V, n° 24). Let $E' = E^N$. Then, in view of 9.2 $E'$ is the smallest strongly normal extension of $F$ in $E$ such that $E$ is P.V. over $E'$, and $G(E'/F)$ is an abelian variety.

**9.6.** In [K19] and in the last section of [K29] the realization of strongly normal extensions by function fields of principal homogenous spaces is studied. Let $G$ be a connected group variety over $C$. A variety $V$ over $F$ is a principal homogenous space (over $F$) for $G$ if $G$ acts simply transitively on it. Given $v \in V$, the map $g \mapsto v.g$ is an isomorphism of varieties over $F(v)$. There is always such a $v$ which is algebraic over $F$. Then $G$ operates over $F(V)$ by right translations. It is shown that a strongly normal extension of $F$ with Galois group $G$ may be so realized.

## §5. Foundational work on algebraic sets and groups

**10.1.** We now come back to $(i)$ in the fundamental theorem (8.4) and to the reason why its formulation had been called provisory. It is correct as stated, and references will be provided. Still, it misrepresents an important aspect of Kolchin's approach. The assertion "is isomorphic to an algebraic group over $C$" refers implicitly to the standard notion of algebraic group, namely, an algebraic variety over $C$ endowed with a group structure such that inverse and product are defined by $C$-morphisms of varieties. In the sequel, I shall refer to those as group varieties over $C$. But Kolchin proceeds differently: he first defines a structure on $G(E/F)$ analogous to that of an algebraic group, in which the fields $C(\sigma)$ play the role of residue fields, so that (i) is in fact the combination of two statements:

(a) *The Galois group $G(E/F)$, endowed with the fields $C(\sigma)$, is an algebraic group in Kolchin's sense.*

(b) *$G(E/F)$ may be identified to a group variety over $C_F$ so that $C(\sigma)$ is the residue field at $\sigma$ ($\sigma \in G(E/F)$).*

**10.2.** To describe this, let me first recall, however briefly, some concepts familiar in [W], which play a great role in Kolchin's approach.

Let $K$ be an algebraically closed ground field. All extensions of $K$ are to be contained in a fixed universal field. Let $V$ be a variety over $K$. Then, as already recalled, every point $x \in V$ has a residue field $K(x)$, a finitely generated extension of $K$. The points of $V$ are partially ordered by a notion of specialization over $K$, written $x \underset{K}{\to} x'$ or simply $x \to x'$. [If $x, x'$ are in an open affine subset, every regular function vanishing at $x$ also vanishes at $x'$.] It implies that $\partial^0(K(x)/K) \geq \partial^0(K(x')/K)$. The specialization is *generic* if there is equality, in which case $x' \to x$, *non-generic* otherwise. The locus $Z_x$ of $x$ over $K$ is the set of specializations (over $K$) of $x$. It is the smallest $K$-irreducible subset of $V$ containing $x$, and $x$ is a generic point of $Z_x$ over $K$.

If $W$ is another variety over $K$, a map $f : V \to W$ is a morphism if $f^0(K(f(x)) \subset K(x)$ for all $x \in V$ and if it is compatible with specializations.

**10.3.** We now return to [K16] and the strongly normal extension $E$ of $F$. Let $G^*$ be the group defined by the strong isomorphisms of $E$ over $F$ (see 8.3) and $G$

---

[5]Kolchin never contributed further to the inverse problem, which puts it outside the purview of this chapter. For a historical survey, see the paper by M. Singer in [K], 527-554.

the subgroup of automorphisms of $E$ over $F$. By definition, $G^*$ is the Galois group of $E$ over $F$, but the main results of [K16] pertain to $G$.

Kolchin introduces the notion of specialization (over $C$) for isomorphisms of $E$ over $F$, and shows that the specialization of a strong isomorphism $\sigma$ is strong (II, n° 5, Prop. 5, p. 774). The strong isomorphism $\sigma$ is isolated if it not a non-generic specialization of another strong isomorphism. $G^*$ has finitely many isolated elements, up to generic specialization. This and some further properties of strong isomorphisms allow Kolchin to define on $G^*$ a structure of algebraic group and morphisms of such groups with the usual properties. Then $G$ is endowed with the induced structure of algebraic group, in which the algebraic subgroups are the intersections of $G$ with algebraic subgroups of $G^*$ defined over $C$, i.e. in fact the groups of rational $C$-points of such subgroups. It is this notion of algebraic subgroup which underlies the statement of the fundamental theorem in [K16], which deals only with $G$. (See n°ˢ 2,3 in III.) The question of whether $G^*$ may be naturally identified with a group variety over $C$, in Weil's sense, is raised in [K16], p. 759, and positively answered in [K18], II, n° 1, Thm. 1, p. 878. More precisely, the fields $C(\sigma)$ are the residue fields in the group variety structure. This then realizes (a) and (b) in 10.1 and yields a proof of (i) in 8.4.

In [K19] this identification is used to put the discussion of realization via principal homogeneous spaces (see 9.6) in the framework of the usual algebraic group theory. But Kolchin comes back in [K29] to his original point of view and develops it much more systematically.

**10.4.** Kolchin's treatment of "algebraic sets and groups" is the subject matter of Chapter V in [K29]. The references in this section and in 10.5 and 10.6 are to that chapter.

Let $K$ be as in 10.2. The starting notion is that of pre-$K$ set (n° 2). The set $V$ is a pre-$K$ set if to each point $x \in V$ is assigned a finitely generated extension $K(x)$ of $K$, subject to relations of specializations, satisfying certain natural conditions. $V$ is a $K$-group if moreover it is a group and we have

$$K(x.y) \subset K(x).K(y), \quad K(x^{-1}.y) \subset K(x).K(y) \qquad (x,y \in V)$$

and some further conditions relative to specializations. $V$ is a homogeneous $K$-space if it is a pre-$K$-set acted upon transitively by a $K$-group $G$, again with certain natural conditions (p. 220). A $K$-set is a pre-$K$-subset of a homogeneous $K$-space. Thus, in this algebraic geometry, the algebraic sets (the $K$-sets) are all contained in homogeneous spaces. Kolchin then develops his theory along lines familiar in other contexts: effect of extensions of the ground field, Zariski topology, regular extensions, and also, for $K$-groups: connected components, products, morphisms, quotients (to be discussed below), linear groups, and Galois cohomology, ending up with a proof of the Chevalley-Barsotti theorem mentioned in 9.5.

**10.5.** We now want to relate those concepts to the corresponding ones in algebraic geometry. There, in any treatment familiar to me, one starts with a notion of affine variety (or scheme), and the varieties (or schemes) are by definition locally affine. Kolchin's approach is *a priori* very different, as we saw. However, a product of $\mathbf{G}_a$'s, i.e. the additive group of a vector space over $K$, is homogeneous, and its $K$-subsets are included in Kolchin's setup. They are essentially affine varieties in disguise, and provide the link between the two points of view.

A $K$-affine set is by definition a $K$-set which is isomorphic to a $K$-subset of some $\mathbf{G}_a^n$. The main point is then to show that in a $K$-set $A$, any finite set of points is contained in an open affine $K$-subset, from which it follows that $A$ has a finite open cover by affine $K$-subsets (n° 16, Thm. 4 and Cor., p. 311). This then implies that Kolchin's $K$-groups or homogeneous $K$-spaces may be naturally identified with group or homogeneous varieties (*loc. cit.*, Remark on pp. 311–312). [In view of this identification, Thm.4 quoted above is not surprising, since it is known that any homogeneous variety is quasi-projective (Chow's theorem).]

**10.6.** In this way Kolchin achieves autonomy, at the cost of a huge (his word) chapter on algebraic groups. Whether it is worth the effort or whether a shorter treatment of the Galois theory using the standard theory of algebraic groups could be given, I will not try to assess, but I would like to mention two points, one minor, one major, in which Kolchin's approach has paid off. The former is the construction of the quotient space $G/H$ of a $K$-group $G$ by a closed $K$-subgroup $H$. One wants a structure of $K$-variety on $G/H$ with a universal property, such that $\pi : G \to G/H$ is a separable $K$-morphism commuting with the action of $G$. For linear groups, there is a simple trick to achieve this (see [B]). For non-linear groups the original approach of Weil [W1] is rather awkward. The usual method nowadays, which also proceeds from the local to the global, is to start from the existence of the quotient of some invariant Zariski-open set, proved by M. Rosenlicht, and to use translations (see [Ss]). From Kolchin's point of view, this existence proof is straightforward and directly global. The starting point is Theorem 4 on p. 240, according to which, given a closed subset $A$ of a homogeneous $K$-space, there is a smallest field $L \supset K$ such that $A$ is an $L$-subset. It is finitely generated over $K$ and denoted $K(A)$. Then, Kolchin assigns to $x \in G/H$ the field $K(\pi^{-1}(x))$. If $x' \in G/H$, then $x \to x'$ if there exist $y \in \pi^{-1}(x), y' \in \pi^{-1}(x')$ such that $y \to y'$. It is then rather easily seen that his axioms for a homogeneous $K$-space are satisfied and that $\pi$ is separable (n° 11).

**10.7.** The second point reaches further, to the last major project of Kolchin: the foundation of a theory of differential algebraic groups, i.e., roughly speaking, groups locally defined by algebraic differential equations [K35]. In the seventies, a theory of affine or linear differential algebraic groups had been developed by P.J. Cassidy, but Kolchin decided to supply the foundations for a general theory. In spite of many technical complications, he could use Chapter V of [K29] almost as a blueprint: his axiomatic treatment there could be adapted to this more general situation, and, in fact, a number of sections could be taken almost verbatim, with minor changes in notation. This initiates more broadly a "differential algebraic geometry", at any rate for subsets of homogeneous spaces, as in [K29].

In the linear case, the theory was pursued by P.J. Cassidy and led to a classification of semisimple differential groups and Lie algebras [Ca]. On the other hand, A. Buium applied similar ideas to the study of diophantine problems over function fields, thus establishing completely new connections with other topics in mathematics. This whole area is in full development and well worth a report, which I have unfortunately neither the time nor the competence to give.[6] Still, I wanted to mention it as a counterpart to the main topic of this lecture: the latter is well-rounded, with statements of considerable aesthetic beauty, seemingly in final form. The former, on the other hand opens up new directions and points to work for the future.

---

[6]I can now refer to the paper by A. Buium and P. Cassidy in [K], 567-636

# References for Chapter VIII

## References to E. R. Kolchin's Work*

[K]     Ellis R. Kolchin, Selected Works with Commentary (H. Bass, A. Buium, and P. J. Cassidy, eds.), Amer. Math. Soc., 1999.

[K4]    ———, *On the basis theorem for differential systems*, Trans. Amer. Math. Soc. **52** (1942), 115–127; [K], pp. 49–61.

[K10]   ———, *Algebraic matric groups and the Picard-Vessiot theory of homogeneous linear ordinary differential equations*, Ann. of Math. (2) **49** (1948), 1–42; [K], pp. 87–128.

[K11]   ———, *On certain concepts in the theory of algebraic groups*, Ann. of Math. (2) **49** (1948), 774–789; [K], pp. 129–144.

[K12]   ———, *Existence theorems connected with the Picard-Vessiot theory of homogeneous linear ordinary differential equations*, Bull. Amer. Math. Soc. **54** (1948), 927–932; [K], pp. 145–150.

[K14]   ———, *Two proofs of a theorem on algebraic groups* (with C. Chevalley), Proc. Amer. Math. Soc. **2** (1951), 126–134; [K], pp. 159–167.

[K15]   ———, *Picard-Vessiot theory of partial differential fields*, Proc. Amer. Math. Soc. **3** (1952), 596-603; [K], pp. 169–176.

[K16]   ———, *Galois theory of differential fields*, Amer. J. Math. **75** (1953), 753–824; [K], pp. 177–248.

[K18]   ———, *On the Galois theory of differential fields*, Amer. J. Math. **77** (1955), 868–894; [K], pp. 261–287.

[K19]   ———, *Algebraic groups and the Galois theory of differential fields* (with S. Lang), Amer. J. Math. **80** (1958), 103–110; [K], pp. 289–296.

[K29]   ———, Differential Algebra and Algebraic Groups, Academic Press, 1973.

[K35]   ———, Differentiable Algebraic Groups, Academic Press, 1985.

## References by Others

[Bk]    F. Beukers, *Differential Galois theory*, From Number Theory to Physics (Les Houches, 1989; M. Waldschmidt et al., eds.), Springer, 1992. 413–439.

[B]     A. Borel, Linear Algebraic Groups, 2nd enlarged edition, GTM **126**, Springer, 1991.

[Ca]    P. J. Cassidy, *The classification of the semisimple differential algebraic groups and the linear semisimple differential algebraic Lie algebras*, J. Algebra **121** (1988), 169–238.

[C]     C. Chevalley, *Une démonstration d'un théorème sur les groupes algébriques*, J. Math. Pures Appl. (9) **39** (1960), 307–317.

[L]     A. Lowey, *Die Rationalitätsgruppe einer linearen homogenen Differentialgleichung*, Math. Annalen **65** (1908), 129–160.

[P1]    E. Picard, Traité d'Analyse, t. III, Gauthier-Villars, Paris 1896; 2nd ed., 1908.

[P2]    ———, Oeuvres, t.2, Éditions du CNRS, Paris, 1979.

[Se]    A. Seidenberg, *Contribution to the Picard-Vessiot theory of homogeneous linear differential equations*, Amer. J. Math. **78** (1956), 808–817.

[Ss]    C. S. Seshadri, *Some results on the quotient space by an algebraic group of automorphisms*, Math. Annalen **149** (1963), 286–301.

[v]     E. Vessiot, *Méthodes d'intégration élémentaires*, Encycl. Sci. Math. Pures Appl., Tome II, Vol. 3 (1910), pp. 58–170.

[W1]    A. Weil, Foundations of Algebraic Geometry, Amer. Math. Soc., 1946.

[W2]    ———, *On algebraic groups and homogeneous spaces*, Amer. J. Math. **77** (1955), 493–512; Oeuvres Sci. II, 235–254.

---

*The numbering follows the preliminary version of [K] that I had to work with when I first wrote this chapter. The final version of [K] uses a different system.

# Name Index

Araki, S., IV, $2.5.3^{13)}$ (78)
Aronhold, S., V, 2,1, 4.17 (97, 112)
Artin, M., III, 13 (44)

Barsotti, I., VI, 2.5 (123)
Bergmann, S., IV, 6.1 (88)
Bochner, S., III, 8 (38)
Bohr, H., III, 8 (38)
Borel, A., VI, 3.2; 4.4, 4.6; VII, $5^{5)}$, 6 (124, 128, 129, 154)
Bourbaki, N., II, 10 (17)
Brauer, R., II, 10; III, 9, 12; IV, 5.3; V, 4.17; VII, 5 (17, 39, 43, 87, 113, 153)
Brown, K., VI, 7.3 (143)
Bruhat, F., VI, 2.3, 7.3; VII, $14^{9)}$ (121, 143, 161)
Buium, A., VIII, 10.7 (178)
Burnside, W., II, 7 (15)

Capelli, A., I, 2.2; III, 11 (6, 42)
Carda, K., V, §6 (115–116)
Cassidy, P.J., VIII, 10.7 (178)
Cartan, É., I, 2.1, 2.4, 3.3; II, 3, 7, 12; III, 1, 2, 4, 6, 7; IV; V, §5; VI, 4.8, 6.3; VII, 4, 11 (5–7, 10, 15, 20; III–IV almost *passim*; 114, 130, 136, 150, 157)
Cartan, H., IV, 6.1; VII, 5 (87, 153)
Casimir, H.B.G., II, 8; III, 9 (16, 39)
Chevalley, C., I, 2.1; II, 9; VI, 2.3, 2.5, 3.2, 3.3, 3.5, 4.4, 6.5; VII, 5 (5, 17; VI–VII almost *passim*)
Chow, W.L., VI, 6.9 (141)
Clebsch, A., V, 2.4 (99)
de Concini, C., III, 13; IV, 4.5 (44, 84)

Demazure, M., VII, 13 (160)
Deruyts, J., II, 14 (24)
Dickson, L., VII, 11 (157)

Ehresmann, C., IV, 5.4; VI, 2.3; VII, 1 (87, 121, 147)
Engel, F.,I, §1; II, 2; III, $4^{8)}$; V, §1, 4.6-7, 5.2 (1, 10, 49, 94–96, 106–107, 114)

Fano, G., II, 3, 13 (11, 21–24)
Freudenthal, H., III, 8; VI, 6.5, 6.8 (38, 137–141)
Frobenius, G., V, 4.0, 4.7 (102, 107)

Gantmacher, F., IV, $2.5.3^{13)}$ (78)
Garrett, P., VI, 7.3 (143)
Gordan, P., II, 6; V, 1.3, 2.1–2.3, 4.15, 4.17 (14, 96–98, 111–113)
Goto, M., VI, 1.2, 2.3 (119, 121)
Green, J.A., II, 14 (24)
Grosshans, F., V, 4.17 (114)
Grothendieck, A., VI, 4.4 (127)

Haboush, W., II, 11; III, 13 (18, 44)
Hadžiev, Dž., V, 4.17 (113)
Harish-Chandra I, 2.1; II, 9; III, 8; IV, $6.3^{21)}$; VI, 2.3; VII, 7 (5, 17, 38, 88, 121, 155)
Helgason, S., IV, 4.5 (84)
Hilbert, D., II, 6; V, 4.17 (14, 112)
Hopf, H., IV, $2.5.2^{11)}$; VI, 3.2; VII, 5 (76, 124, 153)
Howe, R., III, 14 (45)
Humt, G., IV, $2.4.2^{6)}$ (69)
Hurwitz, A., I, 3.3; II, 5, 6; III, 3, 4, 10 (7, 12–15, 33–34, 40)

Iwahori, N., VI, 7.3 (143)
Iwasawa, K., VII, 1 (147)

Jacobson, N., III, 9; VI, 4.4 (39, 128)

Killing, W., I, 2.1; III, 4; V, 5.2 (5, 34, 114)
Klein, F., I, 1.3; II, 4; VI, 5.1 (4, 11, 131)
Kolchin, E.R., VI, 2.1, 2.6, 3.2; VIII (120, 123, 124; VIII *passim*)
Koszul, J.-L., VII, 5 (153)

Lang, S., VI, 4.3 (127)
Lardy, P., IV, $2.5.3^{13)}$ (78)
Lichnerowicz, A., IV, $6.3^{20)}$ (88)
Lie, S., I, §1; V, 1.1, 1.4; VI, 5.1; VIII, 1 (1–4, 93, 96, 131, 165)
Loewy, A., II, 4 (11)

Malcev, A., VII, 1 (147)
Maschke, H., II, 4 (11-12)
Matsumoto, H., VI, 7.3(143)
Matsushima, Y., VI, 1.2 (119)
Maurer, L., V, §4; VI, 1.1, 2.3; VII, 4, $10^{6)}$; VIII, 3 (102–113, 119, 121, 150, 156, 169)
Moore, E.H., II, 4 (11)

# Subject Index

abelian variety VI, 2.5 (122)
absolutely simple group VI, 6.5 (137)
algebraic group VI ff.
   linear V, 3.2, 4.1 (101–103)
algebraic torus V, 4.11; VI, 3.2 (108, 124)
angular parameter IV, 2,2, 2.4.2 (65, 69)
anisotropic quadratic form VI, 6.6 (139)
apartment VI, 5.2 (132)

Barsotti-Chevalley theorem VI, 2.6; VIII, 10.5
   (123, 177)
base change VI, 6.8 (140)
Borel subalgebra VI, 2.2 (120)
Borel subgroups VII, 12 (158)
Bruhat decomposition VI, 2.3 (121–122)

Campbell-Hausdorff formula I, 2.3 (6)
Cartan
   decomposition
      of Lie group IV, 2.4.4, 2.4.5 (71–72)
      of semisimple Lie algebra IV, 1.4 (62)
   polyhedron IV, 2.2 (65)
   subalgebra IV, 2.4.2; VII, 2, 9 (69, 148,
      155)
   subgroup VI, 3.2; VII, 9 (124, 156)
Casimir operator II, 9 (17)
Cayley's operator III, 11 (42)
Clebsch-Gordan series V, 2.4 (99)
complex of flags VI, 5.3 (132)
Coxeter complex VI, 5.4 (133)

diagram VI, 3.3 (125)
differential field
   of constants VIII, 5.1 (170)
   ordinary VIII, 5.1 (170)
   partial VIII, 5.1 (170)
   universal VIII, 8.1 (173)
dual Weyl module VI, 2.4 (122)

extension
   of a differential field VIII, 5.2 (170)
   Picard-Vessiot VIII, 5.3 (171)
   strongly normal VIII, 8.2 (174)
extremal involution VI, 6.6 (139)
extremal pair of involutions VI, 6.6 (139)

finite and continuous group
   in Cartan's sense IV, 3.1 (79)
   In Lie's sense I, 1.3 (4)
flat IV, 2.4.2; VI, 7.1 (68, 142)
FTPG VI, 6.1 (134)

Galois group
   of a P.V.-extension VIII, 5.4 (171)
   of a strongly normal extension VIII, 8.3
      (174)
Gleichzusammensetzung I, 1.3 (4)

Haar measure III, 8 (38)
Heisenberg commutation relations III, 9 (38)
Helmholtz-Lie problem III, 1 (29)

invariant problem II, 1 (9)
isogeny
   purely inseparable of height one VI, 4.4
      (128)
   special VI, 3.3 (126)
isotropic
   quadratic form VI, 6.6 (139)
   subspace VI, 5.3 (132)

Jacobi identity I, 1.2 (3)

$k$-closed VI, 4.1 (126)
$k$-form VI, 4.8 (130)
$k$-variety VI, 4.1 (126)
Killing form I, 2.1 (5)

Lie algebra
   absolutely simple (over $\mathbb{R}$) IV, 2.4 (67)
   of $\mathbf{SL}_2(\mathbb{C})$ II (9–25)
      irreducible representations of II, 2 (10)
      linear representations of II, 1 (9)
      proof of full reducibility of representa-
      tions of
         by Cartan II, 12 (20–21)
         by Casimir II, 8 (16–17)
         by Fano II, 13 (21–24)
         by Schur II, 14 (24–25)
         by Study II, 2 (10)
   of $\mathbf{SO}_3$ II, 8 (16)
   semisimple I, 2.1 (5)
   unitary IV, 1.4 (62)

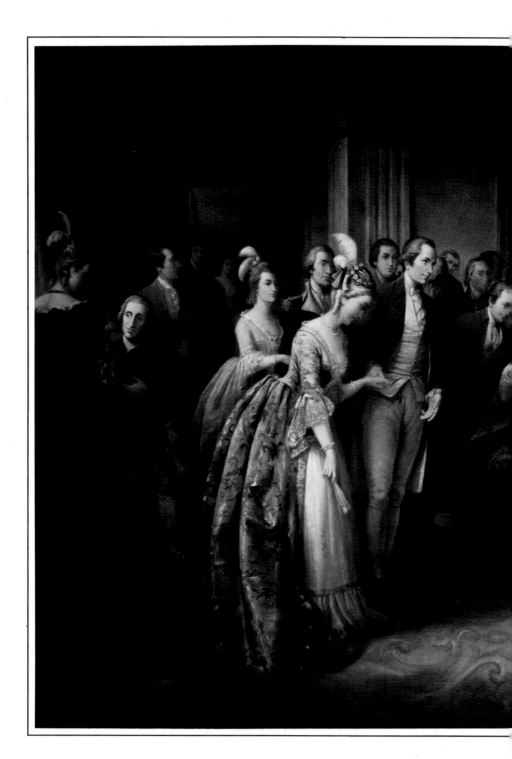

Published by the WHITE HOUSE HISTORICAL ASSOCIATION
With the cooperation of the NATIONAL GEOGRAPHIC SOCIETY, Washington, D. C.

# The First Ladies

By Margaret Brown Klapthor

### THE FIRST LADIES

By *Margaret Brown Klapthor,* Curator Emeritus, Division of Political History, National Museum of American History, Smithsonian Institution

PRODUCED BY THE NATIONAL GEOGRAPHIC SOCIETY AS A PUBLIC SERVICE
*Gilbert M. Grosvenor,* President and Chairman of the Board
*Melvin M. Payne,* Chairman Emeritus
*Owen R. Anderson,* Executive Vice President
*Robert L. Breeden,* Senior Vice President, Publications and Educational Media

PREPARED BY THE SPECIAL PUBLICATIONS DIVISION
*Donald J. Crump,* Director
*Philip B. Silcott,* Associate Director
*Bonnie S. Lawrence,* Assistant Director
*Mary Ann Harrell,* Managing Editor
*Susan C. Burns, Jennifer Urquhart,* Researchers
*Geraldine Linder,* Picture Editor
*Ursula Perrin Vosseler,* Art Director
*Marie Bradby, Louis de la Haba, P. Tyrus Harrington, Margaret McKelway,* Picture Legend Writers
PHOTOGRAPHY: *Steve Adams, Joseph H. Bailey, Sean Baldwin, J. Bruce Baumann, Victor R. Boswell, Jr., David S. Boyer, Sisse Brimberg, Nelson H. Brown, Dan J. Dry, Thomas Hooper, Larry D. Kinney, Erik Kvalsvik, Bates Littlehales, George F. Mobley, Robert S. Oakes, Martin Rogers, James E. Russell, Joseph J. Scherschel, David Valdez, Volkmar Wentzel*
STAFF FOR SIXTH EDITION: *Jane H. Buxton,* Managing Editor; *Lise S. Sajewski,* Researcher; *Viviane Y. Silverman,* Art Director; *Robert W. Messer,* Director of Manufacturing; *David V. Showers,* Production Manager; *Carol R. Curtis, Marisa J. Farabelli, Lisa A. LaFuria, Sandra F. Lotterman, Jennie H. Proctor, Dru McLoud Stancampiano, Marilyn Williams,* Staff Assistants

White House staff members who assisted in the preparation of this edition: *Rex Scouten,* Curator; *Betty Monkman,* Associate Curator; *William Allman,* Assistant Curator.

*Visitors' view of the Executive Mansion in winter—a steel engraving—illustrates a noteworthy venture in biography: Laura C. Holloway's* The Ladies of the White House, *first published in 1870 and apparently the first book on First Ladies.*

COVER: *Portrait of Martha Washington, painted by Eliphalet F. Andrews in 1878, hangs in the East Room of the White House as a companion piece for the Gilbert Stuart portrait of her husband.*
PRECEDING PAGES: *George and Martha Washington appear at right in an unknown artist's painting "Reception at Mount Vernon." Its style suggests a fictionalized illustration for a 19th-century popular magazine. Previously unpublished, the painting hangs in the headquarters of the National Society of the Daughters of the American Revolution, in Washington, D. C., on loan from the Rhode Island Society.*

# Foreword

I feel privileged to have taken my place in the impressive parade of women that began when Martha Washington became the first First Lady by virtue of her husband's election to the Presidency. Two hundred years of our country's history are reflected in the lives of the women who have served in this position. They have come from every part of our nation, from every economic and social level, from varying backgrounds; each has responded in her own way to the demands and opportunities of her role.

Some of the earliest wives had a keen interest in politics and judged political issues astutely. There were women like the brilliant Abigail Adams, whose husband shared with her his political ideas and who listened to her ideas with great respect—so much so that her husband's political enemies called her "Mrs. President."

By the mid-19th century we see the true Victorian woman, whose interest was centered on her home and family; her influence was felt primarily in moral issues rather than in political causes. But the First Lady still had influence. As President Hayes once remarked, "Mrs. Hayes may not have much influence with Congress, but she has great influence with me." The emergence of a First Lady with an awareness of the opportunity she has to bring public attention to social problems began in the 20th century with the gentle Ellen Wilson, President Wilson's first wife. She was so shocked by the squalor of Washington's back alleys that she threw the weight of her position behind one of the first slum clearance bills to be proposed to Congress. She learned just before her untimely death in 1914 that the bill she supported would pass because of her personal interest in it.

But it was Eleanor Roosevelt, with her natural inclination for public service and her personal appreciation of politics, who made the position of First Lady a national force. While she traveled as her husband's extra eyes and ears, her own lectures, writings, work on welfare committees, and visits to troops during World War II added up to a major career of her own.

Since the middle of the 20th century, each First Lady from Jacqueline Kennedy on has had special projects to which she has devoted her energy. This book describes these women, beginning with Martha Washington. We may all be called "The First Lady," but each of us has a story of her own, which appears in this book.

President Bush and I convey our gratitude to the White House Historical Association for publishing this book, now in its sixth edition, and to the National Geographic Society for providing, as a public service, the editorial staff who produced it. We also appreciate the scholarship of the author, Margaret Brown Klapthor, who for many years was curator of the First Ladies collection at the Smithsonian Institution. I hope that everyone with an interest in the White House and its residents will read and enjoy these stories of the First Ladies.

*Barbara Bush*

*"Lincoln's Last Reception,"* a lithograph hastily published after his assassination, portrays him and Mrs. Lincoln among heroes of the Union cause on the night of his second inauguration, March 4, 1865. The artist places Vice President Andrew Johnson, who was ill, at Lincoln's side as he greets Mrs. Ulysses S. Grant, although neither she nor the general attended.

# Contents

# Martha Dandridge Custis Washington
## 1731-1802

"I think I am more like a state prisoner than anything else, there is certain bounds set for me which I must not depart from. . . ." So in one of her few surviving letters, Martha Washington confided to a niece that she did not entirely enjoy her role as first of First Ladies. She once conceded that "many younger and gayer women would be extremely pleased" in her place; she would "much rather be at home."

But when George Washington took his oath of office in New York City on April 30, 1789, and assumed the new duties of President of the United States, his wife brought to their position a tact and discretion developed over 58 years of life in Tidewater Virginia society.

Oldest daughter of John and Frances Dandridge, she was born June 2, 1731, on a plantation near Williamsburg. Typical for a girl in an 18th-century family, her education was almost negligible except in domestic and social skills, but she learned all the arts of a well-ordered household and how to keep a family contented.

As a girl of 18—about five feet tall, dark-haired, gentle of manner—she married the wealthy Daniel Parke Custis. Two babies died; two were hardly past infancy when her husband died in 1757.

From the day Martha married George Washington in 1759, her great concern was the comfort and happiness of her husband and her children. When his career led him to the battlegrounds of the Revolutionary War and finally to the Presidency, she followed him bravely. Her love of private life equaled her husband's; but, as she wrote to her friend Mercy Otis Warren, "I cannot blame him for having acted according to his ideas of duty in obeying the voice of his country." As for herself, "I am still determined to be cheerful and to be happy, in whatever situation I may be; for I have also learned from experience that the greater part of our happiness or misery depends upon our dispositions, and not upon our circumstances."

At the President's House in temporary capitals, New York and Philadelphia, the Washingtons chose to entertain in formal style, deliberately emphasizing the new republic's wish to be accepted as the equal of the established governments of Europe. Still, Martha's warm hospitality made her guests feel welcome and put strangers at ease. She took little satisfaction in "formal compliments and empty ceremonies," and declared that "I am fond only of what comes from the heart." Abigail Adams, who sat at her right during parties and receptions, praised her as "one of those unassuming characters which create Love & Esteem."

In 1797 the Washingtons said farewell to public life and returned to their beloved Mount Vernon, to live surrounded by kinfolk, friends, and a constant stream of guests eager to pay their respects to the celebrated couple. Martha's daughter Patsy had died at 17, her son Jack at 26, but Jack's children figured in the household. After George Washington died in 1799, Martha assured a final privacy by burning their letters; she died of a "severe fever" on May 22, 1802. Both lie buried at Mount Vernon, where Washington himself had planned an unpretentious tomb for them.

*"Dear Patsy" to her husband, Martha Washington had this miniature painted by Charles Will-
son Peale about 1776; through the rest of the Revolutionary War, the general wore it in a
gold locket. "Lady Washington," some admiring Americans called her; in a modest reference
to her domestic skill, she described herself as an "old-fashioned Virginia house-keeper." "A
most becoming pleasentness sits upon her countanance...," wrote Abigail Adams in 1789.*

# Abigail Smith Adams
## 1744-1818

Inheriting New England's strongest traditions, Abigail Smith was born in 1744 at Weymouth, Massachusetts. On her mother's side she was descended from the Quincys, a family of great prestige in the colony; her father and other forebears were Congregational ministers, leaders in a society that held its clergy in high esteem.

Like other women of the time, Abigail lacked formal education; but her curiosity spurred her keen intelligence, and she read avidly the books at hand. Reading created a bond between her and young John Adams, Harvard graduate launched on a career in law, and they were married in 1764. It was a marriage of the mind and of the heart, enduring for more than half a century, enriched by time.

The young couple lived on John's small farm at Braintree or in Boston as his practice expanded. In ten years she bore three sons and two daughters; she looked after family and home when he went traveling as circuit judge. "Alass!" she wrote in December 1773, "How many snow banks devide thee and me. . . ."

Long separations kept Abigail from her husband while he served the country they loved, as delegate to the Continental Congress, envoy abroad, elected officer under the Constitution. Her letters—pungent, witty, and vivid, spelled just as she spoke—detail her life in times of revolution. They tell the story of the woman who stayed at home to struggle with wartime shortages and inflation; to run the farm with a minimum of help; to teach four children when formal education was interrupted. Most of all, they tell of her loneliness without her "dearest Friend." That "one single expression," she said, "dwelt upon my mind and playd about my Heart. . . ."

In 1784, she joined him at his diplomatic post in Paris, and observed with interest the manners of the French. After 1785, she filled the difficult role of wife of the first United States Minister to Great Britain, and did so with dignity and tact. They returned happily in 1788 to Massachusetts and the handsome house they had just acquired in Braintree, later called Quincy, home for the rest of their lives.

As wife of the first Vice President, Abigail became a good friend to Mrs. Washington and a valued help in official entertaining, drawing on her experience of courts and society abroad. After 1791, however, poor health forced her to spend as much time as possible in Quincy. Illness or trouble found her resolute; as she once declared, she would "not forget the blessings which sweeten life."

When John Adams was elected President, she continued a formal pattern of entertaining—even in the primitive conditions she found at the new capital in November 1800. The city was wilderness, the President's House far from completion. Her private complaints to her family provide blunt accounts of both, but for her three months in Washington she duly held her dinners and receptions.

The Adamses retired to Quincy in 1801, and for 17 years enjoyed the companionship that public life had long denied them. Abigail died in 1818, and is buried beside her husband in United First Parish Church. She leaves her country a most remarkable record as patriot and First Lady, wife of one President and mother of another.

*Proudly appearing in London as a citizen of her newly independent country, wife of its first diplomat to the court of King George III, Abigail Adams sat in 1785 for her portrait by a young American artist and family friend, Mather Brown. Some scholars identify the painting above as Brown's, which passed out of family hands; others attribute it to Ralph Earl, another artist from Massachusetts who spent part of 1785 in London. It came to light in the 1930's as "Portrait of a Lady." Well-authenticated portraits by other painters confirm the likeness to Abigail Adams at about the age of 40.*

*Jefferson's daughter Martha ("Patsy") posed for this oil-on-ivory miniature in 1789, at age 17; "a delicate likeness of her father," a friend called her during the White House years when she sometimes served as his hostess. Of her mother, whose name she bore, no portrait of any kind has survived.*

# Martha Wayles Skelton Jefferson
## 1748-1782

When Thomas Jefferson came courting, Martha Wayles Skelton at 22 was already a widow, an heiress, and a mother whose firstborn son would die in early childhood. Family tradition says that she was accomplished and beautiful—with slender figure, hazel eyes, and auburn hair—and wooed by many. Perhaps a mutual love of music cemented the romance; Jefferson played the violin, and one of the furnishings he ordered for the home he was building at Monticello was a "forte-piano" for his bride.

They were married on New Year's Day, 1772, at the bride's plantation home "The Forest," near Williamsburg. When they finally reached Monticello in a late January snowstorm to find no fire, no food, and the servants asleep, they toasted their new home with a leftover half-bottle of wine and "song and merriment and laughter." That night, on their own mountaintop, the love of Thomas Jefferson and his bride seemed strong enough to endure any adversity.

The birth of their daughter Martha in September increased their happiness. Within ten years the family gained five more children. Of them all, only two lived to grow up: Martha, called Patsy, and Mary, called Maria or Polly.

The physical strain of frequent pregnancies weakened Martha Jefferson so gravely that her husband curtailed his political activities to stay near her. He served in Virginia's House of Delegates and as governor, but he refused an appointment by the Continental Congress as a commissioner to France. Just after New Year's Day, 1781, a British invasion forced Martha to flee the capital in Richmond with a baby girl a few weeks old—who died in April. In June the family barely escaped an enemy raid on Monticello. She bore another daughter the following May, and never regained a fair measure of strength. Jefferson wrote on May 20 that her condition was dangerous. After months of tending her devotedly, he noted in his account book for September 6, "My dear wife died this day at 11-45 A.M."

Apparently he never brought himself to record their life together; in a memoir he referred to ten years "in unchequered happiness." Half a century later his daughter Martha remembered his sorrow: "the violence of his emotion . . . to this day I dare not describe to myself." For three weeks he had shut himself in his room, pacing back and forth until exhausted. Slowly that first anguish spent itself. In November he agreed to serve as commissioner to France, eventually taking "Patsy" with him in 1784 and sending for "Polly" later.

When Jefferson became President in 1801, he had been a widower for 19 years. He had become as capable of handling social affairs as political matters. Occasionally he called on Dolley Madison for assistance. And it was Patsy—now Mrs. Thomas Mann Randolph, Jr.—who appeared as the lady of the President's House in the winter of 1802-1803, when she spent seven weeks there. She was there again in 1805-1806, and gave birth to a son named for James Madison, the first child born in the White House. It was Martha Randolph with her family who shared Jefferson's retirement at Monticello until he died there in 1826.

# Dolley Payne Todd Madison
## 1768-1849

For half a century she was the most important woman in the social circles of America. To this day she remains one of the best known and best loved ladies of the White House—though often referred to, mistakenly, as Dorothy or Dorothea.

She always called herself Dolley; and by that name the New Garden Monthly Meeting of the Society of Friends, in piedmont North Carolina, recorded her birth to John and Mary Coles Payne, settlers from Virginia. In 1769 John Payne took his family back to his home colony, and in 1783 he moved them to Philadelphia, city of the Quakers. Dolley grew up in the strict discipline of the Society, but nothing muted her happy personality and her warm heart.

John Todd, Jr., a lawyer, exchanged marriage vows with Dolley in 1790. Just three years later he died in a yellow-fever epidemic, leaving his wife with a small son.

By this time Philadelphia had become the capital city. With her charm and her laughing blue eyes, fair skin, and black curls, the young widow attracted distinguished attention. Before long Dolley was reporting to her best friend that "the great little Madison has asked. . . . to see me this evening."

Although Representative James Madison of Virginia was 17 years her senior, and Episcopalian in background, they were married in September 1794. The marriage, though childless, was notably happy; "our hearts understand each other," she assured him. He could even be patient with Dolley's son, Payne, who mishandled his own affairs—and, eventually, mismanaged Madison's estate.

Discarding the somber Quaker dress after her second marriage, Dolley chose the finest of fashions. Margaret Bayard Smith, chronicler of early Washington social life, wrote: "She looked a Queen. . . . It would be *absolutely impossible* for any one to behave with more perfect propriety than she did."

Blessed with a desire to please and a willingness to be pleased, Dolley made her home the center of society when Madison began, in 1801, his eight years as Jefferson's Secretary of State. She assisted at the White House when the President asked her help in receiving ladies, and presided at the first inaugural ball in Washington when her husband became Chief Executive in 1809.

Dolley's social graces made her famous. Her political acumen, prized by her husband, is less renowned, though her gracious tact smoothed many a quarrel. Hostile statesmen, difficult envoys from Spain or Tunisia, warrior chiefs from the west, flustered youngsters—she always welcomed everyone. Forced to flee from the White House by a British army during the War of 1812, she returned to find the mansion in ruins. Undaunted by temporary quarters, she entertained as skillfully as ever.

At their plantation Montpelier in Virginia, the Madisons lived in pleasant retirement until he died in 1836. She returned to the capital in the autumn of 1837, and friends found tactful ways to supplement her diminished income. She remained in Washington until her death in 1849, honored and loved by all. The delightful personality of this unusual woman is a cherished part of her country's history.

*Already a hostess of outstanding success, Dolley Madison sat for this portrait by Gilbert Stuart in 1804. As wife of the Secretary of State, she began her long career of official hospitality; she sustained it for eight years as First Lady. Her manners, said a contemporary, "would disarm envy itself." Widowed, impoverished, and old, she never lost her charm or dignity; she enjoyed the deepening respect of her friends and country to the last.*

*Invariably elegant, Elizabeth Monroe chose an ermine scarf to complement her black velvet Empire-style gown for this portrait by an unknown artist. Although she had gone abroad with her husband on diplomatic missions, delicate health limited her activities—and enjoyment—in her eight years as First Lady.*

# Elizabeth Kortright Monroe
## 1768-1830

Romance glints from the little that is known of Elizabeth Kortright's early life. She was born in New York City in 1768, daughter of an old New York family. Her father, Lawrence, had served the Crown by privateering during the French and Indian War and made a fortune. He took no active part in the War of Independence; and James Monroe wrote to his friend Thomas Jefferson in Paris in 1786 that he had married the daughter of a gentleman "injured in his fortunes" by the Revolution.

Strange choice, perhaps, for a patriot veteran with political ambitions and little money of his own; but Elizabeth was beautiful, and love was decisive. They were married in February 1786, when the bride was not yet 18.

The young couple planned to live in Fredericksburg, Virginia, where Monroe began his practice of law. His political career, however, kept them on the move as the family increased by two daughters and a son who died in infancy.

In 1794, Elizabeth Monroe accompanied her husband to France when President Washington appointed him United States Minister. Arriving in Paris in the midst of the French Revolution, she took a dramatic part in saving Lafayette's wife, imprisoned and expecting death on the guillotine. With only her servants in her carriage, the American Minister's wife went to the prison and asked to see Madame Lafayette. Soon after this hint of American interest, the prisoner was set free. The Monroes became very popular in France, where the diplomat's lady received the affectionate name of *la belle Américaine*.

For 17 years Monroe, his wife at his side, alternated between foreign missions and service as governor or legislator of Virginia. They made the plantation of Oak Hill their home after he inherited it from an uncle, and appeared on the Washington scene in 1811 when he became Madison's Secretary of State.

Elizabeth Monroe was an accomplished hostess when her husband took the Presidential oath in 1817. Through much of the administration, however, she was in poor health and curtailed her activities. Wives of the diplomatic corps and other dignitaries took it amiss when she decided to pay no calls—an arduous social duty in a city of widely scattered dwellings and unpaved streets.

Moreover, she and her daughter Eliza changed White House customs to create the formal atmosphere of European courts. Even the White House wedding of her daughter Maria was private, in "the New York style" rather than the expansive Virginia social style made popular by Dolley Madison. A guest at the Monroes' last levee, on New Year's Day in 1825, described the First Lady as "regal-looking" and noted details of interest: "Her dress was superb black velvet; neck and arms bare and beautifully formed; her hair in puffs and dressed high on the head and ornamented with white ostrich plumes; around her neck an elegant pearl necklace. Though no longer young, she is still a very handsome woman."

In retirement at Oak Hill, Elizabeth Monroe died on September 23, 1830; and family tradition says that her husband burned the letters of their life together.

# Louisa Catherine Johnson Adams
## 1775-1852

Only First Lady born outside the United States, Louisa Catherine Adams did not come to this country until four years after she had married John Quincy Adams. Political enemies sometimes called her English. She was born in London to an English mother, Catherine Nuth Johnson, but her father was American—Joshua Johnson, of Maryland—and he served as United States consul after 1790.

A career diplomat at 27, accredited to the Netherlands, John Quincy developed his interest in charming 19-year-old Louisa when they met in London in 1794. Three years later they were married, and went to Berlin in course of duty. At the Prussian court she displayed the style and grace of a diplomat's lady; the ways of a Yankee farm community seemed strange indeed in 1801 when she first reached the country of which she was a citizen. Then began years divided among the family home in Quincy, Massachusetts, their house in Boston, and a political home in Washington, D. C. When the Johnsons had settled in the capital, Louisa felt more at home there than she ever did in New England.

She left her two older sons in Massachusetts for education in 1809 when she took two-year-old Charles Francis to Russia, where Adams served as Minister. Despite the glamour of the tsar's court, she had to struggle with cold winters, strange customs, limited funds, and poor health; an infant daughter born in 1811 died the next year. Peace negotiations called Adams to Ghent in 1814 and then to London. To join him, Louisa had to make a forty-day journey across war-ravaged Europe by coach in winter; roving bands of stragglers and highwaymen filled her with "unspeakable terrors" for her son. Happily, the next two years gave her an interlude of family life in the country of her birth.

Appointment of John Quincy as Monroe's Secretary of State brought the Adamses to Washington in 1817, and Louisa's drawing room became a center for the diplomatic corps and other notables. Good music enhanced her Tuesday evenings at home, and theater parties contributed to her reputation as an outstanding hostess.

But the pleasure of moving to the White House in 1825 was dimmed by the bitter politics of the election and by her own poor health. She suffered from deep depression. Though she continued her weekly "drawing rooms," she preferred quiet evenings—reading, composing music and verse, playing her harp. The necessary entertainments were always elegant, however; and her cordial hospitality made the last official reception a gracious occasion although her husband had lost his bid for re-election and partisan feeling still ran high.

Louisa thought she was retiring to Massachusetts permanently, but in 1831 her husband began 17 years of notable service in the House of Representatives. The Adamses could look back on a secure happiness as well as many trials when they celebrated their fiftieth wedding anniversary at Quincy in 1847. He was fatally stricken at the Capitol the following year; she died in Washington in 1852, and today lies buried at his side in the family church at Quincy.

*"Try as she might, the Madam could never be Bostonian, and it was her cross in life," wrote Henry Adams of his London-born grandmother. Nor did Louisa Johnson Adams feel at ease in what she called the "Bull Bait" of Washington's political life. Gilbert Stuart painted her portrait in 1821.*

*With the devotion that always marked their marriage, Andrew Jackson once assured Rachel: "recollection never fails me of your likeness." Nearly heartbroken by her death, he wore a miniature of her daily and kept it on his bedside table every night. Family tradition held that the portrait above was the very memento he carried, and attributed it to Anna C. Peale. Current scholarship identifies it as the work of Louisa Catherine Strobel, who was in France from 1815 to 1830; she probably copied a study from life, for one of "Aunt Rachel's" relatives.*

# Rachel Donelson Jackson
## 1767-1828

Wearing the white dress she had purchased for her husband's inaugural ceremonies in March 1829, Rachel Donelson Jackson was buried in the garden at The Hermitage, her home near Nashville, Tennessee, on Christmas Eve in 1828. Lines from her epitaph —"A being so gentle and so virtuous slander might wound, but could not dishonor" —reflected his bitterness at campaign slurs that seemed to precipitate her death.

Rachel Donelson was a child of the frontier. Born in Virginia, she journeyed to the Tennessee wilderness with her parents when only 12. At 17, while living in Kentucky, she married Lewis Robards, of a prominent Mercer County family. His unreasoning jealousy made it impossible for her to live with him; in 1790 they separated, and she heard that he was filing a petition for divorce.

Andrew Jackson married her in 1791; and after two happy years they learned to their dismay that Robards had not obtained a divorce, only permission to file for one. Now he brought suit on grounds of adultery. After the divorce was granted, the Jacksons quietly remarried in 1794. They had made an honest mistake, as friends well understood, but whispers of adultery and bigamy followed Rachel as Jackson's career advanced in both politics and war. He was quick to take offense at, and ready to avenge, any slight to her.

Scandal aside, Rachel's unpretentious kindness won the respect of all who knew her—including innumerable visitors who found a comfortable welcome at The Hermitage. Although the Jacksons never had children of their own, they gladly opened their home to the children of Rachel's many relatives. In 1809 they adopted a nephew and named him Andrew Jackson, Jr. They also reared other nephews; one, Andrew Jackson Donelson, eventually married his cousin Emily, one of Rachel's favorite nieces.

*Jackson's hostess Emily Donelson: by his friend and portraitist Ralph E. W. Earl.*

When Jackson was elected President, he planned to have young Donelson for private secretary, with Emily as company for Rachel. After losing his beloved wife he asked Emily to serve as his hostess.

Though only 21 when she entered the White House, she skillfully cared for her uncle, her husband, four children (three born at the mansion), many visiting relatives, and official guests. Praised by contemporaries for her wonderful tact, she had the courage to differ with the President on issues of principle. Frail throughout her lifetime, Emily died of tuberculosis in 1836.

During the last months of the administration, Sarah Yorke Jackson, wife of Andrew Jackson, Jr., presided at the mansion in her stead.

# Hannah Hoes Van Buren
## 1783-1819

Cousins in a close-knit Dutch community, Hannah Hoes and Martin Van Buren grew up together in Kinderhook, New York. Evidently he wanted to establish his law practice before marrying his sweetheart — they were not wed until 1807, when he was 24 and his bride just three months younger. Apparently their marriage was a happy one, though little is known of Hannah as a person.

Van Buren omitted even her name from his autobiography; a gentleman of that day would not shame a lady by public references. A niece who remembered "her loving, gentle disposition" emphasized "her modest, even timid manner." Church records preserve some details of her life; she seems to have considered formal church affiliation a matter of importance.

*Angelica Singleton Van Buren: daughter-in-law and hostess of the widower President.*

She bore a son in Kinderhook, three others in Hudson, where Martin served as county surrogate; but the fourth son died in infancy. In 1816 the family moved to the state capital in Albany. Soon the household included Martin's law partner and three apprentices; relatives came and went constantly, and Hannah could return their visits. Contemporary letters indicate that she was busy, sociable, and happy. She gave birth to a fifth boy in January 1817.

But by the following winter her health was obviously failing, apparently from tuberculosis. Not yet 36, she died on February 5, 1819. The Albany *Argus* called her "an ornament of the Christian faith."

Her husband never remarried; he moved into the White House in 1837 as a widower with four bachelor sons. Now accustomed to living in elegant style, he immediately began to refurbish a mansion shabby from public use under Jackson. Across Lafayette Square, Dolley Madison reigned as matriarch of Washington society; when her young relative-by-marriage Angelica Singleton came up from South Carolina for a visit, Dolley took her to the White House to pay a call.

Angelica's aristocratic manners, excellent education, and handsome face won the heart of the President's eldest son, Abraham. They were married in November 1838; next spring a honeymoon abroad polished her social experience. Thereafter, while Abraham served as the President's private secretary, Angelica presided as lady of the house. The only flaw in her pleasure in this role was the loss of a baby girl. Born at the White House, she lived only a few hours. In later years, though spending much time in South Carolina and in Europe, Angelica and her husband made their home in New York City; she died there in 1878.

"Jannetje," Martin Van Buren called his wife, in the Dutch language of their ancestors. An unknown artist painted two portraits of her: one for her husband, one for her niece Maria Hoes Cantine. Initials on the back of the gold brooch indicate that this belonged to the niece; an inset holds a plaited lock of reddish-brown hair, probably a memento of "Aunt Hannah." Prolonged illness, almost certainly tuberculosis, claimed Hannah Van Buren at age 35: years later, Maria remembered "the perfect composure" of her deathbed farewell to her children, and her wish that money usually spent on gifts to pallbearers "should be given to the poor."

# Anna Tuthill Symmes Harrison
## 1775-1864

Anna Harrison was too ill to travel when her husband set out from Ohio in 1841 for his inauguration. It was a long trip and a difficult one even by steamboat and railroad, with February weather uncertain at best, and she at age 65 was well acquainted with the rigors of frontier journeys.

As a girl of 19, bringing pretty clothes and dainty manners, she went out to Ohio with her father, Judge John Cleves Symmes, who had taken up land for settlement on the "north bend" of the Ohio River. She had grown up a young lady of the East, completing her education at a boarding school in New York City.

A clandestine marriage on November 25, 1795, united Anna Symmes and Lt. William Henry Harrison, an experienced soldier at 22. Though the young man came from one of the best families of Virginia, Judge Symmes did not want his daughter to face the hard life of frontier forts; but eventually, seeing her happiness, he accepted her choice.

*Hostess for a single month: Jane Irwin Harrison, portrayed by an unknown artist.*

Though Harrison won fame as an Indian fighter and hero of the War of 1812, he spent much of his life in a civilian career. His service in Congress as territorial delegate from Ohio gave Anna and their two children a chance to visit his family at Berkeley, their plantation on the James River. Her third child was born on that trip, at Richmond in September 1800. Harrison's appointment as governor of Indiana Territory took them even farther into the wilderness; he built a handsome house at Vincennes that blended fortress and plantation mansion. Five more children were born to Anna.

Facing war in 1812, the family went to the farm at North Bend. Before peace was assured, she had borne two more children. There, at news of her husband's landslide electoral victory in 1840, home-loving Anna said simply: "I wish that my husband's friends had left him where he is, happy and contented in retirement."

When she decided not to go to Washington with him, the President-elect asked his daughter-in-law Jane Irwin Harrison, widow of his namesake son, to accompany him and act as hostess until Anna's proposed arrival in May. Half a dozen other relatives happily went with them. On April 4, exactly one month after his inauguration, he died, so Anna never made the journey. She had already begun her packing when she learned of her loss.

Accepting grief with admirable dignity, she stayed at her home in North Bend until the house burned in 1858; she lived nearby with her last surviving child, John Scott Harrison, until she died in February 1864 at the age of 88.

*Dressed in mourning, Anna Harrison posed for Cornelia Stuart Cassady in 1843—possibly at the Ohio home she had not left during her husband's Presidency. Her health precarious in 1841, she had temporarily entrusted the duties of White House hostess to her widowed daughter-in-law, Jane Irwin Harrison; before she could assume them herself, her husband died.*

*Heirloom portrait survived Civil War hazards to record the tranquil charm of Letitia Tyler for her descendants. An unknown artist painted it sometime before a paralytic stroke crippled her in 1839. At the White House she lived in seclusion. Her first daughter-in-law, Priscilla Cooper Tyler, assumed the duties of hostess until 1844, when she moved to Philadelphia. Between Letitia Tyler's death in 1842 and her husband's second marriage, the role of First Lady passed briefly to her second daughter, Letitia Tyler Semple.*

# Letitia Christian Tyler
## 1790-1842

Letitia Tyler had been confined to an invalid's chair for two years when her husband unexpectedly became President. Nobody had thought of that possibility when he took his oath of office as Vice President on March 4, 1841; indeed, he had planned to fill his undemanding duties from his home in Williamsburg where his wife was most comfortable, her Bible, prayer book, and knitting at her side.

Born on a Tidewater Virginia plantation in the 18th century, Letitia was spiritually akin to Martha Washington and Martha Jefferson. Formal education was no part of this pattern of life, but Letitia learned all the skills of managing a plantation, rearing a family, and presiding over a home that would be John Tyler's refuge during an active political life. They were married on March 29, 1813 — his twenty-third birthday. Thereafter, whether he served in Congress or as Governor of Virginia, she attended to domestic duties. Only once did she join him for the winter social season in Washington. Of the eight children she bore, seven survived; but after 1839 she was a cripple, though "still beautiful now in her declining years."

So her admiring new daughter-in-law, Priscilla Cooper Tyler, described her — "the most entirely unselfish person you can imagine. . . . Notwithstanding her very delicate health, mother attends to and regulates all the household affairs and all so quietly that you can't tell when she does it."

In a second-floor room at the White House, Letitia Tyler kept her quiet but pivotal role in family activities. She did not attempt to take part in the social affairs of the administration. Her married daughters had their own homes; the others were too young for the full responsibility of official entertaining; Priscilla at age 24 assumed the position of White House hostess, met its demands with spirit and success, and enjoyed it.

Daughter of a well-known tragedian, Priscilla Cooper had gone on the stage herself at 17. Playing Desdemona to her father's Othello in Richmond, she won the instant interest of Robert Tyler, whom she married in 1839. Intelligent and beautiful, with dark brown hair, she charmed the President's guests — from visiting celebrities like Charles Dickens to enthusiastic countrymen. Once she noted ruefully: "such hearty shakes as they gave my poor little hand too!" She enjoyed the expert advice of Dolley Madison, and the companionship of her young sister-in-law Elizabeth until she married William N. Waller in 1842.

For this wedding Letitia made her only appearance at a White House social function. "Lizzie looked surpassingly lovely," said Priscilla, and "our dear mother" was "far more attractive to me . . . than any other lady in the room," greeting her guests "in her sweet, gentle, self-possessed manner."

The first President's wife to die in the White House, Letitia Tyler ended her days peacefully on September 10, 1842, holding a damask rose in her hand. She was taken to Virginia for burial at the plantation of her birth, deeply mourned by her family. "She had everything about her," said Priscilla, "to awaken love. . . ."

# Julia Gardiner Tyler
## 1820-1889

"I grieve my love a belle should be," sighed one of Julia Gardiner's innumerable admirers in 1840; at the age of 20 she was already famous as the "Rose of Long Island."

Daughter of Juliana McLachlan and David Gardiner, descendant of prominent and wealthy New York families, Julia was trained from earliest childhood for a life in society; she made her debut at 15. A European tour with her family gave her new glimpses of social splendors. Late in 1842 the Gardiners went to Washington for the winter social season, and Julia became the undisputed darling of the capital. Her beauty and her practiced charm attracted the most eminent men in the city, among them President Tyler, a widower since September.

Tragedy brought his courtship poignant success the next winter. Julia, her sister Margaret, and her father joined a Presidential excursion on the new steam frigate *Princeton;* and David Gardiner lost his life in the explosion of a huge naval gun. Tyler comforted Julia in her grief and won her consent to a secret engagement.

The first President to marry in office took his vows in New York on June 26, 1844. The news was then broken to the American people, who greeted it with keen interest, much publicity, and some criticism about the couple's difference in age: 30 years.

As young Mrs. Tyler said herself, she "reigned" as First Lady for the last eight months of her husband's term. Wearing white satin or black lace to obey the conventions of mourning, she presided with vivacity and animation at a series of parties. She enjoyed her position immensely, and filled it with grace. For receptions she revived the formality of the Van Buren administration; she welcomed her guests with plumes in her hair, attended by maids of honor dressed in white. She once declared, with truth: "Nothing appears to delight the President more than . . . to hear people sing my praises."

The Tylers' happiness was unshaken when they retired to their home at Sherwood Forest in Virginia. There Julia bore five of her seven children; and she acted as mistress of the plantation until the Civil War. As such, she defended both states' rights and the institution of slavery. She championed the political views of her husband, who remained for her "the President" until the end of his life.

His death in 1862 came as a severe blow to her. In a poem composed for his sixty-second birthday she had assured him that "what e'er changes time may bring, I'll love thee as thou art!"

Even as a refugee in New York, she devoted herself to volunteer work for the Confederacy. Its defeat found her impoverished. Not until 1958 would federal law provide automatic pensions for Presidential widows; but Congress in 1870 voted a pension for Mary Lincoln, and Julia Tyler used this precedent in seeking help. In December 1880 Congress voted her $1,200 a year—and after Garfield's assassination it passed bills to grant uniform amounts of $5,000 annually to Mrs. Garfield, Mrs. Lincoln, Mrs. Polk, and Mrs. Tyler. Living out her last years comfortably in Richmond, Julia died there in 1889 and was buried there at her husband's side.

*"Most beautiful woman of the age and . . . most accomplished,"* President Tyler called his new bride, who keenly enjoyed her success as hostess at the White House. She posed for this portrait three years after leaving it.

*"Time has dealt kindly with her personal charms. . . ," declared one of many admiring citizens when Sarah Polk became First Lady. In 1846 both she and her husband posed at the White House for George P. A. Healy; three years later they took his fine portraits home to "Polk Place" in Nashville, Tennessee. In her 80th year, ladies of the state commissioned a copy of hers for the Executive Mansion; they entrusted the work to George Dury, who had already painted her at 75—still handsome and alert. Revered throughout the land, she kept her dignity of manner and her clarity of mind to the end.*

# Sarah Childress Polk
## 1803-1891

Silks and satins little Sarah took for granted, growing up on a plantation near Murfreesboro, Tennessee. Elder daughter of Captain Joel and Elizabeth Childress, she gained something rarer from her father's wealth. He sent her and her sister away to school, first to Nashville, then to the Moravians' "female academy" at Salem, North Carolina, one of the very few institutions of higher learning available to women in the early 19th century. So she acquired an education that made her especially fitted to assist a man with a political career.

James K. Polk was laying the foundation for that career when he met her. He had begun his first year's service in the Tennessee legislature when they were married on New Year's Day, 1824; he was 28, she 20. The story goes that Andrew Jackson had encouraged their romance; he certainly made Polk a political protégé, and as such Polk represented a district in Congress for 14 sessions.

In an age when motherhood gave a woman her only acknowledged career, Sarah Polk had to resign herself to childlessness. Moreover, no lady would admit to a political role of her own, but Mrs. Polk found scope for her astute mind as well as her social skills. She accompanied her husband to Washington whenever she could, and they soon won a place in its most select social circles. Constantly—but privately—Sarah was helping him with his speeches, copying his correspondence, giving him advice. Much as she enjoyed politics, she would warn him against overwork. He would hand her a newspaper—"Sarah, here is something I wish you to read. . . ."—and she would set to work as well.

A devout Presbyterian, she refused to attend horse races or the theater; but she always maintained social contacts of value to James. When he returned to Washington as President in 1845, she stepped to her high position with ease and evident pleasure. She appeared at the inaugural ball, but did not dance.

Contrasted with Julia Tyler's waltzes, her entertainments have become famous for sedateness and sobriety. Some later accounts say that the Polks never served wine, but in December 1845 a Congressman's wife recorded in her diary details of a four-hour dinner for forty at the White House—glasses for six different wines, from pink champagne to ruby port and sauterne, "formed a rainbow around each plate." Skilled in tactful conversation, Mrs. Polk enjoyed wide popularity as well as deep respect.

Only three months after retirement to their fine new home "Polk Place" in Nashville, he died, worn out by years of public service. Clad always in black, Sarah Polk lived on in that home for 42 years, guarding the memory of her husband and accepting honors paid to her as honors due to him. The house became a place of pilgrimage.

During the Civil War, Mrs. Polk held herself above sectional strife and received with dignity leaders of both Confederate and Union armies; all respected Polk Place as neutral ground. She presided over her house until her death in her 88th year. Buried beside her husband, she was mourned by a nation that had come to regard her as a precious link to the past.

# Margaret Mackall Smith Taylor
## 1788-1852

After the election of 1848, a passenger on a Mississippi riverboat struck up a conversation with easy-mannered Gen. Zachary Taylor, not knowing his identity. The passenger remarked that he didn't think the general qualified for the Presidency—was the stranger "a Taylor man"? "Not much of a one," came the reply. The general went on to say that he hadn't voted for Taylor, partly because his wife was opposed to sending "Old Zack" to Washington, "where she would be obliged to go with him!" It was a truthful answer.

Moreover, the story goes that Margaret Taylor had taken a vow during the Mexican War: If her husband returned safely, she would never go into society again. In fact she never did, though prepared for it by genteel upbringing.

"Peggy" Smith was born in Calvert County, Maryland, daughter of Ann Mackall and Walter Smith, a major in the Revolutionary War according to family tradition. In 1809, visiting a sister in Kentucky, she met young Lieutenant Taylor. They were married the following June, and for a while the young wife stayed on the farm given them as a wedding present by Zachary's father. She bore her first baby there, but cheerfully followed her husband from one remote garrison to another along the western frontier of civilization. An admiring civilian official cited her as one of the "delicate females . . . reared in tenderness" who had to educate "worthy and most interesting" children at a fort in Indian country.

Two small girls died in 1820 of what Taylor called "a violent bilious fever," which left their mother's health impaired; three girls and a boy grew up. Knowing the hardships of a military wife, Taylor opposed his daughters' marrying career soldiers— but each eventually married into the Army.

The second daughter, Knox, married Lt. Jefferson Davis in gentle defiance of her parents. In a loving letter home, she imagined her mother skimming milk in the cellar or going out to feed the chickens. Within three months of her wedding, Knox died of malaria. Taylor was not reconciled to Davis until they fought together in Mexico; in Washington the second Mrs. Davis became a good friend of Mrs. Taylor's, often calling on her at the White House.

Though Peggy Taylor welcomed friends and kinfolk in her upstairs sitting room, presided at the family table, met special groups at her husband's side, and worshiped regularly at St. John's Episcopal Church, she took no part in formal social functions. She relegated all the duties of official hostess to her youngest daughter, Mary Elizabeth, then 25 and recent bride of Lt. Col. William W. S. Bliss, adjutant and secretary to the President. Betty Bliss filled her role admirably. One observer thought that her manner blended "the artlessness of a rustic belle and the grace of a duchess."

For Mrs. Taylor, her husband's death—on July 9, 1850—was an appalling blow. Never again did she speak of the White House. She spent her last days with the Blisses, dying on August 18, 1852.

*Hostess in her mother's stead, Betty Taylor Bliss probably posed for this daguerreotype about a decade after her White House days. Of Mrs. Taylor, a semi-invalid for years, no authentic likeness survives; contemporaries described her as slender and stately, a brunette in youth.*

*Dignity of a matron in middle life marks this likeness of Abigail Fillmore, rendered in oils by an unknown artist—possibly copying a photograph. By contemporary descriptions she had been fair of coloring and taller at 5 feet 6 inches than the average woman of her time. A snippet of Washington gossip suggests her manner as First Lady, accusing her of lacking "good form" in seeming "so motherly to her guests." In the duties of hostess her daughter Mary Abigail gave her welcome help. "Abby," a talented young woman of poise and polish, managed the former President's household from her mother's death until her own shockingly unexpected death at age 22. After five years Fillmore found escape from loneliness in a second happy marriage: to Caroline Carmichael McIntosh, a wealthy and childless widow who survived him.*

# Abigail Powers Fillmore
## 1798-1853

First of First Ladies to hold a job after marriage, Abigail Fillmore was helping her husband's career. She was also revealing her most striking personal characteristic: eagerness to learn and pleasure in teaching others.

She was born in Saratoga County, New York, in 1798, while it was still on the fringe of civilization. Her father, a locally prominent Baptist preacher named Lemuel Powers, died shortly thereafter. Courageously, her mother moved on westward, thinking her scanty funds would go further in a less settled region, and ably educated her small son and daughter beyond the usual frontier level with the help of her husband's library.

Shared eagerness for schooling formed a bond when Abigail Powers at 21 met Millard Fillmore at 19, both students at a recently opened academy in the village of New Hope. Although she soon became young Fillmore's inspiration, his struggle to make his way as a lawyer was so long and ill paid that they were not married until February 1826. She even resumed teaching school after the marriage. And then her only son, Millard Powers, was born in 1828.

Attaining prosperity at last, Fillmore bought his family a six-room house in Buffalo, where little Mary Abigail was born in 1832. Enjoying comparative luxury, Abigail learned the ways of society as the wife of a Congressman. She cultivated a noted flower garden; but much of her time, as always, she spent in reading. In 1847, Fillmore was elected state comptroller; with the children away in boarding school and college, the parents moved temporarily to Albany.

In 1849, Abigail Fillmore came to Washington as wife of the Vice President; 16 months later, after Zachary Taylor's death at a height of sectional crisis, the Fillmores moved into the White House.

Even after the period of official mourning the social life of the Fillmore administration remained subdued. The First Lady presided with grace at state dinners and receptions; but a permanently injured ankle made her Friday-evening levees an ordeal—two hours of standing at her husband's side to greet the public. In any case, she preferred reading or music in private. Pleading her delicate health, she entrusted many routine social duties to her attractive daughter, "Abby." With a special appropriation from Congress, she spent contented hours selecting books for a White House library and arranging them in the oval room upstairs, where Abby had her piano, harp, and guitar. Here, wrote a friend, Mrs. Fillmore "could enjoy the music she so much loved, and the conversation of . . . cultivated society. . . ."

Despite chronic poor health, Mrs. Fillmore stayed near her husband through the outdoor ceremonies of President Pierce's inauguration while a raw northeast wind whipped snow over the crowd. Returning chilled to the Willard Hotel, she developed pneumonia; she died there on March 30, 1853. The House of Representatives and the Senate adjourned, and public offices closed in respect, as her family took her body home to Buffalo for burial.

# Jane Means Appleton Pierce
## 1806-1863

In looks and in pathetic destiny young Jane Means Appleton resembled the heroine of a Victorian novel. The gentle dignity of her face reflected her sensitive, retiring personality and physical weakness. Her father had died—he was a Congregational minister, the Reverend Jesse Appleton, president of Bowdoin College—and her mother had taken the family to Amherst, New Hampshire. And Jane met a Bowdoin graduate, a young lawyer with political ambitions, Franklin Pierce.

Although he was immediately devoted to Jane, they did not marry until she was 28—surprising in that day of early marriages. Her family opposed the match; moreover, she always did her best to discourage his interest in politics. The death of a three-day-old son, the arrival of a new baby, and Jane's dislike of Washington counted heavily in his decision to retire at the apparent height of his career, as United States Senator, in 1842. Little Frank Robert, the second son, died the next year of typhus.

Service in the Mexican War brought Pierce the rank of brigadier and local fame as a hero. He returned home safely, and for four years the Pierces lived quietly at Concord, New Hampshire, in the happiest period of their lives. With attentive pleasure Jane watched her son Benjamin growing up.

Then, in 1852, the Democratic Party made Pierce their candidate for President. His wife fainted at the news. When he took her to Newport for a respite, Benny wrote to her: "I hope he won't be elected for I should not like to be at Washington and I know you would not either." But the President-elect convinced Jane that his office would be an asset for Benny's success in life.

On a journey by train, January 6, 1853, their car was derailed and Benny killed before their eyes. The whole nation shared the parents' grief. The inauguration on March 4 took place without an inaugural ball and without the presence of Mrs. Pierce. She joined her husband later that month, but any pleasure the White House might have brought her was gone. From this loss she never recovered fully. Other events deepened the somber mood of the new administration: Mrs. Fillmore's death in March, that of Vice President Rufus King in April.

Always devout, Jane Pierce turned for solace to prayer. She had to force herself to meet the social obligations inherent in the role of First Lady. Fortunately she had the companionship and help of a girlhood friend, now her aunt by marriage, Abigail Kent Means. Mrs. Robert E. Lee wrote in a private letter: "I have known many of the ladies of the White House, none more truly excellent than the afflicted wife of President Pierce. Her health was a bar to any great effort on her part to meet the expectations of the public in her high position but she was a refined, extremely religious and well educated lady."

With retirement, the Pierces made a prolonged trip abroad in search of health for the invalid—she carried Benny's Bible throughout the journey. The quest was unsuccessful, so the couple came home to New Hampshire to be near family and friends until Jane's death in 1863. She was buried near Benny's grave.

*Treasure for a family's annals, this likeness — a photograph from about 1850 — captures the mutual devotion of Jane Pierce and Benny, last of her three sons. Two months before Pierce's inauguration, a railroad accident killed 11-year-old Benny before his parents' eyes. When his stricken mother finally brought herself to appear at White House functions, the guests who acknowledged her "winning smile" could not fail to recognize the "traces of bereavement . . . on a countenance too ingenuous for concealment. . . ."*

# Harriet Lane
## 1830-1903

Unique among First Ladies, Harriet Lane acted as hostess for the only President who never married: James Buchanan, her favorite uncle and her guardian after she was orphaned at the age of eleven. And of all the ladies of the White House, few achieved such great success in deeply troubled times as this polished young woman in her twenties.

In the rich farming country of Franklin County, Pennsylvania, her family had prospered as merchants. Her uncle supervised her sound education in private school, completed by two years at the Visitation Convent in Georgetown. By this time "Nunc" was Secretary of State, and he introduced her to fashionable circles as he had promised, "in the best manner." In 1854 she joined him in London, where he was minister to the Court of St. James's. Queen Victoria gave "dear Miss Lane" the rank of ambassador's wife; admiring suitors gave her the fame of a beauty.

In appearance "Hal" Lane was of medium height, with masses of light hair almost golden. In manner she enlivened social gatherings with a captivating mixture of spontaneity and poise.

After the sadness of the Pierce administration, the capital eagerly welcomed its new "Democratic Queen" in 1857. Harriet Lane filled the White House with gaiety and flowers, and guided its social life with enthusiasm and discretion, winning national popularity.

As sectional tensions increased, she worked out seating arrangements for her weekly formal dinner parties with special care, to give dignitaries their proper precedence and still keep political foes apart. Her tact did not falter, but her task became impossible—as did her uncle's. Seven states had seceded by the time Buchanan retired from office and thankfully returned with his niece to his spacious country home, Wheatland, near Lancaster, Pennsylvania.

From her teenage years, the popular Miss Lane flirted happily with numerous beaux, calling them "pleasant but dreadfully troublesome." Buchanan often warned her against "rushing precipitately into matrimonial connexions," and she waited until she was almost 36 to marry. She chose, with her uncle's approval, Henry Elliott Johnston, a Baltimore banker. Within the next 18 years she faced one sorrow after another: the loss of her uncle, her two fine young sons, and her husband.

Thereafter she decided to live in Washington, among friends made during years of happiness. She had acquired a sizable art collection, largely of European works, which she bequeathed to the government. Accepted after her death in 1903, it inspired an official of the Smithsonian Institution to call her "First Lady of the National Collection of Fine Arts." In addition, she had dedicated a generous sum to endow a home for invalid children at the Johns Hopkins Hospital in Baltimore. It became an outstanding pediatric facility, and its national reputation is a fitting memorial to the young lady who presided at the White House with such dignity and charm. The Harriet Lane Outpatient Clinics serve thousands of children today.

*"Mischievous romp of a niece,"* James Buchanan once called the little tomboy who at age 26 became White House hostess for the bachelor President. Artist John Henry Brown captured her queenly beauty 21 years after Harriet Lane assumed the role of First Lady. Fun-loving and flirtatious, spending lavishly on parties and clothes, she brightened a gloomy administration as civil war threatened it, warning her guests to refrain from political discussions.

# Mary Todd Lincoln
## 1818-1882

As a girlhood companion remembered her, Mary Todd was vivacious and impulsive, with an interesting personality—but "she now and then could not restrain a witty, sarcastic speech that cut deeper than she intended. . . ." A young lawyer summed her up in 1840: "the very creature of excitement." All of these attributes marked her life, bringing her both happiness and tragedy.

Daughter of Eliza Parker and Robert Smith Todd, pioneer settlers of Kentucky, Mary lost her mother before the age of seven. Her father remarried; and Mary remembered her childhood as "desolate" although she belonged to the aristocracy of Lexington, with high-spirited social life and a sound private education.

Just 5 feet 2 inches at maturity, Mary had clear blue eyes, long lashes, light-brown hair with glints of bronze, and a lovely complexion. She danced gracefully, she loved finery, and her crisp intelligence polished the wiles of a Southern coquette.

Nearly 21, she went to Springfield, Illinois, to live with her sister Mrs. Ninian Edwards. Here she met Abraham Lincoln—in his own words, "a poor nobody then." Three years later, after a stormy courtship and broken engagement, they were married. Though opposites in background and temperament, they were united by an enduring love—by Mary's confidence in her husband's ability and his gentle consideration of her excitable ways.

Their years in Springfield brought hard work, a family of boys, and reduced circumstances to the pleasure-loving girl who had never felt responsibility before. Lincoln's single term in Congress, for 1847-1849, gave Mary and the boys a winter in Washington, but scant opportunity for social life. Finally her unwavering faith in her husband won ample justification with his election as President in 1860.

Though her position fulfilled her high social ambitions, Mrs. Lincoln's years in the White House mingled misery with triumph. An orgy of spending stirred resentful comment. While the Civil War dragged on, Southerners scorned her as a traitor to her birth, and citizens loyal to the Union suspected her of treason. When she entertained, critics accused her of unpatriotic extravagance. When, utterly distraught, she curtailed her entertaining after her son Willie's death in 1862, they accused her of shirking her social duties.

Yet Lincoln, watching her put her guests at ease during a White House reception, could say happily: "My wife is as handsome as when she was a girl, and I . . . fell in love with her; and what is more, I have never fallen out."

Her husband's assassination in 1865 shattered Mary Todd Lincoln. The next 17 years held nothing but sorrow. With her son "Tad" she traveled abroad in search of health, tortured by distorted ideas of her financial situation. After Tad died in 1871, she slipped into a world of illusion where poverty and murder pursued her.

A misunderstood and tragic figure, she passed away in 1882 at her sister's home in Springfield—the same house from which she had walked as the bride of Abraham Lincoln, 40 years before.

*Mathew Brady's camera reveals Mary Lincoln before hostile gossip, private grief, and national tragedy had marked her face—probably in 1861. Of six portraits he made that year, she found only a profile "passable."*

Married at 16, Eliza McCardle Johnson taught her husband writing and arithmetic. In poor health when she went to the White House, she left the duties of First Lady to her daughter Martha Patterson (opposite), who modestly said of her family: "We are plain people, from the mountains of Tennessee, called here for a short time by a national calamity. I trust too much will not be expected of us." Byrd Venable Farioletti painted Mrs. Johnson's portrait in 1961, from a photograph taken during the White House years.

# Eliza McCardle Johnson
## 1810-1876

"I knew he'd be acquitted; I knew it," declared Eliza McCardle Johnson, told how the Senate had voted in her husband's impeachment trial. Her faith in him had never wavered during those difficult days in 1868, when her courage dictated that all White House social events should continue as usual.

That faith began to develop many years before in east Tennessee, when Andrew Johnson first came to Greeneville, across the mountains from North Carolina, and established a tailor shop. Eliza was almost 16 then and Andrew only 17; and local tradition tells of the day she first saw him. He was driving a blind pony hitched to a small cart, and she said to a girl friend, "There goes my beau!" She married him within a year, on May 17, 1827.

Eliza was the daughter of Sarah Phillips and John McCardle, a shoemaker. Fortunately she had received a good basic education that she was delighted to share with her new husband. He already knew his letters and could read a bit, so she taught him writing and arithmetic. With their limited means, her skill at keeping a house and bringing up a family—five children, in all—had much to do with Johnson's success.

He rose rapidly, serving in the state and national legislatures and as governor. Like him, when the Civil War came, people of east Tennessee remained loyal to the Union; Lincoln sent him to Nashville as military governor in 1862. Rebel forces caught Eliza at home with part of the family. Only after months of uncertainty did they rejoin Andrew Johnson in Nashville. By 1865 a soldier son and son-in-law had died, and Eliza was an invalid for life.

Quite aside from the tragedy of Lincoln's death, she found little pleasure in her husband's position as President. At the White House, she settled into a second-floor room that became the center of activities for a large

*Martha Johnson Patterson, wearing onyx earrings that Mrs. Polk had given to her.*

family: her two sons, her widowed daughter Mary Stover and her children; her older daughter Martha with her husband, Senator David T. Patterson, and their children. As a schoolgirl Martha had often been the Polks' guest at the mansion; now she took up its social duties. She was a competent, unpretentious, and gracious hostess even during the impeachment crisis.

At the end of Johnson's term, Eliza returned with relief to her home in Tennessee, restored from wartime vandalism. She lived to see the legislature of her state vindicate her husband's career by electing him to the Senate in 1875, and survived him by nearly six months, dying at the Pattersons' home in 1876.

# Julia Dent Grant
## 1826-1902

Quite naturally, shy young Lieutenant Grant lost his heart to friendly Julia; and he made his love known, as he said himself years later, "in the most awkward manner imaginable." She told her side of the story—her father opposed the match, saying, "the boy is too poor," and she answered angrily that she was poor herself. The "poverty" on her part came from a slave-owner's lack of ready cash.

Daughter of Frederick and Ellen Wrenshall Dent, Julia had grown up on a plantation near St. Louis in a typically Southern atmosphere. In memoirs prepared late in life—unpublished until 1975—she pictured her girlhood as an idyll: "one long summer of sunshine, flowers, and smiles. . . ." She attended the Misses Mauros' boarding school in St. Louis for seven years among the daughters of other affluent parents. A social favorite in that circle, she met "Ulys" at her home, where her family welcomed him as a West Point classmate of her brother Frederick; soon she felt lonely without him, dreamed of him, and agreed to wear his West Point ring.

Julia and her handsome lieutenant became engaged in 1844, but the Mexican War deferred the wedding for four long years. Their marriage, often tried by adversity, met every test; they gave each other a life-long loyalty. Like other army wives, "dearest Julia" accompanied her husband to military posts, to pass uneventful days at distant garrisons. Then she returned to his parents' home in 1852 when he was ordered to the West.

Ending that separation, Grant resigned his commission two years later. Farming and business ventures at St. Louis failed, and in 1860 he took his family—four children now—back to his home in Galena, Illinois. He was working in his father's leather goods store when the Civil War called him to a soldier's duty with his state's volunteers. Throughout the war, Julia joined her husband near the scene of action whenever she could.

After so many years of hardship and stress, she rejoiced in his fame as a victorious general, and she entered the White House in 1869 to begin, in her words, "the happiest period" of her life. With Cabinet wives as her allies, she entertained extensively and lavishly. Contemporaries noted her finery, jewels and silks and laces.

Upon leaving the White House in 1877, the Grants made a trip around the world that became a journey of triumphs. Julia proudly recalled details of hospitality and magnificent gifts they received.

But in 1884 Grant suffered yet another business failure and they lost all they had. To provide for his wife, Grant wrote his famous personal memoirs, racing with time and death from cancer. The means thus afforded and her widow's pension enabled her to live in comfort, surrounded by children and grandchildren, till her own death in 1902. She had attended in 1897 the dedication of Grant's monumental tomb in New York City where she was laid to rest. She had ended her own chronicle of their years together with a firm declaration: "the light of his glorious fame still reaches out to me, falls upon me, and warms me."

*At Mathew Brady's studio, Julia Dent Grant adopts a characteristic pose to hide an eye defect. Grant rejected the idea of corrective surgery, teasing her gently: "...I might not like you half so well with any other eyes."*

# Lucy Ware Webb Hayes
## 1831-1889

There was no inaugural ball in 1877—when Rutherford B. Hayes and his wife, Lucy, left Ohio for Washington, the outcome of the election was still in doubt. Public fears had not subsided when it was settled in Hayes' favor; and when Lucy watched her husband take his oath of office at the Capitol, her serene and beautiful face impressed even cynical journalists.

She came to the White House well loved by many. Born in Chillicothe, Ohio, daughter of Maria Cook and Dr. James Webb, she lost her father at age two. She was just entering her teens when Mrs. Webb took her sons to the town of Delaware to enroll in the new Ohio Wesleyan University, but she began studying with its excellent instructors. She graduated from the Wesleyan Female College in Cincinnati at 18, unusually well educated for a young lady of her day.

"Rud" Hayes at 27 had set up a law practice in Cincinnati, and he began paying calls at the Webb home. References to Lucy appeared in his diary: "Her low sweet voice is very winning...a heart as true as steel.... Intellect she has too.... By George! I am in love with her!" Married in 1852, they lived in Cincinnati until the Civil War, and he soon came to share her deeply religious opposition to slavery. Visits to relatives and vacation journeys broke the routine of a happy domestic life in a growing family. Over twenty years Lucy bore eight children, of whom five grew up.

She won the affectionate name of "Mother Lucy" from men of the 23rd Ohio Volunteer Infantry who served under her husband's command in the war. They remembered her visits to camp—to minister to the wounded, cheer the homesick, and comfort the dying. Hayes' distinguished combat record earned him election to Congress, and three postwar terms as governor of Ohio. She not only joined him in Washington for its winter social season, she also accompanied him on visits to state reform schools, prisons, and asylums. As the popular first lady of her state, she gained experience in what a woman of her time aptly called "semi-public life."

Thus she entered the White House with confidence gained from her long and happy married life, her knowledge of political circles, her intelligence and culture, and her cheerful spirit. She enjoyed informal parties, and spared no effort to make official entertaining attractive. Though she was a temperance advocate and liquor was banned at the mansion during this administration, she was a very popular hostess. She took criticism of her views in good humor (the famous nickname "Lemonade Lucy" apparently came into use only after she had left the mansion). She became one of the best-loved women to preside over the White House, where the Hayeses celebrated their silver wedding anniversary in 1877, and an admirer hailed her as representing "the new woman era."

The Hayes term ended in 1881, and the family home was now "Spiegel Grove," an estate at Fremont, Ohio. There husband and wife spent eight active, contented years together until her death in 1889. She was buried in Fremont, mourned by her family and hosts of friends.

*Sunny in temperament, winning in manner, firm in high principle, Lucy
Hayes met both praise and gibes when White House menus omitted liquor.
The Woman's Christian Temperance Union honored her with this portrait.*

*Intellectual Lucretia Garfield—"Crete" to the husband who depended upon her political insights—may have posed for this Brady study while still a Congressman's wife. Illness and turmoil filled her 200 days as First Lady.*

# Lucretia Rudolph Garfield
## 1832-1918

In the fond eyes of her husband, President James A. Garfield, Lucretia "grows up to every new emergency with fine tact and faultless taste." She proved this in the eyes of the nation, though she was always a reserved, self-contained woman. She flatly refused to pose for a campaign photograph, and much preferred a literary circle or informal party to a state reception.

Her love of learning she acquired from her father, Zeb Rudolph, a leading citizen of Hiram, Ohio, and devout member of the Disciples of Christ. She first met "Jim" Garfield when both attended a nearby school, and they renewed their friendship in 1851 as students at the Western Reserve Eclectic Institute, founded by the Disciples.

But "Crete" did not attract his special attention until December 1853, when he began a rather cautious courtship, and they did not marry until November 1858, when he was well launched on his career as a teacher. His service in the Union Army from 1861 to 1863 kept them apart; their first child, a daughter, died in 1863. But after his first lonely winter in Washington as a freshman Representative, the family remained together. With a home in the capital as well as one in Ohio they enjoyed a happy domestic life. A two-year-old son died in 1876, but five children grew up healthy and promising; with the passage of time, Lucretia became more and more her husband's companion.

In Washington they shared intellectual interests with congenial friends; she went with him to meetings of a locally celebrated literary society. They read together, made social calls together, dined with each other and traveled in company until by 1880 they were as nearly inseparable as his career permitted.

Garfield's election to the Presidency brought a cheerful family to the White House in 1881. Though Mrs. Garfield was not particularly interested in a First Lady's social duties, she was deeply conscientious and her genuine hospitality made her dinners and twice-weekly receptions enjoyable. At the age of 49 she was still a slender, graceful little woman with clear dark eyes, her brown hair beginning to show traces of silver.

In May she fell gravely ill, apparently from malaria and nervous exhaustion, to her husband's profound distress. "When you are sick," he had written her seven years earlier, "I am like the inhabitants of countries visited by earthquakes." She was still a convalescent, at a seaside resort in New Jersey, when he was shot by a demented assassin on July 2. She returned to Washington by special train—"frail, fatigued, desperate," reported an eyewitness at the White House, "but firm and quiet and full of purpose to save."

During the three months her husband fought for his life, her grief, devotion, and fortitude won the respect and sympathy of the country. In September, after his death, the bereaved family went home to their farm in Ohio. For another 36 years she led a strictly private but busy and comfortable life, active in preserving the records of her husband's career. She died on March 14, 1918.

*Daily throughout his White House years, according to family tradition, President Chester A. Arthur had a fresh bouquet of flowers placed before this hand-tinted, silver-framed photograph of his wife. Ellen Herndon Arthur had died ten months before his election to the Vice Presidency. A sophisticated host in his own right, he asked his sister Mary McElroy to assume the role of White House hostess for form's sake, and to oversee the care of his motherless daughter, Ellen.*

# Ellen Lewis Herndon Arthur
## 1837-1880

Chester Alan Arthur's beloved "Nell" died of pneumonia on January 12, 1880. That November, when he was elected Vice President, he was still mourning her bitterly. In his own words: "Honors to me now are not what they once were." His grief was the more poignant because she was only 42 and her death sudden. Just two days earlier she had attended a benefit concert in New York City—while he was busy with politics in Albany—and she caught cold that night while waiting for her carriage. She was already unconscious when he reached her side.

Her family connections among distinguished Virginians had shaped her life. She was born at Culpeper Court House, only child of Elizabeth Hansbrough and William Lewis Herndon, U.S.N. They moved to Washington, D. C., when he was assigned to help his brother-in-law Lt. Matthew Fontaine Maury establish the Naval Observatory. While Ellen was still just a girl her beautiful contralto voice attracted attention; she joined the choir at St. John's Episcopal Church on Lafayette Square.

Then her father assumed command of a mail steamer operating from New York; and in 1856 a cousin introduced her to "Chet" Arthur, who was establishing a law practice in the city. By 1857 they were engaged. In a birthday letter that year he reminded her of "the soft, moonlight nights of June, a year ago . . . happy, happy days at Saratoga —the golden, fleeting hours at Lake George." He wished he could hear her singing.

That same year her father died a hero's death at sea, going down with his ship in a gale off Cape Hatteras. The marriage did not take place until October 1859; and a son named for Commander Herndon died when only two. But another boy was born in

*Mary Arthur McElroy: youngest sister of the President, ranking lady of the Arthur administration.*

1864 and a girl, named for her mother, in 1871. Arthur's career brought the family an increasing prosperity; they decorated their home in the latest fashion and entertained prominent friends with elegance. At Christmas there were jewels from Tiffany for Nell, the finest toys for the children.

At the White House, Arthur would not give anyone the place that would have been his wife's. He asked his sister Mary (Mrs. John E. McElroy) to assume certain social duties and help care for his daughter. He presented a stained-glass window to St. John's Church in his wife's memory; it depicted angels of the Resurrection, and at his special request it was placed in the south transept so that he could see it at night from the White House with the lights of the church shining through.

# Frances Folsom Cleveland
## 1864-1947

"I detest him so much that I don't even think his wife is beautiful." So spoke one of President Grover Cleveland's political foes—the only person, it seems, to deny the loveliness of this notable First Lady, first bride of a President to be married in the White House.

She was born in Buffalo, New York, only child of Emma C. Harmon and Oscar Folsom—who became a law partner of Cleveland's. As a devoted family friend Cleveland bought "Frank" her first baby carriage. As administrator of the Folsom

*Rose Elizabeth Cleveland: her bachelor brother's hostess in 15 months of his first term of office.*

estate after his partner's death, though never her legal guardian, he guided her education with sound advice. When she entered Wells College, he asked Mrs. Folsom's permission to correspond with her, and he kept her room bright with flowers.

Though Frank and her mother missed his inauguration in 1885, they visited him at the White House that spring. There affection turned into romance—despite 27 years' difference in age—and there the wedding took place on June 2, 1886.

Cleveland's scholarly sister Rose gladly gave up the duties of hostess for her own career in education; and with a bride as First Lady, state entertainments took on a new interest. Mrs. Cleveland's unaffected charm won her immediate popularity. She held two receptions a week—one on Saturday afternoons, when women with jobs were free to come.

After the President's defeat in 1888, the Clevelands lived in New York City, where baby Ruth was born. With his unprecedented re-election, the First Lady returned to the White House as if she had been gone but a day. Through the political storms of this term she always kept her place in public favor. People took keen interest in the birth of Esther at the mansion in 1893, and of Marion in 1895. When the family left the White House, Mrs. Cleveland had become one of the most popular women ever to serve as hostess for the nation.

She bore two sons while the Clevelands lived in Princeton, New Jersey, and was at her husband's side when he died at their home, "Westland," in 1908.

In 1913 she married Thomas J. Preston, Jr., a professor of archeology, and remained a figure of note in the Princeton community until she died. She had reached her 84th year—nearly the age at which the venerable Mrs. Polk had welcomed her and her husband on a Presidential visit to the South, and chatted of changes in White House life from bygone days.

*Youngest of First Ladies, Frances Folsom Cleveland stirred the public's sentimental admiration as a White House bride at 21 and earned nation-wide respect as a charming hostess, a loyal wife, and a capable mother. As President Cleveland expected, she proved "pretty level-headed." The fashionable Swedish portraitist Anders L. Zorn painted this study from life in 1899.*

# Caroline Lavinia Scott Harrison
## 1832-1892

The centennial of President Washington's inauguration heightened the nation's interest in its heroic past, and in 1890 Caroline Scott Harrison lent her prestige as First Lady to the founding of the National Society of the Daughters of the American Revolution. She served as its first President General. She took a special interest in the history of the White House, and the mature dignity with which she carried out her duties may overshadow the fun-loving nature that had charmed "Ben" Harrison when they met as teenagers.

Born at Oxford, Ohio, in 1832, "Carrie" was the second daughter of Mary Potts Neal and the Reverend Dr. John W. Scott, a Presbyterian minister and founder of the Oxford Female Institute. As her father's pupil—brown-haired, petite, witty—she infatuated the reserved young Ben, then an honor student at Miami University; they were engaged before his graduation and married in 1853.

After early years of struggle while he established a law practice in Indianapolis, they enjoyed a happy family life interrupted only by the Civil War. Then, while General Harrison became a man of note in his profession, his wife cared for their son and daughter, gave active service to the First Presbyterian Church and to an orphans' home, and extended cordial hospitality to her many friends. Church views to the contrary, she saw no harm in private dancing lessons for her daughter—she liked dancing herself. Blessed with considerable artistic talent, she was an accomplished pianist; she especially enjoyed painting for recreation.

Illness repeatedly kept her away from Washington's winter social season during her husband's term in the Senate, 1881-1887, and she welcomed their return to private life; but she moved with poise to the White House in 1889 to continue the gracious way of life she had always created in her own home.

During the administration the Harrisons' daughter, Mary Harrison McKee, her two children, and other relatives lived at the White House. The First Lady tried in vain to have the overcrowded mansion enlarged but managed to assure an extensive renovation with up-to-date improvements. She established the collection of china associated with White House history. She worked for local charities as well. With other ladies of progressive views, she helped raise funds for the Johns Hopkins University medical school on condition that it admit women. She gave elegant receptions and dinners. In the winter of 1891-1892, however, she had to battle illness as she tried to fulfill her social obligations. She died of tuberculosis at the White House in October 1892, and after services in the East Room was buried from her own church in Indianapolis.

When official mourning ended, Mrs. McKee acted as hostess for her father in the last months of his term. (In 1896 he married his first wife's widowed niece and former secretary, Mary Scott Lord Dimmick; she survived him by nearly 47 years, dying in January 1948.)

*To honor the memory of their first President General, Caroline Harrison,*
*the Daughters of the American Revolution commissioned this posthumous*
*portrait by Daniel Huntington and presented it to the White House in 1894.*

# Ida Saxton McKinley
## 1847-1907

There was little resemblance between the vivacious young woman who married William McKinley in January 1871 — a slender bride with sky-blue eyes and fair skin and masses of auburn hair — and the petulant invalid who moved into the White House with him in March 1897. Now her face was pallid and drawn, her close-cropped hair gray; her eyes were glazed with pain or dulled with sedative. Only one thing had remained the same: love which had brightened early years of happiness and endured through more than twenty years of illness.

Ida had been born in Canton, Ohio, in 1847, elder daughter of a socially prominent and well-to-do family. James A. Saxton, a banker, was indulgent to his two daughters. He educated them well in local schools and a finishing school, and then sent them to Europe on the grand tour.

Being pretty, fashionable, and a leader of the younger set in Canton did not satisfy Ida, so her broad-minded father suggested that she work in his bank. As a cashier she caught the attention of Maj. William McKinley, who had come to Canton in 1867 to establish a law practice, and they fell deeply in love. While he advanced in his profession, his young wife devoted her time to home and husband. A daughter, Katherine, was born on Christmas Day, 1871; a second, in April 1873. This time Ida was seriously ill, and the frail baby died in August. Phlebitis and epileptic seizures shattered the mother's health; and even before little Katie died in 1876, she was a confirmed invalid.

As Congressman and then as governor of Ohio, William McKinley was never far from her side. He arranged their life to suit her convenience. She spent most of her waking hours in a small Victorian rocking chair that she had had since childhood; she sat doing fancywork and crocheting bedroom slippers while she waited for her husband, who indulged her every whim.

At the White House, the McKinleys acted as if her health were no great handicap to her role as First Lady. Richly and prettily dressed, she received guests at formal receptions seated in a blue velvet chair. She held a fragrant bouquet to suggest that she would not shake hands. Contrary to protocol, she was seated beside the President at state dinners and he, as always, kept close watch for signs of an impending seizure. If necessary, he would cover her face with a large handkerchief for a moment. The First Lady and her devoted husband seemed oblivious to any social inadequacy. Guests were discreet and newspapers silent on the subject of her "fainting spells." Only in recent years have the facts of her health been revealed.

When the President was shot by an assassin in September 1901, after his second inauguration, he thought primarily of her. He murmured to his secretary: "My wife — be careful, Cortelyou, how you tell her — oh, be careful." After his death she lived in Canton, cared for by her younger sister, visiting her husband's grave almost daily. She died in 1907, and lies entombed beside the President and near their two little daughters in Canton's McKinley Memorial Mausoleum.

*Both Ida McKinley and her devoted husband posed in the private quarters of the White House for watercolor miniatures by Emily Drayton Taylor in April 1899. Despite varied illnesses, including epilepsy, she took up the role of First Lady with an indomitable determination.*

With assured, cultivated taste, Edith Roosevelt approved decor for a renovated White House in 1902; she decided that the new ground-floor corridor, entryway for guests on important occasions, should display likenesses of "all the ladies . . . including myself." That same year Theobald Chartran had painted her confidently posed in the South Grounds. Her firmness and prudent advice earned the deep respect of her exuberant husband, and all who knew her. Ever since her day, the corridor has served as a gallery for portraits of recent First Ladies.

# Edith Kermit Carow Roosevelt
## 1861-1948

Edith Kermit Carow knew Theodore Roosevelt from infancy; as a toddler she became a playmate of his younger sister Corinne. Born in Connecticut in 1861, daughter of Charles and Gertrude Tyler Carow, she grew up in an old New York brownstone on Union Square—an environment of comfort and tradition. Throughout childhood she and "Teedie" were in and out of each other's houses.

Attending Miss Comstock's school, she acquired the proper finishing touch for a young lady of that era. A quiet girl who loved books, she was often Theodore's companion for summer outings at Oyster Bay, Long Island; but this ended when he entered Harvard. Although she attended his wedding to Alice Hathaway Lee in 1880, their lives ran separately until 1885, when he was a young widower with an infant daughter, Alice.

Putting tragedy behind him, he and Edith were married in London in December, 1886. They settled down in a house on Sagamore Hill, at Oyster Bay, headquarters for a family that added five children in ten years: Theodore, Kermit, Ethel, Archibald, and Quentin. Throughout Roosevelt's intensely active career, family life remained close and entirely delightful. A small son remarked one day, "When Mother was a little girl, she must have been a boy!"

Public tragedy brought them into the White House, eleven days after President McKinley succumbed to an assassin's bullet. Assuming her new duties with characteristic dignity, Mrs. Roosevelt meant to guard the privacy of a family that attracted everyone's interest, and she tried to keep reporters outside her domain. The public, in consequence, heard little of the vigor of her character, her sound judgment, her efficient household management.

But in this administration the White House was unmistakably the social center of the land. Beyond the formal occasions, smaller parties brought together distinguished men and women from varied walks of life. Two family events were highlights: the wedding of "Princess Alice" to Nicholas Longworth, and Ethel's debut. A perceptive aide described the First Lady as "always the gentle, high-bred hostess; smiling often at what went on about her, yet never critical of the ignorant and tolerant always of the little insincerities of political life."

T.R. once wrote to Ted Jr. that "if Mother had been a mere unhealthy Patient Griselda I might have grown set in selfish and inconsiderate ways." She continued, with keen humor and unfailing dignity, to balance her husband's exuberance after they retired in 1909.

After his death in 1919, she traveled widely abroad but always returned to Sagamore Hill as her home. Alone much of the time, she never appeared lonely, being still an avid reader—"not only cultured but scholarly," as T.R. had said. She kept till the end her interest in the Needlework Guild, a charity which provided garments for the poor, and in the work of Christ Church at Oyster Bay. She died on September 30, 1948, at the age of 87.

# Helen Herron Taft
## 1861-1943

As "the only unusual incident" of her girlhood, "Nellie" Herron Taft recalled her visit to the White House at 17 as the guest of President and Mrs. Hayes, intimate friends of her parents. Fourth child of Harriet Collins and John W. Herron, born in 1861, she had grown up in Cincinnati, Ohio, attending a private school in the city and studying music with enthusiasm.

The year after this notable visit she met "that adorable Will Taft," a tall young lawyer, at a sledding party. They found intellectual interests in common; friendship matured into love; Helen Herron and William Howard Taft were married in 1886. A "treasure," he called her, "self-contained, independent, and of unusual application." He wondered if they would ever reach Washington "in any official capacity" and suggested to her that they might—when she became Secretary of the Treasury!

No woman could hope for such a career in that day, but Mrs. Taft welcomed each step in her husband's: state judge, Solicitor General of the United States, federal circuit judge. In 1900 he agreed to take charge of American civil government in the Philippines. By now the children numbered three: Robert, Helen, and Charles. The delight with which she undertook the journey, and her willingness to take her children to a country still unsettled by war, were characteristic of this woman who loved a challenge. In Manila she handled a difficult role with enthusiasm and tact; she relished travel to Japan and China, and a special diplomatic mission to the Vatican.

Further travel with her husband, who became Secretary of War in 1904, brought a widened interest in world politics and a cosmopolitan circle of friends. His election to the Presidency in 1908 gave her a position she had long desired.

As First Lady, she still took an interest in politics but concentrated on giving the administration a particular social brilliance. Only two months after the inauguration she suffered a severe stroke. An indomitable will had her back in command again within a year. At the New Year's reception for 1910, she appeared in white crepe embroidered with gold—a graceful figure. Her daughter left college for a year to take part in social life at the White House, and the gaiety of Helen's debut enhanced the 1910 Christmas season.

During four years famous for social events, the most outstanding was an evening garden party for several thousand guests on the Tafts' silver wedding anniversary, June 19, 1911. Mrs. Taft remembered this as "the greatest event" in her White House experience. Her own book, *Recollections of Full Years*, gives her account of a varied life. And the capital's famous Japanese cherry trees, planted around the Tidal Basin at her request, form a notable memorial.

Her public role in Washington did not end when she left the White House. In 1921 her husband was appointed Chief Justice of the United States—the position he had desired most of all—and she continued to live in the capital after his death in 1930. Retaining to the end her love of travel and of classical music, she died at her home on May 22, 1943.

*"My dearest and best critic,"* William Howard Taft *wrote of his wife. She wore her inaugural gown and a tiara to sit for Bror Kronstrand in 1910 at her New England summer home; the artist added a White House setting.*

# Ellen Louise Axson Wilson
## 1860-1914

"I am naturally the most unambitious of women and life in the White House has no attractions for me." Mrs. Wilson was writing to thank President Taft for advice concerning the mansion he was leaving. Two years as first lady of New Jersey had given her valuable experience in the duties of a woman whose time belongs to the people. She always played a public role with dignity and grace but never learned to enjoy it.

Those who knew her in the White House described her as calm and sweet, a motherly woman, pretty and refined. Her soft Southern voice had kept its slow drawl through many changes of residence.

Ellen Louise Axson grew up in Rome, Georgia, where her father, the Reverend S. E. Axson, was a Presbyterian minister. Thomas Woodrow Wilson first saw her when he was about six and she only a baby. In 1883, as a young lawyer from Atlanta, "Tommy" visited Rome and met "Miss Ellie Lou" again—a beautiful girl now, keeping house for a bereaved father. He thought, "what splendid laughing eyes!" Despite their instant attraction they did not marry until 1885, because she was unwilling to leave her heartbroken father.

That same year Bryn Mawr College offered Wilson a teaching position at an annual salary of $1,500. He and his bride lived near the campus, keeping her little brother with them. Humorously insisting that her own children must not be born Yankees, she went to relatives in Georgia for the birth of Margaret in 1886 and Jessie in 1887. But Eleanor was born in Connecticut, while Wilson was teaching at Wesleyan University.

His distinguished career at Princeton began in 1890, bringing his wife new social responsibilities. From such demands she took refuge, as always, in art. She had studied briefly in New York, and the quality of her paintings compares favorably with professional art of the period. She had a studio with a skylight installed at the White House in 1913, and found time for painting despite the weddings of two daughters within six months and the duties of hostess for the nation.

The Wilsons had preferred to begin the administration without an inaugural ball, and the First Lady's entertainments were simple; but her unaffected cordiality made her parties successful. In their first year she convinced her scrupulous husband that it would be perfectly proper to invite influential legislators to a private dinner, and when such an evening led to agreement on a tariff bill, he told a friend, "You see what a wise wife I have!"

Descendant of slave owners, Ellen Wilson lent her prestige to the cause of improving housing in the capital's Negro slums. Visiting dilapidated alleys, she brought them to the attention of debutantes and Congressmen. Her death spurred passage of a remedial bill she had worked for. Her health failing slowly from Bright's disease, she died serenely on August 6, 1914. On the day before her death, she made her physician promise to tell Wilson "later" that she hoped he would marry again; she murmured at the end, "... take good care of my husband." Struggling grimly to control his grief, Wilson took her to Rome for burial among her kin.

*First wife of Woodrow Wilson, Ellen Louise Axson shared his life from his teaching days at Bryn Mawr to the Presidency. In Washington she devoted much time to humanitarian causes, particularly better housing for Negroes. She told a cousin that she wondered "how anyone who reaches middle age can bear it if she cannot feel, on looking back, that whatever mistakes she may have made she has on the whole lived for others and not for herself." In 1913 Robert Vonnoh depicted Mrs. Wilson serving tea to her three daughters, Margaret, Eleanor, and Jessie, at the artist's home in Cornish, New Hampshire. When Ellen Wilson became gravely ill in 1914, Margaret took over as official hostess.*

*Solace and support for a troubled President, Edith Wilson saw him through the agonies of wartime, frustrations of protracted peace negotiations, rejection of his dream for world order, illness, and death. Adolpho Muller-Ury portrayed her soon after she had entered the White House, early in 1916.*

# Edith Bolling Galt Wilson
## 1872-1961

"Secret President," "first woman to run the government"—so legend has labeled a First Lady whose role gained unusual significance when her husband suffered prolonged and disabling illness. A happy, protected childhood and first marriage had prepared Edith Wilson for the duties of helpmate and hostess; widowhood had taught her something of business matters.

Descendant of Virginia aristocracy, she was born in Wytheville in 1872, seventh among eleven children of Sallie White and Judge William Holcombe Bolling. Until the age of 12 she never left the town; at 15 she went to Martha Washington College to study music, with a second year at a smaller school in Richmond.

Visiting a married sister in Washington, pretty young Edith met a businessman named Norman Galt; in 1896 they were married. For 12 years she lived as a contented (though childless) young matron in the capital, with vacations abroad. In 1908 her husband died unexpectedly. Shrewdly, Edith Galt chose a good manager who operated the family's jewelry firm with financial success.

By a quirk of fate and a chain of friendships, Mrs. Galt met the bereaved President, still mourning profoundly for his first wife. A man who depended on feminine companionship, the lonely Wilson took an instant liking to Mrs. Galt, charming and intelligent and unusually pretty. Admiration changed swiftly to love. In proposing to her, he made the poignant statement that "in this place time is not measured by weeks, or months, or years, but by deep human experiences. . . ." They were married privately on December 18, 1915, at her home; and after they returned from a brief honeymoon in Virginia, their happiness made a vivid impression on their friends and White House staff.

Though the new First Lady had sound qualifications for the role of hostess, the social aspect of the administration was overshadowed by the war in Europe and abandoned after the United States entered the conflict in 1917. Edith Wilson submerged her own life in her husband's, trying to keep him fit under tremendous strain. She accompanied him to Europe when the Allies conferred on terms of peace.

Wilson returned to campaign for Senate approval of the peace treaty and the League of Nations Covenant. His health failed in September 1919; a stroke left him partly paralyzed. His constant attendant, Mrs. Wilson took over many routine duties and details of government. But she did not initiate programs or make major decisions, and she did not try to control the executive branch. She selected matters for her husband's attention and let everything else go to the heads of departments or remain in abeyance. Her "stewardship," she called this. And in *My Memoir*, published in 1939, she stated emphatically that her husband's doctors had urged this course upon her.

In 1921, the Wilsons retired to a comfortable home in Washington, where he died three years later. A highly respected figure in the society of the capital, Mrs. Wilson lived on to ride in President Kennedy's inaugural parade. She died later in 1961: on December 28, the anniversary of her famous husband's birth.

# Florence Kling Harding
## 1860-1924

Daughter of the richest man in a small town—Amos Kling, a successful business-man—Florence Mabel Kling was born in Marion, Ohio, in 1860, to grow up in a setting of wealth, position, and privilege. Much like her strong-willed father in temperament, she developed a self-reliance rare in girls of that era.

A music course at the Cincinnati Conservatory completed her education. When only 19, she eloped with Henry De Wolfe, a neighbor two years her senior. He proved a spendthrift and a heavy drinker who soon deserted her, so she returned to Marion with her baby son. Refusing to live at home, she rented rooms and earned her own money by giving piano lessons to children of the neighborhood. She divorced De Wolfe in 1886 and resumed her maiden name; he died at age 35.

Warren G. Harding had come to Marion when only 16 and, showing a flair for newspaper work, had managed to buy the little *Daily Star*. When he met Florence a courtship quickly developed. Over Amos Kling's angry opposition they were mar-ried in 1891, in a house that Harding had planned, and this remained their home for the rest of their lives. (They had no children.)

Mrs. Harding soon took over the *Star*'s circulation department, spanking news-boys when necessary. "No pennies escaped her," a friend recalled, and the paper prospered while its owner's political success increased. As he rose through Ohio politics and became a United States Senator, his wife directed all her acumen to his career. He became Republican nominee for President in 1920 and "the Duchess," as he called her, worked tirelessly for his election. In her own words: "I have only one real hobby—my husband."

She had never been a guest at the White House; and former President Taft, meet-ing the President-elect and Mrs. Harding, discussed its social customs with her and stressed the value of ceremony. Writing to Nellie, he concluded that the new First Lady was "a nice woman" and would "readily adapt herself."

When Mrs. Harding moved into the White House, she opened mansion and grounds to the public again—both had been closed throughout President Wilson's illness. She herself suffered from a chronic kidney ailment, but she threw herself into the job of First Lady with energy and willpower. Garden parties for veterans were regular events on a crowded social calendar. The President and his wife relaxed at poker parties in the White House library, where liquor was available although the Eighteenth Amendment made it illegal.

Mrs. Harding always liked to travel with her husband. She was with him in the summer of 1923 when he died unexpectedly in California, shortly before the public learned of the major scandals facing his administration.

With astonishing fortitude she endured the long train ride to Washington with the President's body, the state funeral at the Capitol, the last service and burial at Marion. She died in Marion on November 21, 1924, surviving Warren Harding by little more than a year of illness and sorrow.

*"The Duchess," her husband called Florence Harding. Her drive and determination helped put the 29th President in the White House in 1921. There, in the Queens' Bedroom, hangs this portrait from life by Philip de László.*

# Grace Anna Goodhue Coolidge
## 1879-1957

For her "fine personal influence exerted as First Lady of the Land," Grace Coolidge received a gold medal from the National Institute of Social Sciences. In 1931 she was voted one of America's twelve greatest living women.

She had grown up in the Green Mountain city of Burlington, Vermont, only child of Andrew and Lemira B. Goodhue, born in 1879. While still a girl she heard of a school for deaf children in Northampton, Massachusetts, and eventually decided to share its challenging work. She graduated from the University of Vermont in 1902 and went to teach at the Clarke School for the Deaf that autumn.

In Northampton she met Calvin Coolidge; they belonged to the same boating, picnicking, whist-club set, composed largely of members of the local Congregational Church. In October 1905 they were married at her parents' home. They lived modestly; they moved into half of a duplex two weeks before their first son was born, and she budgeted expenses well within the income of a struggling small-town lawyer.

To Grace Coolidge may be credited a full share in her husband's rise in politics. She worked hard, kept up appearances, took her part in town activities, attended her church, and offset his shyness with a gay friendliness. She bore a second son in 1908, and it was she who played backyard baseball with the boys. As Coolidge was rising to the rank of governor, the family kept the duplex; he rented a dollar-and-a-half room in Boston and came home on weekends.

In 1921, as wife of the Vice President, Grace Coolidge went from her housewife's routine into Washington society and quickly became the most popular woman in the capital. Her zest for life and her innate simplicity charmed even the most critical. Stylish clothes—a frugal husband's one indulgence—set off her good looks.

After Harding's death, she planned the new administration's social life as her husband wanted it: unpretentious but dignified. Her time and her friendliness now belonged to the nation, and she was generous with both. As she wrote later, she was "I, and yet, not I—this was the wife of the President of the United States and she took precedence over me. . . ." Under the sorrow of her younger son's sudden death at 16, she never let grief interfere with her duties as First Lady. Tact and gaiety made her one of the most popular hostesses of the White House, and she left Washington in 1929 with the country's respect and love.

For greater privacy in Northampton, the Coolidges bought "The Beeches," a large house with spacious grounds. Calvin Coolidge died there in 1933. He had summed up their marriage in his *Autobiography:* "For almost a quarter of a century she has borne with my infirmities, and I have rejoiced in her graces." After his death she sold The Beeches, bought a smaller house, and in time undertook new ventures she had longed to try: her first airplane ride, her first trip to Europe. She kept her aversion to publicity and her sense of fun until her death in 1957. Her chief activity as she grew older was serving as a trustee of the Clarke School; her great pleasure was the family of her surviving son, John.

First as a Vice President's wife, then as First Lady, Grace Goodhue Coolidge brought to Washington society the charm and simplicity of her New England upbringing and her love of people, outdoor activity, and animals. Howard Chandler Christy portrayed her in 1924 with her famous collie Rob Roy.

# Lou Henry Hoover
## 1874-1944

Admirably equipped to preside at the White House, Lou Henry Hoover brought to it long experience as wife of a man eminent in public affairs at home and abroad. She had shared his interests since they met in a geology lab at Leland Stanford University. She was a freshman, he a senior, and he was fascinated, as he declared later, "by her whimsical mind, her blue eyes and a broad grinnish smile."

Born in Iowa, in 1874, she grew up there for ten years. Then her father, Charles D. Henry, decided that the climate of southern California would favor the health of his wife, Florence. He took his daughter on camping trips in the hills—her greatest pleasures in her early teens. Lou became a fine horsewoman; she hunted, and preserved specimens with the skill of a taxidermist; she developed an enthusiam for rocks, minerals, and mining. She entered Stanford in 1894—"slim and supple as a reed," a classmate recalled, with a "wealth of brown hair"—and completed her course before marrying Herbert Hoover in 1899.

The newlyweds left at once for China, where he won quick recognition as a mining engineer. His career took them about the globe—Ceylon, Burma, Siberia, Australia, Egypt, Japan, Europe—while her talent for homemaking eased their time in a dozen foreign lands. Two sons, Herbert and Allan, were born during this adventurous life, which made their father a youthful millionaire.

During World War I, while Hoover earned world fame administering emergency relief programs, she was often with him but spent some time with the boys in California. In 1919 she saw construction begin for a long-planned home in Palo Alto. In 1921, however, his appointment as Secretary of Commerce took the family to Washington. There she spent eight years busy with the social duties of a Cabinet wife and an active participation in the Girl Scout movement, including service as its president.

The Hoovers moved into the White House in 1929, and the First Lady welcomed visitors with poise and dignity throughout the administration. However, when the first day of 1933 dawned, Mr. and Mrs. Hoover were away on holiday. Their absence ended the New Year's Day tradition of the public being greeted personally by the President at a reception in the Executive Mansion.

Mrs. Hoover paid with her own money the cost of reproducing furniture owned by Monroe for a period sitting room in the White House. She also restored Lincoln's study for her husband's use. She dressed handsomely; she "never fitted more perfectly into the White House picture than in her formal evening gown," remarked one secretary. The Hoovers entertained elegantly, using their own private funds for social events while the country suffered worsening economic depression.

In 1933 they retired to Palo Alto, but maintained an apartment in New York. Mr. Hoover learned the full lavishness of his wife's charities only after her death there on January 7, 1944; she had helped the education, he said, "of a multitude of boys and girls." In retrospect he stated her ideal for the position she had held: "a symbol of everything wholesome in American life."

*Perfectionist of hospitality, Lou Hoover as First Lady considered formal entertaining a duty to the American people. For the White House Collection, Richard M. Brown copied a 1932 portrait from life by Philip de László.*

*Douglas Chandor's portrait of Eleanor Roosevelt at age 65 captures a range of moods and suggests her astonishing energy. She devoted a long career to victims of poverty, prejudice, and war. Painfully shy and self-conscious in her youth, often a controversial figure in the role of President's wife, she won international respect in later years as "First Lady of the World."*

# Anna Eleanor Roosevelt Roosevelt
## 1884-1962

A shy, awkward child, starved for recognition and love, Eleanor Roosevelt grew into a woman with great sensitivity to the underprivileged of all creeds, races, and nations. Her constant work to improve their lot made her one of the most loved—and for some years one of the most reviled—women of her generation.

She was born in New York City on October 11, 1884, daughter of lovely Anna Hall and Elliott Roosevelt, younger brother of Theodore. When her mother died in 1892, the children went to live with Grandmother Hall; her adored father died only two years later. Attending a distinguished school in England gave her, at 15, her first chance to develop self-confidence among other girls.

Tall, slender, graceful of figure but apprehensive at the thought of being a wallflower, she returned for a debut that she dreaded. In her circle of friends was a distant cousin, handsome young Franklin Delano Roosevelt. They became engaged in 1903 and were married in 1905, with her uncle the President giving the bride away. Within eleven years Eleanor bore six children; one son died in infancy. "I suppose I was fitting pretty well into the pattern of a fairly conventional, quiet, young society matron," she wrote later in her autobiography.

In Albany, where Franklin served in the state Senate from 1910 to 1913, Eleanor started her long career as political helpmate. She gained a knowledge of Washington and its ways while he served as Assistant Secretary of the Navy. When he was stricken with poliomyelitis in 1921, she tended him devotedly. She became active in the women's division of the State Democratic Committee to keep his interest in politics alive. From his successful campaign for governor in 1928 to the day of his death, she dedicated her life to his purposes. She became eyes and ears for him, a trusted and tireless reporter.

When Mrs. Roosevelt came to the White House in 1933, she understood social conditions better than any of her predecessors and she transformed the role of First Lady accordingly. She never shirked official entertaining; she greeted thousands with charming friendliness. She also broke precedent to hold press conferences, travel to all parts of the country, give lectures and radio broadcasts, and express her opinions candidly in a daily syndicated newspaper column, "My Day."

This made her a tempting target for political enemies but her integrity, her graciousness, and her sincerity of purpose endeared her personally to many—from heads of state to servicemen she visited abroad during World War II. As she had written wistfully at 14: "... no matter how plain a woman may be if truth & loyalty are stamped upon her face all will be attracted to her...."

After the President's death in 1945 she returned to a cottage at his Hyde Park estate; she told reporters: "the story is over." Within a year, however, she began her service as American spokesman in the United Nations. She continued a vigorous career until her strength began to wane in 1962. She died in New York City that November, and was buried at Hyde Park beside her husband.

# Elizabeth Virginia Wallace Truman
## 1885-1982

Whistle-stopping in 1948, President Harry Truman often ended his campaign talk by introducing his wife as "the Boss" and his daughter, Margaret, as "the Boss's Boss," and they smiled and waved as the train picked up steam. The sight of that close-knit family gallantly fighting against such long odds had much to do with his surprise victory at the polls that November.

Strong family ties in the southern tradition had always been important around Independence, Missouri, where a baby girl was born to Margaret ("Madge") Gates and David Wallace on February 13, 1885. Christened Elizabeth Virginia, she grew up as "Bess." Harry Truman, whose family moved to town in 1890, always kept his first impression of her—"golden curls" and "the most beautiful blue eyes." A relative said, "there never was but one girl in the world" for him. They attended the same schools from fifth grade through high school.

In recent years their daughter has written a vivid sketch of Bess as a girl: "a marvelous athlete—the best third baseman in Independence, a superb tennis player, a tireless ice skater—and she was pretty besides." She also had many "strong opinions. . . . and no hesitation about stating them Missouri style—straight from the shoulder."

For Bess and Harry, World War I altered a deliberate courtship. He proposed and they became engaged before Lieutenant Truman left for the battlefields of France in 1918. They were married in June 1919; they lived in Mrs. Wallace's home, where Mary Margaret was born in 1924.

When Harry Truman became active in politics, Mrs. Truman traveled with him and shared his platform appearances as the public had come to expect a candidate's wife to do. His election to the Senate in 1934 took the family to Washington. Reluctant to be a public figure herself, she always shared his thoughts and interests in private. When she joined his office staff as a secretary, he said, she earned "every cent I pay her." His wartime role as chairman of a special committee on defense spending earned him national recognition—and a place on the Democratic ticket as President Roosevelt's fourth-term running mate. Three months after their inauguration Roosevelt was dead. On April 12, 1945, Harry Truman took the President's oath of office—and Bess, who managed to look on with composure, was the new First Lady.

In the White House, its lack of privacy was distasteful to her. As her husband put it later, she was "not especially interested" in the "formalities and pomp or the artificiality which, as we had learned . . . , inevitably surround the family of the President." Though she conscientiously fulfilled the social obligations of her position, she did only what was necessary. While the mansion was rebuilt during the second term, the Trumans lived in Blair House and kept social life to a minimum.

They returned to Independence in 1953. After her husband's death in 1972, Mrs. Truman continued to live in the family home. There she enjoyed visits from Margaret and her husband, Clifton Daniel, and their four sons. She died in 1982 and was buried beside her husband in the courtyard of the Harry S. Truman Library.

*Dignified and witty Bess Truman kept the original painting of this White House replica in her Missouri home. Her daughter, Margaret Truman Daniel, once described her as "a warmhearted, kind lady, with a robust sense of humor."*

# Mamie Geneva Doud Eisenhower
## 1896-1979

Mamie Eisenhower's bangs and sparkling blue eyes were as much trademarks of an administration as the President's famous grin. Her outgoing manner, her feminine love of pretty clothes and jewelry, and her obvious pride in husband and home made her a very popular First Lady.

Born in Boone, Iowa, Mamie Geneva Doud moved with her family to Colorado when she was seven. Her father retired from business, and Mamie and her three sisters grew up in a large house in Denver. During winters the family made long visits to relatives in the milder climate of San Antonio, Texas.

There, in 1915, at Fort Sam Houston, Mamie met Dwight D. Eisenhower, a young second lieutenant on his first tour of duty. She drew his attention instantly, he recalled: "a vivacious and attractive girl, smaller than average, saucy in the look about her face and in her whole attitude." On St. Valentine's Day in 1916 he gave her a miniature of his West Point class ring to seal a formal engagement; they were married at the Doud home in Denver on July 1.

For years Mamie Eisenhower's life followed the pattern of other Army wives: a succession of posts in the United States, in the Panama Canal Zone; duty in France, in the Philippines. She once estimated that in 37 years she had unpacked her household at least 27 times. Each move meant another step up the career ladder for her husband, with increasing responsibilities for her.

The first son Doud Dwight or "Icky," who was born in 1917, died of scarlet fever in 1921. A second child, John, was born in 1922 in Denver. Like his father he had a career in the Army; later he became an author and served as ambassador to Belgium.

During World War II, while promotion and fame came to "Ike," his wife lived in Washington. After he became president of Columbia University in 1948, the Eisenhowers purchased a farm at Gettysburg, Pennsylvania. It was the first home they had ever owned. His duties as commander of North Atlantic Treaty Organization forces— and hers as his hostess at a chateau near Paris—delayed work on their dream home, finally completed in 1955. They celebrated with a housewarming picnic for the staff from their last temporary quarters: the White House.

When Eisenhower had campaigned for President, his wife cheerfully shared his travels; when he was inaugurated in 1953, the American people warmly welcomed her as First Lady. Diplomacy—and air travel—in the postwar world brought changes in their official hospitality. The Eisenhowers entertained an unprecedented number of heads of state and leaders of foreign governments, and Mamie's evident enjoyment of her role endeared her to her guests and to the public.

In 1961 the Eisenhowers returned to Gettysburg for eight years of contented retirement together. After her husband's death in 1969, Mamie continued to live on the farm, devoting more of her time to her family and friends. Mamie Eisenhower died on November 1, 1979. She is buried beside her husband in a small chapel on the grounds of the Eisenhower Library in Abilene, Kansas.

*Her famed and favorite pink casts a soft glow from Mrs. Eisenhower's rhinestone-studded inaugural gown. After many separations in Army life, she and the President happily resided in the White House for eight years.*

*Jacqueline Kennedy, portrayed by Aaron Shikler in 1970, stirred national interest in obtaining antiques and works of art for the White House. She planned the first historical guidebook of the mansion for its many visitors.*

# Jacqueline Lee Bouvier Kennedy Onassis
## 1929-

The inauguration of John F. Kennedy in 1961 brought to the White House and to the heart of the nation a beautiful young wife and the first young children of a President in half a century. Jacqueline Kennedy not only looked like a fairy-tale princess, her family background had permitted her to be reared like one.

She was born Jacqueline Lee Bouvier, daughter of John Vernon Bouvier III and his wife, Janet Lee. Her early years were divided between New York City and East Hampton, Long Island, where she learned to ride almost as soon as she could walk. She was educated at the best of private schools; she wrote poems and stories, drew illustrations for them, and studied ballet.

Her mother, who had obtained a divorce, married Hugh D. Auchincloss in 1942 and brought her two girls to "Merrywood," his home near Washington, D. C., with summers spent at his estate in Newport, Rhode Island. Jacqueline was dubbed "the Debutante of the year" for the 1947-1948 season, but her social success did not keep her from continuing her education. As a Vassar student she traveled extensively, and she spent her junior year in France. These experiences left her with a great empathy for people of foreign countries, especially the French.

Back in Washington, with a job as "inquiring photographer" for a local newspaper, her path soon crossed that of Senator Kennedy, who had the reputation of being the most eligible bachelor in the capital. Their romance progressed slowly and privately, but their wedding at Newport in 1953 attracted nationwide publicity as a grand extravaganza.

With marriage "Jackie" had to adapt herself to the new role of wife to one of the country's most energetic political figures. Her own public appearances were highly successful, but limited in number. After the sadness of a miscarriage and the stillbirth of a daughter, Caroline Bouvier was born in 1957; John Jr. was born between the election of 1960 and Inauguration Day. Patrick Bouvier, born prematurely on August 7, 1963, died two days later.

To the role of First Lady, Jacqueline Kennedy brought beauty, intelligence, and cultivated taste. Her interest in the arts, publicized by press and television, inspired an attention to culture never before evident at a national level. She devoted much time and study to making the White House a museum of decorative arts and American history as well as a family residence of elegance and charm. But she defined her major role as "to take care of the President" and added that "if you bungle raising your children, I don't think whatever else you do well matters very much."

Mrs. Kennedy's gallant courage during the tragedy of her husband's assassination won her the admiration of the world. Thereafter it seemed the public would never allow her the privacy she desired for herself and her children. She moved to New York City; and in 1968 she married the wealthy Greek businessman Aristotle Onassis, 23 years her senior, who died in March 1975. Since 1978 she has worked in New York City as an editor for Doubleday & Company, Inc.

# Claudia Taylor (Lady Bird) Johnson
## 1912-

Christened Claudia Alta Taylor when she was born in a country mansion near Karnack, Texas, she received her nickname "Lady Bird" as a small child; and as Lady Bird she is known and loved throughout America today. Perhaps that name was prophetic, as there has seldom been a First Lady so attuned to nature and the importance of conserving the environment.

Her mother, Minnie Pattillo Taylor, died when Lady Bird was five, so she was reared by her father, her aunt, and family servants. From her father, Thomas Jefferson Taylor, who had prospered, she learned much about the business world. An excellent student, she also learned to love classical literature. At the University of Texas she earned bachelor's degrees in arts and in journalism.

In 1934 Lady Bird met Lyndon Baines Johnson, then a Congressional secretary visiting Austin on official business; he promptly asked her for a date, which she accepted. He courted her from Washington with letters, telegrams, and telephone calls. Seven weeks later he was back in Texas; he proposed to her and she accepted. In her own words: "Sometimes Lyndon simply takes your breath away." They were married in November 1934.

The years that followed were devoted to Lyndon's political career, with "Bird" as partner, confidante, and helpmate. She helped keep his Congressional office open during World War II when he volunteered for naval service; and in 1955, when he had a severe heart attack, she helped his staff keep things running smoothly until he could return to his post as Majority Leader of the Senate. He once remarked that voters "would happily have elected her over me."

After repeated miscarriages, she gave birth to Lynda Bird (now Mrs. Charles S. Robb) in 1944; Luci Baines (Mrs. Ian Turpin) was born three years later.

In the election of 1960, Lady Bird successfully stumped for Democratic candidates across 35,000 miles of campaign trail. As wife of the Vice President, she became an ambassador of goodwill by visiting 33 foreign countries. Moving to the White House after Kennedy's murder, she did her best to ease a painful transition. She soon set her own stamp of Texas hospitality on social events, but these were not her chief concern. She created a First Lady's Committee for a More Beautiful Capital, then expanded her program to include the entire nation. She took a highly active part in her husband's war-on-poverty program, especially the Head Start project for preschool children.

When the Presidential term ended, the Johnsons returned to Texas, where he died in 1973. Mrs. Johnson's *White House Diary*, published in 1970, and a 1981 documentary film, *The First Lady, A Portrait of Lady Bird Johnson*, give sensitive and detailed views of her contributions to the President's Great Society administration. Today Lady Bird leads a life devoted to her husband's memory, her children, and seven grandchildren. She still supports causes dear to her—notably the National Wildflower Research Center, which she founded in 1982, and The Lyndon Baines Johnson Library. She also serves on the Board of the National Geographic Society as a trustee emeritus.

*Flower- and wilderness-lover, Lady Bird Johnson sponsored a beautification project that came to include the nation's cities and highways. "A woman of great depth and excellent judgment," President Johnson described her.*

*While First Lady, Pat Nixon traveled over the world more than any of her predecessors, alone or with her husband. She encouraged volunteer work — "individual attention and love and concern" — to help the less fortunate.*

# Patricia Ryan Nixon
## 1912-

Born Thelma Catherine Ryan on March 16 in Ely, Nevada, "Pat" Nixon acquired her nickname within hours. Her father, William Ryan, called her his "St. Patrick's babe in the morn" when he came home from the mines before dawn.

Soon the family moved to California and settled on a small truck farm near Los Angeles — a life of hard work with few luxuries. Her mother, Kate Halberstadt Bender Ryan, died in 1925; at 13 Pat assumed all the household duties for her father and two older brothers. At 18, she lost her father after nursing him through months of serious illness. Left on her own and determined to continue her education, she worked her way through the University of Southern California. She held part-time jobs on campus, as a sales clerk in a fashionable department store, and as an extra in the movies — and she graduated cum laude in 1937.

She accepted a position as a high-school teacher in Whittier; and there she met Richard Nixon, who had come home from Duke University Law School to establish a practice. They became acquainted at a Little Theater group when they were cast in the same play, and were married on June 21, 1940.

During World War II, she worked as a government economist while he served in the Navy. She campaigned at his side in 1946 when he entered politics, running successfully for Congress, and afterward. Within six years she saw him elected to the House, the Senate, and the Vice Presidency on the ticket with Dwight D. Eisenhower. Despite the demands of official life, the Nixons were devoted parents to their two daughters, Tricia (now Mrs. Edward Cox), and Julie (now Mrs. David Eisenhower).

A tireless campaigner when he ran unsuccessfully for President in 1960, she was at his side when he ran again in 1968 — and won. She had once remarked succinctly, "It takes heart to be in political life."

A gracious hostess as First Lady, she used her position to encourage volunteer service — "the spirit of people helping people." She invited hundreds of families to nondenominational Sunday services in the East Room, unprecedented in White House history. She instituted a series of performances by artists in varied American traditions — from opera to bluegrass — and took pride in the number of fine antiques and significant paintings added to the White House Collection.

She had shared her husband's journeys abroad in the Vice Presidential years and continued this during his Presidency. Official travel took them to most of the nations of the world, including the historic visit to the People's Republic of China and the summit meeting in the Soviet Union. On her first solo trip, the First Lady made a journey of compassion to take relief supplies to earthquake victims in Peru. Later she visited Africa and South America with the unique diplomatic standing of Personal Representative of the President. Always she was a charming envoy for the nation.

Mrs. Nixon returned to private life with her husband upon his resignation in 1974. She had met the troubled days of Watergate with dignity. She said simply, "I love my husband, I believe in him, and I am proud of his accomplishments."

*Once a professional dancer with Martha Graham and a teacher of dance to underprivileged youngsters, Betty Ford continues her support of dance and encourages the other arts. She staunchly advocates women's rights as well.*

# Elizabeth Bloomer Ford
## 1918-

In 25 years of political life, Betty Bloomer Ford did not expect to become First Lady. As wife of Representative Gerald R. Ford, she looked forward to his retirement and more time together. In late 1973 his selection as Vice President was a surprise to her. She was just becoming accustomed to their new roles when he became President upon Mr. Nixon's resignation in August 1974.

Born Elizabeth Anne Bloomer in Chicago, she grew up in Grand Rapids, Michigan, and graduated from high school there. She studied modern dance at Bennington College in Vermont, decided to make it a career, and became a member of Martha Graham's noted concert group in New York City, supporting herself as a fashion model for the John Robert Powers firm.

Close ties with her family and her hometown took her back to Grand Rapids, where she became fashion coordinator for a department store. She also organized her own dance group and taught dance to handicapped children.

Her first marriage, at age 24, ended in divorce five years later on the grounds of incompatibility. Not long afterward she began dating Jerry Ford, football hero, graduate of the University of Michigan and Yale Law School, and soon a candidate for Congress. They were married during the 1948 campaign; he won his election; and the Fords lived in the Washington area for nearly three decades thereafter.

Their four children—Michael, Jack, Steven, and Susan—were born in the next ten years. As her husband's political career became more demanding, Betty Ford found herself shouldering many of the family responsibilities. She supervised the home, did the cooking, undertook volunteer work, and took part in the activities of "House wives" and "Senate wives" for Congressional and Republican clubs. In addition, she was an effective campaigner for her husband.

Betty Ford faced her new life as First Lady with dignity and serenity. She accepted it as a challenge. "I like challenges very much," she said. She had the self-confidence to express herself with humor and forthrightness whether speaking to friends or to the public. Forced to undergo radical surgery for breast cancer in 1974, she reassured many troubled women by discussing her ordeal openly. She explained that "maybe if I as First Lady could talk about it candidly and without embarrassment, many other people would be able to as well." As soon as possible, she resumed her duties as hostess at the Executive Mansion and her role as a public-spirited citizen. She did not hesitate to state her views on controversial issues such as the Equal Rights Amendment, which she strongly supported.

From their home in California, she was equally frank about her successful battle against dependency on drugs and alcohol. She helped establish the Betty Ford Center for treatment of this problem at the Eisenhower Medical Center in Rancho Mirage.

She has described the role of First Lady as "much more of a 24-hour job than anyone would guess" and says of her predecessors: "Now that I realize what they've had to put up with, I have new respect and admiration for every one of them."

For many years, Rosalynn Carter has taken an active interest in volunteer work and in helping the mentally ill, the elderly, the handicapped, and the poor. The shy girl from Plains became a determined campaigner for her husband and called herself "more a political partner than a political wife." She used her influence as First Lady in behalf of women's rights. In 1980 the District of Columbia honored Mrs. Carter for her "commitment . . . to build a more caring society."

# Rosalynn Smith Carter
## 1927-

"She's the girl I want to marry," Jimmy Carter told his mother after his first date with 17-year-old Rosalynn Smith, who had grown up as a friend and neighbor of the Carter family in Plains, Georgia.

Born in Plains on August 18, 1927, Rosalynn was the first of four children in the family of Allethea Murray Smith and Wilburn Edgar Smith. The future First Lady grew up in a small-town atmosphere where strong family ties were the norm and where dedication to church and community was expected. When she was 13, her father died and her mother became a dressmaker to help support the family. As the oldest child, Rosalynn worked beside her mother, helping with the sewing, the housekeeping, and the other children.

Times were difficult, but Rosalynn completed high school and enrolled in Georgia Southwestern College at Americus. In 1945, after her freshman year, she first dated Jimmy Carter, who was home from the U. S. Naval Academy at Annapolis. Their romance progressed, and in 1946 they were married.

The young couple went to Norfolk, Virginia, Ensign Carter's first duty station after graduation. The Navy kept them on the move. Their sons were born in different places: John William in Virginia, James Earl III in Hawaii, and Donnel Jeffrey in Connecticut. The Carters' only daughter, Amy Lynn, was born in Georgia in 1967.

When his father died in 1953, Jimmy left the service, and the Carters returned to Plains to run the family business. Managing the accounts of the peanut, fertilizer, and seed enterprise, Rosalynn soon found herself working full time.

Jimmy entered politics in 1962, winning a seat in the Georgia Senate. Rosalynn, an important member of his campaign team, helped develop support for her husband's successful bid for the governorship of Georgia in 1970. During his Presidential campaigns, Rosalynn traveled independently throughout the United States. Her belief in her husband's ability to lead the nation was communicated in a quiet, friendly manner that made her an effective campaigner.

A skillful speaker, Rosalynn was a serious, hardworking First Lady who went regularly to her own office in the East Wing to manage routine duties and special projects. She attended Cabinet meetings and high-level briefings. Repeatedly during the administration, she represented the Chief Executive at ceremonial occasions, and she served as the President's personal emissary to Latin American countries.

As First Lady, she focused national attention on the performing arts. She invited to the White House leading classical artists from around the world, as well as traditional American artists. She also took a strong interest in programs to aid mental health, the community, and the elderly. During 1977-1978, she served as the Honorary Chairperson of the President's Commission on Mental Health.

Now settled once again back home in Plains, Rosalynn continues to be active in promoting mental health programs. Her autobiography *First Lady from Plains* was published in 1984.

# Nancy Davis Reagan
## 1923-

"My life really began when I married my husband," says Nancy Reagan, who in the 1950's happily gave up an acting career for a permanent role as the wife of Ronald Reagan and mother to their children. Her story actually begins in New York City, her birthplace. She was born on July 6, 1923, according to her autobiography *Nancy*, published in 1980. When the future First Lady was six, her mother, Edith—a stage actress—married Dr. Loyal Davis, a neurosurgeon. Dr. Davis adopted Nancy, and she grew up in Chicago. It was a happy time: summer camp, tennis, swimming, dancing. She received her formal education at Girls' Latin School and at Smith College in Massachusetts, where she majored in theater.

Soon after graduation she became a professional actress. She toured with a road company, then landed a role on Broadway in the hit musical *Lute Song*. More parts followed. One performance drew an offer from Hollywood. Billed as Nancy Davis, she performed in 11 films from 1949 to 1956. Her first screen role was in *Shadow on the Wall*. Other releases included *The Next Voice You Hear* and *East Side, West Side*. In her last movie, *Hellcats of the Navy*, she played opposite her husband.

She had met Ronald Reagan in 1951, when he was president of the Screen Actors Guild. The following year they were married in a simple ceremony in Los Angeles in the Little Brown Church in the Valley. Mrs. Reagan soon retired from making movies so she "could be the wife I wanted to be. . . . A woman's real happiness and real fulfillment come from within the home with her husband and children," she says. President and Mrs. Reagan have a daughter, Patricia Ann, and a son, Ronald Prescott.

While her husband was Governor of California from 1967 to 1975, she worked with numerous charitable groups. She spent many hours visiting veterans, the elderly, and the emotionally and physically handicapped. These people continued to interest her as First Lady. She gave her support to the Foster Grandparent Program, the subject of her 1982 book, *To Love A Child*. Increasingly, she has concentrated on the fight against drug and alcohol abuse among young people. She visited prevention and rehabilitation centers, and in 1985 she held a conference at the White House for First Ladies of 17 countries to focus international attention on this problem.

Mrs. Reagan shared her lifelong interest in the arts with the nation by using the Executive Mansion as a showcase for talented young performers in the PBS television series "In Performance at the White House." In her first year in the mansion she directed a major renovation of the second- and third-floor quarters.

Now living in retirement with her husband in California, she continues to work on her campaign to teach children to "just say no" to drugs, though her husband and her home remain her first priority. In her book *My Turn*, published in 1989, she gives her own account of her life in the White House. Through the joys and sorrows of those days, including the assassination attempt on her husband, Nancy Reagan held fast to her belief in love, honesty, and selflessness. "The ideals have endured because they are right and are no less right today than yesterday."

*Nancy Davis Reagan focused national attention on the problem of drug abuse in young people. Artist Aaron Shikler portrays her in the Red Room, her favorite of the White House state rooms, by a door leading into the State Dining Room.*

# Barbara Pierce Bush
## 1925-

Rarely has a First Lady been greeted by the American people and the press with the approbation and warmth accorded to Barbara Pierce Bush. Perhaps this is prompted by the image she calls "everybody's grandmother." People are comfortable with her white hair, her warm, relaxed manner, and her keen wit. With characteristic directness, she says people like her because they know "I'm fair and I like children and I adore my husband."

Barbara was born in 1925 to Marvin Pierce, president of McCall Corporation, and his wife, Pauline. In the suburban town of Rye, New York, she had a happy childhood. She went to boarding school at Ashley Hall in South Carolina, and it was at a dance during Christmas vacation when she was only 16 that she met George Bush, a senior at Phillips Academy in Andover, Massachusetts. They became engaged a year and a half later, just before he went off to war as a Navy fighter pilot. By the time George returned on leave, Barbara had dropped out of Smith College. Two weeks later, on January 6, 1945, they were married.

After the war, George graduated from Yale, and they set out for Texas to start their lives together. Six children were born to them: George, Robin, Jeb, Neil, Marvin, and Dorothy. Meanwhile, George built a business in the oil industry. With Texas as home base, he then turned to politics and public service, serving as a Member of Congress, U. S. Ambassador to the United Nations, Chairman of the Republican National Committee, Chief of the U. S. Liaison Office in the People's Republic of China, Director of the Central Intelligence Agency, and later as Vice President. During 44 years of marriage, Mrs. Bush managed 29 moves of the family.

When her husband was away, she became the family linchpin, providing everything from discipline to carpools. The death of their daughter Robin from leukemia when she was not quite four left George and Barbara Bush with a lifelong compassion. She says, "Because of Robin, George and I love every living human more."

Barbara Bush was always an asset to her husband during his campaigns for public office. Her friendly, forthright manner won her high marks from the voters and the press. As wife of the Vice President, she selected the promotion of literacy as her special cause. Now, as First Lady, she calls working for a more literate America the "most important issue we have." She is involved with many organizations devoted to this cause and is honorary Chairman of the Barbara Bush Foundation for Family Literacy. A strong advocate of volunteerism, Mrs. Bush helps many causes—including the homeless, AIDS, the elderly, and school volunteer programs.

The First Lady keeps in close touch with their five grown children and their spouses and eleven grandchildren. She still finds time these days to pick up her needlepoint or to read, and she continues to enjoy a game of tennis, a swim, or a walk. The family gatherings at the White House and at the Bushes' summer home in Kennebunkport, Maine, are times of relaxation and refreshment in a busy schedule in which she tries "to do some good every day."

*An active volunteer for many years for the cause of literacy, Barbara Bush feels everybody should help some charitable cause. "If it worries you, then you've got to do something about it." Such sincerity has won the hearts of the nation.*

# Illustrations Credits

A photographic portrait of Mrs. Bush appears because an official painting has not yet been acquired.

Composition for *The First Ladies* by the Typographic section of National Geographic Production Services, Pre-Press Division. Printed and bound by United Color Press, Monroe, Ohio.

*Host and hostess of the White House, President Grover Cleveland and First Lady Frances Folsom Cleveland welcome officers of the Army and Navy at a formal reception in the Blue Room in 1888. Young Mrs. Cleveland greets a lady escorted by Maj. Gen. John M. Schofield, USA, of Civil War renown.*